INTERNATIONAL SERIES OF MONOGRAPHS IN

ANALYTICAL CHEMISTRY

GENERAL EDITORS: R. BELCHER AND H. FREISER

VOLUME 56

STATIONARY PHASES
IN
GAS CHROMATOGRAPHY

D1545612

STATIONARY PHASES
IN
GAS CHROMATOGRAPHY

G. E. BAIULESCU

*Department of Analytical Chemistry,
University of Bucharest*

and

V. A. ILIE

*Institute of Chemical Research,
Bucharest, Romania*

PERGAMON PRESS

OXFORD · NEW YORK · TORONTO
SYDNEY · BRAUNSCHWEIG

Pergamon Press Offices:

U. K.	Pergamon Press Ltd., Headington Hill Hall, Oxford, England
U. S. A.	Pergamon Press Inc., Maxwell House, Fairview Park, Elmsford, New York 10523, U. S. A.
C A N A D A	Pergamon of Canada Ltd., 207 Queen's Quay West, Toronto 1, Canada
A U S T R A L I A	Pergamon Press (Aust.) Pty. Ltd., 19a Boundary Street, Rushcutters Bay, N.S.W. 2011, Australia
F R A N C E	Pergamon Press SARL, 24 rue des Ecoles, 75240 Paris, Cedex 05, France
W E S T G E R M A N Y	Pergamon Press GmbH, 3300 Braunschweig, Postfach 2923, Burgplatz 1, West Germany

First edition 1975

Library of Congress Cataloging in Publication Data

Baiulescu, George.
Stationary phases in gas chromatography.
(International series of monographs in analytical chemistry; v. 56)
Includes bibliographical references and indexes.
1. Gas chromatography. I. Ilie, V. A., joint author. II. Title.
QD79.C45B3413 1975   543'.08   74-32148 ISBN 0-08-018075-2

859942

CONTENTS

CHAPTER 1

THE CHROMATOGRAPHIC COLUMN

1.1 RETENTION PARAMETERS

A chromatographic separation of mixtures of solutes arises from the different distribution of the solutes between the stationary phase and the mobile phase. The distribution of the components between the two phases determines their different retention on the chromatographic column. In a chromatographic experiment, the quantity which is actually observed is the peak retention time.

Compounds that are not retained by the stationary phase leave the chromatographic column at time t_0 — the so-called dead time (the time needed by the front of the carrier gas to go through the column). The solutes that are retained by the stationary phase elute after a time t. A measure of the interaction of a solute with the stationary phase is given by the value of the net retention time t' which is proportional to the partition coefficient K:

$$t' = t - t_0 = fK. \tag{1}$$

The exact determination of the dead time t_0 is of the utmost importance in order precisely to evaluate the retention data. A thorough discussion of the calculation of dead times is available.[1–4]

Other authors[5–8] have described methods by which dead times can be calculated from retention data of n-alkanes on a chromatographic column. They have devised a linear relation between the logarithm of the

1

net retention time and the number of carbon atoms n in the molecule of the n-alkane:

$$\log t'_n = a + bn, \qquad (2)$$

where a and b are constants. The hypothesis which states that the ratio $\log (t'_{n+1}/t'_n)$ should be a constant, and independent of n, has not been completely verified.[9, 10] Taking into consideration the activity coefficients and the vapour pressures, it can be concluded that b is not constant.

The calculation of dead times from the retention of C_1—C_5 alkanes does not constitute an exact measure of b. The exact dead time is hard to measure. Among the usual methods is that using the air peak. The determinations using methane give less errors than the extrapolation of data obtained from alkanes.[10] In spite of that, Kaiser[1] holds that we, nevertheless, need a "synthetic" dead time for calculations. The best "synthetic" dead time is obtained when linearizing the logarithms of the retention times of the n-alkanes. This value may differ from the effective dead time, but it is practical and can be used for exact retention calculations. Kaiser's suggestion is especially valid for high molecular weight alkanes.

The ratio of the net retention time t' and the retention time of an inert gas t_0 is called the *capacity ratio* (k):

$$k = t'/t_0. \qquad (3)$$

For an equilibrium between the mobile phase and the stationary phase:

$$k = KV_{st}/V_m, \qquad (4)$$

where K is the partition coefficient (the ratio of the concentration of the component in the stationary phase to its concentration in the mobile phase), and V_{st} and V_m are the volumes of the stationary and mobile phases, respectively.

Knowing the net retention time t', one can calculate the specific retention volume per gram of stationary phase or per square metre of specific surface. The retention volume represents the quantity of carrier gas that passes through the column from the moment of introduction of the sample until the maximum elution of the component.

In order to calculate the absolute volume of gas that flows through the column, it is first necessary to calculate the average pressure in the column, correcting for the compressibility j :[11]

$$j = \frac{3}{2} \frac{(p_i/p_o)^2 - 1}{(p_i/p_o)^3 - 1},$$ (5)

where p_i and p_0 are the inlet and the outlet column pressures, respectively, of the carrier gas. Thus the net retention volume V_N is:

$$V_N = t'F_0 j \quad \text{(ml)},$$ (6)

where F_o is the flow rate of the eluent at the outlet pressure p_o and at the temperature of the column.

The net retention volume, divided by the quantity of stationary phase G_L (in grams) from the column, gives the specific retention volume V_g^T at the column temperature T_c,

$$V_g^T = V_N/G_L \quad \text{(ml/g)}.$$ (7)

To convert the results to 0°C,

$$V_g = \frac{V_N}{G_L} \frac{273}{T_c} \quad \text{(ml/g)}.$$ (8)

For gas–solid elution chromatography the specific retention volume at the column temperature V_S^T may be obtained by dividing the net retention volume by the specific surface area A of the adsorbent:

$$V_S^T = V_N/A \quad \text{(ml/m}^2\text{)}.$$ (9)

The specific retention volume is an absolute retention parameter, independent of the instrumental and working conditions. It is characteristic of the chemical nature of the component and of the stationary phase used. From the theoretical point of view it is an ideal parameter, but unfortunately the exact determination of specific retention volumes is difficult, being very susceptible to error especially when measuring the pressure, the temperature, or the amount of the stationary phase.[12-15]

For this reason, the relative retention parameters have proved to be more useful in practice. Such a relative parameter is the relative retention r_{23}:

$$r_{23} = V_{g_2}/V_{g_3} = V_{N_2}/V_{N_3} = t_2'/t_3' = k_2/k_3 = K_2/K_3. \tag{10}$$

The indices 2 and 3 refer to two components from the same chromatogram. Rohrschneider[16] has recently described how specific retention volumes in gas–liquid chromatography can be estimated from relative retention data. He formulated the following expression:

$$\ln V_g(\text{n-octane}) = A + B(t_{N_x}/t_{N_{\text{n-octane}}}) + C \ln(t_{N_y}/t_{N_{\text{n-octane}}})$$
$$+ D \ln(t_{N_z}/t_{N_{\text{n-octane}}}) + E \ln(t_{N_u}/t_{N_{\text{n-octane}}}). \tag{11}$$

A through E are constants, and t_N is the net retention time of a solute. Using data from the compilation of McReynolds[17], Rohrschneider fitted eqn. (11) to data for various sets of solutes. He found that with the solutes he chose, going beyond a second relative retention term did not significantly improve the accuracy of the equation. With $x =$ n-decane and $y =$ benzene he obtained the following constants: $A = -0.733$, $B = +2.083$, and $C = +0.623$.

The resulting equation predicts V_g for n-octane on seventy-two liquid phases at 120°C with an average error of 14%. The specific retention volume of any other solute S can be calculated if its retention relative to n-octane is known, because

$$(t_{N_S}/t_{N_{\text{n-octane}}}) V_g(\text{n-octane}) = V_g(S). \tag{12}$$

The relative retention data required for the above equations can be obtained from any isothermal gas chromatogram which has an adsorbed solute peak. The anomalous behaviour of two of the liquid phases examined by Rohrschneider (sorbitol and diglycerol) is easily explained by consideration of the extent of gas–liquid interfacial adsorption on these liquid phases.[18] The best correlation result was obtained with $x =$ 2,5-dimethyltetrahydrofuran, $y =$ 2-methyltetrahydrofuran, and $z =$ n-decane. The following constants were obtained: $A = -0.239$, $B = -0.873$, $C = -0.220$, and $D = 2.733$. This equation predicts the specific retention volume of n-octane on sixty-five liquid phases with an average error of 9% and a maximum error of 21%. In gas chromatography retention data obtained by comparing the retention of a compound RX with the retention of two similar reference compounds can also be used. Thus for the determination of retention indices[19] and the number of methylene units (MU),[17] the n-alkanes may be used as reference compounds.

In the determination of the equivalent chain length (ECL)[20] and the steroidic number (SN),[21] methyl esters of fatty acids and steroids, respectively, may be used as reference substances. All the above discussed quantities are logarithmic retention ratios:

$$(I/100, MU, ECL, SN) = N + Z \frac{\log r_{RX/N}}{\log t'_{N+Z} - \log t'_N}, \qquad (13)$$

where N and $N+Z$ represent the number of carbon atoms in the reference materials.

Rohrschneider[10] showed that the presentation of retention data in this form contains more information, and their determination can be made with the greatest precision and accuracy.

The retention index is a parameter very frequently used for the identification of compounds. If the esters of saturated fatty acids are used as reference materials in the ECL system, the same results for the ECL of the methyl esters of saturated linear fatty acids are obtained no matter what stationary phase is used. The same thing is valid for the steroidic system, in which androstane and cholestane have a steroidic number of 19 and 27, respectively. The disadvantage of using such systems is that it is difficult to transform the data obtained into thermodynamic quantities.

1.2 MOBILE AND STATIONARY PHASE CONTRIBUTIONS TO PEAK DISPERSION

In general, peak dispersion arises from the following factors:

(1) axial molecular diffusion;
(2) a finite rate of equilibration of solute between the mobile phase and the stationary phase in the column;

(3) non-equilibrium within the mobile phase which is continuously generated by the complex flow and velocity pattern within the column.

For unsorbed solutes the second factor is absent.

According to the "classical" theories of peak dispersion in chromatographic columns,[22-24] contributions to the height equivalent of a theoretical plate (HETP) from various processes in the column are regarded as additive in the form of a linear equation.

Giddings,[25] however, has pointed out that eddy diffusion and lateral mass transfer in the mobile phase co-operate in reducing the dispersion caused by uneven flow patterns across and along the column. Therefore their contribution to plate height should be expressed by a velocity-dependent, coupled eddy diffusion term. This coupling term is always less than the sum of the classical eddy diffusion and mobile phase mass transfer terms. The coupling effect is particularly important at high flow velocities, and results in significantly less peak broadening in the mobile phase than is predicted by the classical theory.

When the carrier gas behaves like an ideal gas it is considered to have a viscosity η which is independent of the pressure. The diffusion coefficient of the component in the gas phase $D_g = D'_g/p$ (diffusion coefficient of the component at 1 atm pressure, D'_g = constant at any temperature). Taking into consideration the above definition, the equation for the experimental plate height $\hat{H}$ is obtained from the elution time and the standard deviation of the peak in time units $(\hat{H} = L(\sigma_t/t)^2)$ in the form[26]

$$\hat{H} = H_g f_1 + C_s v_o f_2, \tag{14}$$

where H_g is the total gas phase contribution found locally within the column, C_s is the stationary phase mass transfer coefficient, v_o is the outlet velocity, and f_1 and f_2 are the pressure correction terms given by

$$f_1 = \frac{9}{8} \frac{(P^4-1)(P^2-1)}{(P^3-1)^2} \tag{15}$$

and

$$f_2 = \frac{3}{2} \frac{(P^2-1)}{(P^3-1)}, \tag{16}$$

where P is the inlet/outlet pressure ratio p_i/p_o.

Equations (14)–(16) are valid if the contributions of the stationary ($H_{s(z)}$) and mobile ($H_{g(z)}$) phases to the plate height at distance z from the inlet is given by eqns. (17):

$$H_{s(z)} = C_s v_{(z)},\tag{17a}$$

$$H_{g(z)} = H_g = \text{const., independent of } z.\tag{17b}$$

All chromatographic theories agree that eqn. (17a) correctly represents the contribution of the stationary phase to the plate height. It is very important, however, that this theoretical conclusion should be verified with experiments carried out over a wide range of carrier gas velocities. Dispersion of the bands due to processes that occur in the gas phase are considered to be caused by variations in the velocity of the mobile phase over the column cross-section.

The contribution of the gas phase to the plate height H_g can be expressed in the form of a general equation:

$$H_g/d_p = h_g = f[(vd_p/D_g), k', r_1, r_2, \ldots] = f(v, k', r_1, r_2, \ldots),\quad(18)$$

where d_p is the particle diameter, $r_1, r_2, \ldots$, are geometric parameters that reflect the general nature of the packing, k' is the column capacity ratio and is directly proportional to the ratio of the stationary phase volume and the mobile phase volume, $h_g = H_g/d_p$ is the reduced plate height, and $v = vd_p/D_m$ is the reduced velocity.

The main advantage that is offered by formulating the plate height in terms of reduced parameters is that experimental data obtained using different carrier gases and with different particle sizes are comparable.

Independent determination of H_g and $C_s v_o$ can be made as follows[26–28]: A parameter $X = v_o/D_{g_o} = v_o p_o/D_g'$ is defined, where D_{g_o} is the diffusion coefficient of the solute at the outlet pressure, D_g' is the diffusion coefficient of the solute in the carrier gas at the standard pressure (1 atm), and p_o is the outlet pressure.

In this instance, eqn. (14) can be written

$$\hat{H}/f_1 = H_g(X, d_p, k', \ldots,) + f_2/f_1 C_s X D_{g_o}.\tag{19}$$

As $f_2/f_1 D_{g_o} = f_2 D_g'/f_1 p_o = Y$,

$$\hat{H}/f_1 = H_g(X, d_p, k') + C_s X Y.\tag{20}$$

The two terms of eqn. (20) may be separated by varying Y, while X, d_p, and k' are maintained constant. Y can be varied either by changing the working pressure, which modifies the ratio f_2/f_1[29], or by changing the carrier gas, which modifies f_2/f_1 as well as D'_g.[27, 30, 31]

In practice, $\hat{H}/f_1$ and Y are represented as a function of X for two groups of working conditions 1 and 2, e.g. using nitrogen or helium as carrier gases. The experimental data should obey the equations:

$$(\hat{H}/f_1)_1 = H_g[X, \ldots,]+C_sXY_1, \tag{21a}$$

$$(\hat{H}/f_1)_2 = H_g[X, \ldots,]+C_sXY_2. \tag{21b}$$

Substracting these equations for the same value of X,

$$C_s = \frac{\Delta(H/f_1)}{X \Delta Y}, \tag{22}$$

where $\Delta(\hat{H}/f_1) = (\hat{H}/f_1)_2-(\hat{H}/f_1)_1$ and $\Delta Y = Y_2-Y_1$.

Equation (22) is valid only if $\hat{H}$ is the true column plate height and does not include contributions from equipment time constants and dead volume. An approximate correction for the latter is given by Saha and Giddings.[27]

The contribution H_g from the gas phase can be obtained from eqn. (20) by substracting C_sXY from $(\hat{H}/f_1)$:

$$H_g = \hat{H}/f_1-C_sXY. \tag{23}$$

Figure 1 shows the results obtained by Giddings and Schettler[26] with a column packed with Chromosorb W, a material that has wide pores. The figure shows the great distances that separate the $\hat{H}/f_1$ curves, indicating that C_s has a large value.

The value of C_s is equal to $1/X$ times the ratio of the gap in the $\hat{H}/f_1$ curves and the gap in Y at the same X-value, as can be seen from eqn. (22). The general form of the equation that gives us the mass transfer coefficient in the stationary phase C_s is

$$C_s = q\frac{k'}{(1+k')^2}\frac{d_s^2}{D_s}, \tag{24}$$

where q is a configuration factor that has the value $\frac{2}{3}$ for a uniform film (for a uniform deposition), d_s is the thickness of the stationary phase film, and D_s is the diffusion coefficient of the solute in the stationary phase.

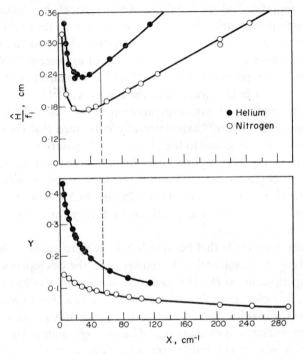

FIG. 1. Plate height data for a column of Chromosorb W (60–80 mesh) containing 20% by weight of dinonyl phthalate, in a copper column (395 cm ×4.8 mm i.d.) at 50°C; 1–5 μl of isobutylene injected. C_s is determined from the ratio of the two heavy lines. [26]

C_s will have its maximum value at $k' = 1$ [32] (assuming that the diffusion coefficient in the stationary phase does not differ too much from one solution to another). The establishment of the dependence between the thickness of the film and the degree of loading is not a simple problem. The good agreement between theory and experimental data (assuming that d_s is constant at low loadings and d_s is proportional to k' at high loadings) is hard to take into account when establishing the dependence of C_s on k' in the absence of complicating factors.

Saha and Giddings[27] tried to correlate the mass transfer coefficient in the liquid stationary phase (DNP) determined experimentally with physical properties and the structure of the support, in particular with the support's pore distribution. Using experimentally determined values for D_s, the results were satisfactory for Chromosorb W and Gas Chrom S but not for Chromosorb P and G. Such lack of agreement for Chromosorb P has been reported elsewhere[33, 34] and is explained by the discontinuities that exist in the structure of these supports.[27]

Using the method of pressure variation as well as different carrier gases, Knox and Saleem[28] experimentally confirmed that the contribution of the stationary phase to the plate height is governed by the simple equation $H_s = C_s v$. The C_s quantities for n-butane and cyclohexane on Chromosorb G (235 μm) with 1 and 6% squalane, have values between 2 and 5 ms (for temperatures between 25 and 39.5°C). But C_s does not decrease with the retention as predicted by theory for column capacity ratio k' greater than unity.

The authors conclude that because it is difficult to see how the theory can be wrong, the explanation is obscure, and the discrepancy calls for further experimental work. The most satisfactory test would be the one in which a homologous series was used which gave k' values over a wide range from about 0.1 to at least 10, and preferably 100. At the same time it would be necessary to measure the diffusion coefficients of the members of the series with considerable accuracy. Tests over a wide range of temperatures would also be most useful.

In gas–solid chromatography, when using solid stationary phases with a uniform surface, the mass transfer coefficient C_s is less than 10^{-6} s because of the rapid adsorptive–desorptive exchange.[26] Nevertheless, for solid stationary phases with a non-uniform surface, this value is multiplied by a heterogeneity factor whose value in turn depends on the distribution of energy centres and can be obtained only if the details of the surface are accessible.

The first quantitative treatment of mass transfer in the mobile phase was by van Deemter et al.,[22] who presumed that mass transfer in the gas phase occurs in pores or channels with a size one-fifth of the particle diameter. The contribution of such a small diffusion distance was assumed to be negligible. In contrast, Jones[24] concluded that the gas-phase contribution was dominant. The present author believes that this contribution comes from two terms: one describes the role of the diffusion in the moving gas, the other accounts for the stagnant gas presumably found within the support particles.

An approximate chromatographic theory was developed in order to explain these effects. The theory concerning mass transfer in the gas phase in capillary columns was outlined by Golay.[23] Applied to packed columns, the above theory is applicable only to flow in the single channels within the support which are comparable to open capillary tubes.[35]

The theoretical treatment of the contribution of non-equilibrium in the gas phase in packed columns to the plate height is very complicated. Difficulties arise because of the complex interstitial geometry of the packing material. But, in spite of that, there is a possibility of estimating this value by taking into account the structural characteristics of the porous media.

Based on the observation that large flow channels occur with a spacing of several particle diameters, a repeating unit of flow was chosen by Giddings[36] which consists of an inner core of circular cross-section where the gas flow is rapid and an outer annulus in which a moderate flow rate occurs through the interstices between the particles (Fig. 2).

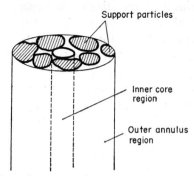

FIG. 2. Model used by Giddings[36] to approximate a repeating flow unit in chromatographic packing.

Giddings considers that for porous supports the major contribution of the gas phase to the plate height arises from interchannel velocity variation, and from non-equilibrium between the streaming interparticle carrier gas and the stagnant interparticle carrier gas. The expression given by Giddings for H_g is

$$H_g = \omega d_p^2 v / D_g, \tag{25}$$

where $\qquad \omega = 0.62 + 0.15[k'/(1+k')] + 0.025[k'/(1+k')]^2. \tag{26}$

The "effective" value of ω, especially applicable at the minimum near $v = d_q v / D_g \cong 1$, can be calculated by an equation proposed by Giddings[36] on the basis of the work of Dal Nogare and Chiu:[32]

$$\omega = 0.18(H_{g\,min}/d_p)^2 = 0.18(h_g)^2_{min}. \tag{27}$$

This shows that ω and the minimum reduced plate height $(h_g)_{min}$ both stem from the same processes. It was deduced theoretically, and later confirmed experimentally, that ω increases as the quantity of stationary phase deposited on the support increases.[36-38]

At higher temperatures (for columns packed with DNP on Chromosorb W and Gas Chrom S) a decrease of h_g (and obviously of the coefficient ω as well) was observed.[37] Experiments carried out on a column which contained 3% DNP deposited on Chromosorb G 40-60, 60-80, and 100-120 mesh, respectively, have demonstrated that $(h_g)_{min}$ and ω decrease with the decrease of the particle diameter of the support.[37]

In connection with the forecasting of the dependence upon fluid velocity of the contribution to the plate height of the gas phase, it can be concluded that the experimental data do not always agree with theory. Thus, in practice, some empirical equations that show the dependence of h as a function of v are much more useful.

Knox et al.[28, 39] showed that the empirical equations of Huber,[40] with slight generalization, could be used to represent the plate height dependence with a high degree of accuracy:

$$h = 2\gamma/v + (a + c/v^n)^{-1}, \tag{28}$$

where γ is an obstructive factor for the molecular diffusion, a and c are constants, and n has a value between 0.3 and 0.7. The second term can be simplified over small velocity ranges by a simple power term, giving

$$h = 2\gamma/v + c'v^{n'}. \tag{29}$$

The value of n' for Chromosorb G increases from 0.45 ($k' = 0$) to 0.65 ($k' = 30$) and is *ca.* 0.3 for glass bead columns.[28]

Experimental data of Knox and Saleem[28] confirm that slow diffusion within the stagnant intraparticle carrier fluid is the major cause of peak dispersion at high velocities, while the effects of uneven distribution of he stationary phase are probably more important at low velocities.

1.3 OPTIMIZATION OF SEPARATION PROCESSES

The efficiency of a chromatographic column is usually given in terms of the number of theoretical plates N or effective plates N_e:

$$N = 16(t/w)^2 = L/H, \tag{30}$$

$$N_e = 16(t'/w)^2 = L/H_e = [k/(1+k)]^2 N, \tag{31}$$

where t and t' are the retention and net retention times of the solute respectively, w is the width of the peak (the intersection of the tangents to the inflection points with the base line), k is the capacity ratio ($= t'/t_0$), and H and H_e are the HETP[41, 42] and the height equivalent of an effective theoretical plate (HEETP)[43], respectively:

$$H = (L/16)(w/t)^2, \tag{32}$$

$$H_e = (L/16)(w/t')^2 = [(1+k)/k]^2 H. \tag{33}$$

H and H_e are empirical constants that are more or less independent of the column length.[44] Although use of H_e has the theoretical disadvantage that it has the value zero, for an inert gas the disadvantage turns into an advantage as H_e is linked to a most important practical notion—the resolution.[43]

The dependence of the theoretical plate height on the velocity of the carrier gas is given in the form of a plot of H vs. v_o.

Van Deemter *et al.*[22] attempted to describe this dependence with an equation

$$H = A + B/v_o + Cv_o, \tag{34}$$

where A, B, and C are pressure independent coefficients (A = eddy diffusion, B = longitudinal diffusion, and $C = C_g + C_s$ represents the mass transfer term). The symbol A is in general use for discussing flow-independent contributions to H. Though this term has no theoretical foundation, it may arise from extra-column broadening.[45] This contribution should obviously be reduced to its lowest possible level. Besides the fact that the A, B, and C parameters do not have a completely true, physical significance, eqn. (34) does not verify the experimental data for greater intervals of velocity of the carrier gas. In all instances where a detailed study is needed, e.g. when it becomes necessary to compare the coefficients appearing with the plate height equations, a more sophisticated expression is used.

Precise experimental data and theory more or less agree:[25, 46]

$$H = \left(\frac{2D_g}{v_o} + C_g v_o + \frac{K' v_o}{D_g + e d_p v_o} \right) f_1 + C_s v_o f_2, \tag{35}$$

where f_1 and f_2 are pressure correction factors and are given in eqns. (15) and (16), K' is a constant that depends on the regularity of the column packing, and γ and e are subunitary numerical constants (for packed columns they have values of 0.1–0.3).[46]

In open tubular columns the flow pattern is uniform when the flow is laminar and there is neither eddy diffusion nor coupling; consequently $A = 0$.

The plate height for open tubular columns is given by the Golay equation (23), which is in good agreement with experimental data:

$$H = \left(\frac{2D_g}{v_o} + \frac{1 + 6k + 11k^2}{24(1+k)^2} \frac{r_o^2}{D_g} v_o \right) f_1 + \frac{k}{6(1+k)} \frac{d_s^2}{D_s} v_o f_2. \tag{36}$$

The main characteristic of eqns. (34)–(36) is the indication of the magnitude of the plate height in the sense that it has the minimal value for a given optimal value of v_o. At the optimal velocity of the carrier gas the efficiency of the column is maximal.

In gas chromatography it is very important that:

(1) the separation should be the most complete possible;

(2) the analysis time should be short;

(3) the technical difficulties should be minimal (temperatures and pressures over 30°C and 5 atm introduce experimental difficulties).

The degree of separation of two solutes (Fig. 3) is expressed by the resolution which is defined as[47]

$$R = 2[(t_2' - t_1')/(w_2 + w_1)] = 2[(t_2 - t_1)/(w_2 + w_1)]. \qquad (37)$$

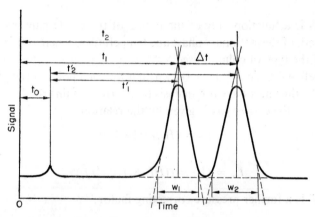

FIG. 3. Separation of a solute pair.

For two peaks with a small relative retention ($r_{12} < 1.1$) the efficiency of the column (N or N_e) is practically the same for both solutes; in this instance one may write[43]

$$R = \frac{(r_{12}-1)}{r_{12}} \sqrt{\left(\frac{N_e}{16}\right)} = \frac{(r_{12}-1)}{r_{12}} \sqrt{\left(\frac{L}{16N_e}\right)} = \frac{(r_{12}-1)}{r_{12}} \frac{t'}{w} \qquad (38)$$

or

$$R = \frac{(r_{12}-1)}{r_{12}} \frac{k}{(1+k)} \sqrt{\left(\frac{N}{16}\right)} = \frac{(r_{12}-1)}{r_{12}} \frac{k}{(1+k)} \sqrt{\left(\frac{L}{16H}\right)}. \qquad (39)$$

Equation (38) shows that the resolution of a pair of solutes with the same relative retention r_{12} is proportional only to $\sqrt{N_e}$. Besides, the right-hand side of this equation shows also that, from the point of view of resolution only, it is the net retention time t' of the solute that is of interest (and not the retention time t).

All this sustains the opinion of Purnell[43] who holds that for practical reasons it is advisable to use the H_e and N_e quantities instead of H and N.

In accordance with eqn. (38) the minimal value necessary for N (the separation factor S) for resolution is

$$N_{ne} = S = [4Rr_{12}/(r_{12}-1)]^2, \tag{40}$$

where S is a function only of the nature of the solute pair that is to be separated, of the stationary phase and temperature, whereas it is independent of the type of column used.

Purnell and Quinn[48] have shown that for each point in the plot of H_e vs. $\bar{v}$, the time necessary for analysis (t_{ne}, corresponding to N_{ne}) at a given velocity of the carrier gas is given by the relation

$$t_{ne} = (H_e/\bar{v})(1+k_2)S \tag{41}$$

or

$$t_{ne} = (1+k_2)(HL/K_e)(\eta/\Delta Pj')S, \tag{42}$$

where j' is a further pressure correcting factor:[49]

$$j' = \left(\frac{p_i/p_o+1}{2}\right)j. \tag{43}$$

Thus the time necessary for analysis depends on:

(a) the identity and the velocity of the carrier gas;
(b) the properties of the column, K_c (the permeability of the column) L, ΔP, and k;
(c) the temperature that is implied in the determination of S, k, H_e, η, and ΔP.

Corresponding to each t_{ne} from the H vs. $\bar{v}$ plot, a L_{ne} column length and a ΔP_{ne} (pressure drop) is necessary. These are given by eqns. (44) and (45):[49]

$$L_{ne} = H_e S, \tag{44}$$

$$\Delta P_{ne} = (\eta K_e j')H_e \bar{v}S. \tag{45}$$

From the H_e vs. $\bar{v}$ plot and using eqns. (41) to (45) the t_{ne} vs. ΔP_{ne} plot for a given temperature can be drawn. For each point on this plot there is a corresponding column length L_{ne} defined by eqn. (44); the minimum of the curve represents the optimal (minimal) time for analysis. From eqns. (40) and (41), it follows that

$$N/t = \bar{v}/H_e(1+k). \tag{46}$$

The factor N/t is being used in the expression for the speed of the analysis and it may be calculated from the H_e vs. $\bar{v}$ plot or even from the chromatogram. For this, one must assume that the analysis is finished at the maximum of the last peak.

Karger and Cooke,[50] in what they called time normalization chromatography, have shown experimentally that a two-parameter operation can achieve improved resolution in a constant analysis time. For the normalized time condition the effect of a parameter on the resolution of a given pair of solutes is studied by modifying this parameter and at the same time altering the column temperature so that the retention time of the second compound of the pair remains constant. Karger and Cooke have measured the variation of resolution of a given pair of solutes (mostly n-dodecane/n-tetradecane), under a normalized time condition, with the column length (for a given analysis time, the longer the column the higher the temperature), the carrier gas outlet velocity, and the support particle diameter. Experimentally they have shown that in length–temperature time normalization chromatography the resolution between two components reaches a maximum as the length varies. The variation of the resolution obtained for the pair n-dodecane/n-tetradecane was measured on a classical packed column and an open tubular column (0.25 mm i.d.). The maximum resolution obtained on the former column was 14 and the analysis time was 1200 s. The open tubular column gives a maximum resolution of 27 with an analysis time of 405 s. This means that with a capillary column a resolution twice as great could be obtained in one-third of the time that was needed for the packed column.

Guiochon[51, 52] has studied time normalization chromatography from a theoretical point of view and has recommended a set of rules that should be obeyed in order to obtain better separations. In general, the best performances (the highest resolution in the least time) will be given by a column when it is operating at a value of 3 for the capacity ratio k for a packed column and 1.5 for a capillary column, at a carrier gas flow rate between 1.4 and 2 times the optimum efficiency. Under these conditions it is advisable to use a support of 80–100 or 100–120 mesh.

Grushka[53] has shown that time normalization can be utilized to improve the peak capacity (the maximum number of solutes which can be resolved between the air peak and the last component) of a chromatographic system. Presuming that the number of the theoretical plates of the column are the same for all the components, the relation for the peak capacity n[53–55] can be written

$$n = 1 + \frac{\sqrt{N}}{H} \ln k_n. \tag{47}$$

From this relation it can be concluded that n depends on the number of plates N and upon the ratio of the retention time of the last component to the inert component ($k_n = t'_n/t_0$). Also n can be increased by careful choice of the column length, the velocity of the gas phase, the column temperature, and the amount of stationary phase.[53]

Presuming that (a) the relation between the relative retention and temperature obeys eqn. (48),

$$\frac{r-1}{r} = \frac{a}{T} - b, \tag{48}$$

where a and b are positive constants for a particular system; (b) the expression of the resolution is given by eqn. (39); and (c) the independence of H of the length and temperature of the column (this is for operation in the vicinity of the optimum velocity), Grushka[56] deduces eqn. (49), which features the dependence of k_{opt} on certain experimentally determinable parameters:

$$\frac{\Delta H}{R_g a} \left(\frac{r-1}{r} \right) + \ln k_0 = \frac{2(k_{opt}+1)}{2 - k_{opt}} + \ln k_{opt}, \tag{49}$$

where ΔH is the heat of solution of the normalized component, R_g is the gas constant, k_0 is the original capacity ratio, and r is the relative retention. It should be stressed that all the parameters required to solve eqn. (49) can be determined experimentally with one column operated at two different temperatures.

The optimum length of the column L_{opt} in the condition of time normalization is[56]

$$L_{opt} = \frac{L_0(1+k_0)}{(1+k_{opt})} , \qquad (50)$$

where L_0 is the initial length of the column.

The present author has shown that eqns. (49) and (50) agree with some experimental data obtained by Karger and Cooke[50] and by Grushka et al.[57]

Unfortunately, most of the theoretical studies concerning the optimization of experimental conditions in analytical gas chromatography were carried out for mixtures of two solutes (binary mixtures). But in practice such simple conditions are seldom met. Usually complex mixtures must be separated. In any complex mixture there is bound to be a pair of compounds which is more difficult to resolve than the others. However, quite often, this pair is not eluted last. It may happen that the most difficultly resolvable pair of compounds is not the same in all temperature or phase ratio ranges used in practice. Under these conditions the experimental solution of the optimization of an analysis may be extremely difficult, the use of a normalized time condition being no longer practical. Ideally, the optimization should be carried out on all the experimental parameters, the column design parameters (column type, length and diameter, nature of the support and the liquid phase, particle diameter, phase ratio) as well as the operational parameters (temperature, nature of carrier gas, flow rate, inlet pressure). Restricting the study to the optimization of only five of these parameters, namely column length, phase ratio, temperature, carrier gas, and flow rate, Guiochon[58] obtained the following relation for the analysis time:

$$t_{R,n} = 128R_{j,l}^3\left(\frac{r_{j,l}}{r_{j,l}-1}\right)^3 \left(\frac{1+k_l}{k_l}\right)^3 (1+k_n) \sqrt{\left(\frac{8\eta H_l^3}{3K_c p_o v_o}\right)}, \qquad (51)$$

where indices j and l refer to two components from the chromatogram, while n corresponds to the last component of the mixture.

According to eqn. (51) the analysis time is proportional to the cube of the resolution of the pair of compounds which are the most difficult to separate. This resolution should be kept at the minimum value which is necessary to obtain the desired results. For quantitative analysis the required resolution usually depends on the concentration ratio of the two solutes, especially if the major one is eluted first, and may vary between 1.2 and 3.0. The choice of the stationary phase is very important because the retention time is proportional to $(r_{j,r} - 1)^{-3}$ and becomes very large when r is near to unity.

Column permeability K_c is also very important. Packed and unpacked capillary columns are much faster than conventionally packed columns simply because they are more permeable.

The carrier gas flow rate should be optimized so that H_l^3/v_o is a minimum. This requirement usually results in a value of v_o which is about 1.5-fold larger than that for which H would be a minimum. If all the k_i values are independent of the phase ratio ($\beta = V_s/V_g$) the relative retention depends only on the temperature. In this instance it is possible to choose this temperature first to maximize r and then to optimize the phase ratio. This last step is easy if H_l is independent of k_l. Let $r_{l,n} = k_l/k_n$. The analysis time is proportional to $(1+k_l)^3(1+r_{l,n}k_l)/k_l^3$ which is minimal for

$$k_l = 1 + \sqrt{(1+3/r_{l,n})}. \tag{52}$$

The validity of the approximation in the sense that the capacity ratio is independent of the phase ratio might be subject to discussions as in some instances it may introduce significant errors. Besides, eqn. (51) does not explain the influence of certain column parameters such as the nature of the stationary phase and the support, the diameter of the column and the support. So an equation for the minimum analysis time that would take into account all experimental parameters is difficult to use in practice, mainly because of the lack of certain data such as diffusivity of the solute in the gas phase, diffusivity of the solute in the stationary phase and the distribution coefficient of the substance between the two phases.

It is obvious that, depending on the nature of the mixture to be separated, the right column material and shape, the nature of the support and its loading[59, 60] has to be decided.

Scott[61] proposed an experimental method for the determination of the concentration of stationary phase deposited on the solid support and the temperature and velocity of the carrier gas that are necessary for an optimum separation. The author used his method for the separation of a mixture of methyl-, ethyl-, and propylacetates with a resolution of 1.45. Glass columns were used (i.d. 3.5 mm, 39 cm long) packed with various concentrations of diethyleneglycol adipate (7.5, 10.0, 12.5, 15.0, and 17.5% w/w) deposited on a support with a particle size of 150–170 BS mesh.

The experimental method proposed by Scott has the following stages:

(a) Plotting resolution vs. gas velocity at different column temperatures for each column. From these graphs the velocity of the carrier gas that gives a resolution of 1.45 may be found for any working temperature and for every concentration of stationary phase.

(b) Plotting analysis time vs. gas velocity at different operating temperatures for columns carrying the various concentrations of stationary phase (one may thus determine the time that corresponds to the gas velocity that gives a 1.45 resolution at any temperature).

(c) Plotting the analysis time vs. temperature for columns carrying different amounts of stationary phase (every curve has a minimum that gives the optimum working temperature) in order to effect the analysis in minimum time.

(d) Plotting the minimum analysis time at optimum temperature vs. the percentage of stationary phase on the support (the minimum on this curve corresponds to the optimum concentration of stationary phase necessary for the analysis).

(e) Plotting the optimum temperature for minimum analysis time vs. percentage by weight of stationary phase on the support, in order to determine the optimum working temperature for the optimum concentration of stationary phase.

(f) Plotting gas velocity vs. column temperature required to give a resolution of 1.45 for different loadings of stationary phase. Using the optimum temperature for every degree of loading, a linear graph may be plotted of the gas velocity against percentage of stationary phase at optimum column temperature to give a resolution of 1.45 in the minimum time. From this last plot, using the optimum concentration of stationary phase (d), the optimum velocity of the carrier gas may be calculated.

Certainly the method described is quite laborious, but as soon as a method of analysis is perfected it can be applied to a wide range of samples.

In this chapter where the optimization of separation parameters was discussed, the possibilities that are offered by temperature programming or the variation of the pressure of the carrier gas were not considered. Unfortunately it is not yet possible to determine from theoretical considerations the instances in which temperature or pressure programming will give a practical reduction of the analysis time, that is, in which the gain in analysis time obtained by using these techniques is not over-compensated by the time lost at the end of the analysis to re-equilibrate the column under the starting conditions.

1.4 TYPES OF COLUMN USED IN GAS CHROMATOGRAPHY

Halász et al.[49, 62] were the first to discuss this problem. They used new types of columns. The columns used in gas chromatography can be classified into two groups: (I) packed columns; (II) open tubular columns.

Packed columns are distinguished from open tubular columns by the fact that in the former the solid support may itself be a stationary phase or it may be coated with a liquid stationary phase. The carrier gas thus cannot pass through the column axially in a direct line for more than a short distance. Only the wall of **open tubes** is coated with a solid or a liquid stationary phase, and so the shape of the free gas cross-section is (more or less) unchanged over the length of the column.

The packed columns comprise: (A) the well-known conventional packed columns, (B) packed capillary columns.

The packing material may be classified into:

(1) Porous particles such as Chromosorb, crushed firebrick, or adsorbents.
(2) Impenetrable core particles, e.g. glass beads coated with a thin film of liquid phase.
(3) An impenetrable core coated with a thin layer of a porous solid impregnated with a liquid phase or a thin layer of adsorbent.

The combination of these classifications leads to a more complete classification of packed columns. It has six classes:

I A1. Conventional packed columns (CP) made with porous particles. Although from the point of view of analysis time these columns do not represent the best means of separation, they are still used for the majority of analyses.

I A2. Glass-bead columns (B) were prepared for the first time by Callear and Cvetanovits.[63] Such columns proved to be very useful for the analysis of compounds with very high boiling point or which were very sensitive to the catalytic activity of the support surface. The difficulties that arose in their preparation as well as the appearance of capillary columns and deactivated supports has reduced somewhat the use of the glass-bead columns. Recent progress in the treatment of the surface of glass beads[64, 65] has brought about an even distribution of the stationary phase, which has made possible very good separations of high boiling point polar compounds.

I A3. Porous-layer bead columns (PLB) were first prepared by Halász and Horváth[66] in order to obtain columns that showed low resistance to mass transfer in the mobile phase and in the stationary phase. They deposited various materials of particle diameter 1 μm or less from solvents onto the surface of the non-porous glass or metal beads. The coating procedures are simple and the powdered particles are apparently held to the surface by van der Waals forces.

The advantage of porous-layer columns is their small phase ratio, and hence the prospect of lower temperatures of analysis than in conventional packed columns. On a column of glass beads coated first with zirconium dioxide and then with triethylene glycol, five components (methane, diethylether, acetone, benzene, and ethyl acetate)were separated in 30 s.[66] Interestingly, the stationary phase has an non-polar character although both components of the adsorbtion layer are polar. Excellent results have been obtained by Halász and Holdinghausen[67] using glass beads coated with 10% of a mixture of highly dispersed iron(III) oxide, 10% Carbowax 20M, and 1% potassium hydroxide. Preparation of this packing is highly reproducible, and the bleeding rate of the column is negligible at temperatures below 240°C. The high rate of mass transfer results in a very high efficiency (theoretical plate height as little as 0.2 mm) which allows very short analysis times.

I B1. Packed capillary columns (PC) were first proposed by Halász and Heine in 1962.[68] As their name indicates, they are packed columns, usually with an inner diameter comparable to that of open capillary tubes (0.25–0.50 mm). The particle diameter of the support is greater than one-third to one-fifth of the column inner diameter. These columns always allow faster analysis than CP. This improvement is mainly a result of their higher permeability. A number of very fast analyses of various mixtures such as light petroleum gas,[68, 69] light paraffins,[70] phenol and polymethylphenols,[71] and amino-acid derivatives[72] have been published. They match in resolution or time of analysis the performances obtained with open tubular columns.[71, 73, 74]

I B2. Glass-bead packed capillary columns (BPC) were first used by Guiochon.[52] Their performance is intermediate between packed capillary and bead columns.

I B3. Porous-layer bead capillary columns have never been reported. As the preceding column type, they are not listed in Halász's classification.[62]

Recently Cramers et al.[75] prepared packed columns of 0.6–0.8 mm inner diameter and up to 15 m long. Theoretical plate numbers of the order of 50,000 at moderate inlet pressures were obtained. The columns are coiled and subsequently packed under pressure and vibration, and all types of support, coated with any stationary phase, may be used.

These columns are known as (C) micro-packed columns (MCP).

Such columns have also been reported by others.[76-79] The relative peak broadening is remarkably low and the speed of analysis appears far better than that attainable on conventional packed columns.

The **open tubular** columns can be divided into two categories:

II 1. Conventional open tubular columns (COT). The tubes can be made of metal,[23] plastic,[80] or glass.[81] These columns yield analysis times one or two orders of magnitude smaller than CP columns. COT are the only columns which supply a practical analysis where more than a few tens of thousands of plates are needed. The analysis of complex mixtures such as crude or essential oils would be impossible on other types of columns.

II 2. Porous-layer open tubular columns (PLOT) were first reported by Halász and Horváth.[82] The tube wall is coated with a thin layer of a porous solid impregnated with a liquid phase. This allows the use of much smaller gas phase/liquid phase ratios than with conventional open tubular column with the same high permeability.

All the types of columns described above may be used in gas–liquid chromatography, but only CP, MCP, PLB, PC, and PLOT columns can be used in gas–solid chromatography, with suitable adsorbents.

The main operational parameter of the chromatographic columns is the specific permeability coefficient K_c, which is independent of the carrier gas, temperature, and column length. For ideal open tubes,

$$K_c = d^2/32, \tag{53}$$

where d is the internal diameter of the column.

For CP, K_c is given by[83]

$$K_c = (d_p^2/180)\varepsilon^2/(1-\varepsilon)^2, \tag{54}$$

where d_p is the average particle diameter and ε the interparticle porosity (for conventional packed columns, $\varepsilon \cong 0.4$). For such columns $K_c = d_p^2/1000$ is a good empirical approximation.

A measure of the performance of chromatographic columns is given by the performance index, IP:[23]

$$IP = 30.7 H_{min}^2 \frac{\eta}{K_c} \left[\frac{k^4}{(1+k)^3 \left(k+\frac{1}{16}\right)} \right]. \tag{55}$$

In the CGS system, IP is measured in poise (dyne/cm² s). The performance index is a function of k and it has a minimum at $k = 3$, the other parameters being constant.

As may be seen from eqn. (55), the relative retention or the separation factor does not appear in the definition of the performance index as this is strictly a column parameter. For this it was considered advisable to introduce a new parameter,[49] the performance parameter (PP) which is defined as:

$$PP = H_{opt}^2(1+k)(\eta/K_cj')S^2. \tag{56}$$

The separation factor S from eqn. (56) shows that the PP depends on the nature of the mixture to be separated.

Recently, Rohrschneider[84] proposed another quantity, Pf, the performance value, defined as

$$Pf = K_c\frac{N}{t_o} = K_cQ = K_c\frac{N}{t}(k+1) \cong \frac{K_c}{C}, \tag{57}$$

where Q is the so-called column quality $(= N/t_o)$, and for rapid separations is approximately inversely proportional to the C constant of the van Deemter equation. A large Pf value corresponds to a rapid separation.

In Table 1 a few properties of the main types of columns is presented.[52] It is always possible to prepare a column that has at least one parameter outside the range specified in Table 1. But 90% of the columns currently in use will meet the requirements specified in Table 1.

TABLE 1. SOME PROPERTIES OF THE MAIN TYPES OF COLUMN[52]

Column type	Permeability $(10^{-7}\,cm^2)$	Phase ratio V_g/V_l	Minimum plate height H_{min} (cm)	Optimum carrier velocity v_{opt} (cm/s)	Sample size (μg)
CP	1–10	4–200	0.05–0.2	5–20	10–1000
B	1.5–15	50–500	0.1–0.3	10–20	1–100
PLB	1.5–15	50–400	0.05–0.2	20–60	1–100
PC	5–40	10–300	0.05–0.2	10–40	1–50
COT	50–800	100–1000	0.03–0.2	10–100	0.1–50
PLOT	200–1000	20–300	0.06–0.2	20–160	1–50

CHAPTER 2

RETENTION DATA AND
THERMODYNAMIC VARIABLES

MARTIN[85] was the first to show the utility of gas chromatography in the study of the thermodynamics of the interaction of volatile compounds with a non-volatile solvent. This technique was further used in order to obtain physical data such as partition coefficients, vapour pressure, vaporization heat, free energy, enthalpy and entropy of organic compounds.[86-89]

2.1 PARTITION COEFFICIENT

The value of the partition coefficient is given by the ratio

$$K = \frac{\text{concn. of the solute per cm}^3 \text{ (or per g) of stationary phase}}{\text{concn. of the solute per cm}^3 \text{ of gas at the column temperature}}.$$

(58)

When the solute forms an ideal solution in the stationary phase, so that Raoult's law is obeyed, i.e. $p_2 = p_2^0 x_2$, where p_2 is the vapour pressure of the compound above the solution, p_2^0 is the vapour pressure of the pure compound, and x_2 is the mole fraction of the solute in the solution, then the equation for the partition coefficient can be written:

$$K = \frac{RT_c}{M_1 p_2^0},$$

(59)

where M_1 is the molecular weight of the stationary phase. If p_2^0 is measured in torr and T in °K, then the units of R are cm³ (torr)/(mole °K) and its value is 62,370.

In practice the solution formed by a volatile component and a non-volatile solvent does not behave ideally. Departure from ideality is expressed quantitatively as the ratio of the actual value of the vapour pressure above the solution to the ideal value of the vapour pressure. This ratio is the activity coefficient of the solute γ. The equation for the partition coefficient can then be written

$$K = \frac{RT_c}{M_1 \gamma p_2^0} = \frac{62{,}370 T_c}{M_1 \gamma p_2^0} . \tag{60}$$

Taking into account that $V_g = 273.16/T_c K$,

$$V_g = \frac{273.16R}{M_1 \gamma p_2^0} . \tag{61}$$

It is obvious that when the activity coefficients are smaller than unity, the vapour pressures are smaller than for ideal behaviour, so that the partition coefficients are greater than for ideal behaviour. When the activity coefficient is greater than unity, the interpretation correspondingly changes.

Equations (60) and (61) are very important for gas chromatography because they link the partition coefficients or the specific retention volumes with certain parameters that can easily be evaluated.

The practical importance of these equations is that by using the values of the activity coefficients and the vapour pressures of a given mixtures of compounds, the stationary phase that will give maximum relative retention can be chosen.

Equations (60) and (61) consider the non-ideality of the liquid phase, but in the meantime the ideality of the gas phase is assumed. The imperfections of gases or gas mixtures can be described in terms of virial coefficients, the non-zero values of which describe the departure of the gas from Boyle's law in terms of a virial series.

It can be shown that the equation for the partition coefficient K corrected for gas phase non-ideality is given by[91, 92]

$$K = \frac{RT}{M_1 p_2^0} \exp\left[\frac{(V_2^0 - B_{11})p_2^0}{RT} + \frac{p^0(2B_{12} - V_2^\infty)}{RT}\right], \tag{62}$$

where V_2^0 is the molar volume of the pure liquid solute, V_2^∞ is the partial molar volume at infinite dilution in the solvent, p^0 is the total pressure, B_{11} is the second virial coefficient of the pure solute vapour, and B_{12} is the second virial coefficient of interaction between the solute vapour and the carrier gas.

Adlard *et al.*[93] took into account the non-ideality of the gas phase for the accurate determination of thermodynamic properties by gas chromatography, and reached the conclusion that the practicality of the method is not affected if the gas phase is considered as ideal.

As can be seen from eqn. (60), the partition coefficient is affected by temperature because of the variation of three parameters: temperature itself, the vapour pressure of the pure compound p_2^0, and the activity coefficient.

Thus it may be concluded that the greater the vapour pressure of the compound at the column temperature, the smaller the partition coefficient. The separation of different solutes with very similar values of their activity coefficients on a given stationary phase should be accomplished using the differences in their vapour pressures, as the solutes will elute from the column in order of their increasing boiling point. There is no need to emphasize the importance of the knowledge of the exact vapour pressures from the point of view of gas chromatography. Some useful data may be found in the references.[94-98] For the compounds whose vapour pressure is not given in the references, vapour pressures can be calculated if the boiling points are known.[99]

When two components have the same vapour pressure, the difference in the partition coefficients will be determined by the difference in their activity coefficients in a given stationary phase. Table 2 gives the values of the partition coefficients of some compounds at 70°, 90°, and 110°C when using di-2-ethylhexylphosphoric acid as the stationary phase (stain-

TABLE 2. PARTITION COEFFICIENTS OF SOME COMPOUNDS ON DI-2-ETHYLHEXYLPHOSPHORIC ACID[100] AT DIFFERENT TEMPERATURES

Compound	70°C	90°C	110°C
Benzene	382.9	200.4	112.5
Cyclohexane	262.8	145.0	83.3
Propanone	78.4	45.4	28.5
Butanone	221.8	121.7	71.8
Methylene chloride	112.1	64.8	40.2
Trichloromethane	287.9	155.2	92.1
Tetrachloromethane	291.7	161.0	97.4

less steel column, 2 m long, 6 mm i.d.; stationary phase loaded 15% on a 60–80 mesh Chromosorb W support.[100]

As can be seen from Table 2, the compounds that have very similar vapour pressures (e.g. the boiling point for benzene is 80.1°C and for cyclohexane is 80.8°C) have different partition coefficients because of their different interactions with the stationary phase. The very similar values of the partition coefficients that have been obtained for trichlormethane (b.p. 61.2°C) and tetrachlormethane (b.p. 76.8°C) may easily be explained by the different shape of the two molecules and thus their different interaction with the stationary phase that compensates for their different behaviour caused by their different boiling points. The trichlorethane may interact with the stationary phase by hydrogen bond formation.

It is very important that the partition coefficient can be related to equilibrium processes. This was first demonstrated by Porter et al.[101] The possibility of calculating the stability constants of olefin–silver ion complexes in ethylene glycol solutions from determinations of partition coefficients has been described.[102–104] Making the plausible assumption that the ratio of olefin (B) to silver ion is 1 : 1, the relationship between the stability constant $K_s = [BAg^+]/[B][Ag^+]$ and the partition coefficient is expressed by:

$$K_s = \frac{(K_1 - K_2)}{K_2[Ag^+]}, \qquad (63)$$

where K_1 is the partition coefficient of the olefin between the silver nitrate-glycol solution and the gas phase, and K_2 is the partition coefficient of the olefin between that solution and the gas phase. Equation (63) is valid only for low concentrations of silver nitrate, where the "salting out" effect is negligible, and for a fast reaction between the silver nitrate and the olefin.

An excellent general discussion concerning the study of chemical equilibria through gas chromatography was given by Purnell.[105] He shows that the simplest gas chromatographic system in which complexing can occur is the two-component system, solute (X) and solvent (S). An additive (A) may be added to the solvent (S) that may react with the solute (X) or the solvent (S). In the meantime, solute (X) may polymerize or depolymerize in the solution.

Thus four main types of interactions exist (the interactions of the A–A or S–S type are not considered because they they are either trivial or too complex).

Class A. Solute X reacts with additive A to give complexes of the type:

$$\text{(i)} \quad AX_n; \qquad \text{(ii)} \quad XA_m; \qquad \text{(iii)} \quad A_mX_n,$$

where both n and $m \geqslant 1$. In each type of system, several complexes may coexist, e.g., AX_n and AX_{n+1}.

Class B. Solute X reacts with solvent S to give complexes of the type:

$$\text{(i)} \quad SX_n; \qquad \text{(ii)} \quad XS_p; \qquad \text{(iii)} \quad S_pX_n,$$

where both n and $p \geqslant 1$. Again several complexes may coexist.

Class C. Solute X reacts with itself:

(i) X polymerizes in solution; (ii) X depolymerizes in solution.

Class D. Additive A reacts with solvent S to give complexes of the type:

$$\text{(i)} \quad SA_{m,\,m+1,\,\text{etc.}}; \qquad \text{(ii)} \quad AS_{p,\,p+1,\,\text{etc.}}$$

This classification is independent of the type of interaction occurring, and the theoretical treatment in each instance is common to reactions such as conventional cation–ligand complexing, hydrogen bonding or the charge-transfer interaction of organic molecules,

Purnell derived the equations that give the values of the stability constants for each system represented and discusses the technical difficulties that arise in the determination of the partition coefficients for the more complicated examples (especially for values of n, m, and p greater than unity). For class A, when $m = n = 1$, the expression obtained for the stability constant of the complex is the same as eqn. (63). It is obvious that from eqn. (63) as well as from the equations obtained by Purnell[105] the so-called stoichiometric stability constants may be obtained.[106]

For purely analytical purposes, such constants are of prime value, as the partition coefficient as used in chromatography is itself stoichiometric rather than thermodynamic. Thus it may be shown that for the separation of two solutes, when using a stationary phase with a relative retention $r_{12} = 1$, a column with an infinite number of theoretical plates is necessary. If an additive is introduced into the stationary phase that allows a ratio of stoichiometric stability constants $K_{s_1}/K_{s_2} = 1.25$ to be obtained, only 900 plates are needed for the separation of the two compounds. For comparison purposes, however, thermodynamic constants are required. The true thermodynamic constant K_t is defined in terms of activities a. For the olefin–silver ion complex,

$$K_t = \frac{a_{\text{complex}}}{a_{\text{Ag}^+}\, a_{\text{olefin}}} = K_s\left[\frac{\gamma_{\text{complex}}}{\gamma_{\text{Ag}^+}\, \gamma_{\text{olefin}}}\right]. \tag{64}$$

Gil-Av and Herling[103] showed that, as the activity coefficient of a solute (γ_{olefin}) can easily be determined by gas chromatography, it is possible to calculate from the experimental data the value K_t' which is proportional to the thermodynamic equilibrium constant

$$K_t = K_s/\gamma_{\text{olefin}} = K_t(\gamma_{\text{Ag}^+}/\gamma_{\text{complex}}). \tag{65}$$

In eqn. (65) the only unknown is the ratio $\gamma_{\text{complex}}/\gamma_{\text{Ag}^+}$. To a first approximation this ratio can be considered as being constant for a given ionic strength, and to change relatively little with the nature of the olefin. K_t is sometimes known as the mixed stability constant, and as it is a function of the activity coefficient value of the olefin, it may differ significantly from the stoichiometric stability constant K_s.[103]

De Ligny[107] recently suggested a method for the determination of complex association constants from gas chromatographic data in terms of the activities of the reactants. The method has been applied to silver ion complexes with substituted alkenes in ethylene glycol at 40°C.[108] The novel feature of this method is that the equilibria involved are represented in terms of activities (rather than concentrations, as is usually done), and that a reference solute is used. This has the advantage that the activity coefficients partially cancel out and only relative retention volumes are required. The equation proposed by de Ligny is

$$\frac{\left[\left(\dfrac{V_x}{V_{x^*}}\right)_{S+A}\left(\dfrac{V_{x^*}}{V_x}\right)_{S+A'}-1\right]}{X_A} = K_a\left(\frac{\gamma_x\gamma_A}{\gamma_{Ax}}\right)_{S+A'}, \tag{66}$$

where V is the retention volume, corrected for the gas hold-up in the apparatus at the mean column pressure and the column temperature;
x is a complex-forming solute (e.g. an alkene);
x^* is a solute that is closely related to x but does not form a complex (e.g. an alkane);
S is a solvent (such as ethylene glycol);
A is a complex-forming species, dissolved in S (e.g. silver ions);
A' is a compound that is similar to A but does not form a complex (such as lithium or sodium ions);
X_A is the mole fraction of A;
K_a is the association constant of the complex Ax; and
γ is the activity coefficient.

From eqn. (66), it follows that, apart from X_A, only the ratios of the retention volumes of x and x^* on columns containing the complex-forming stationary phase (S+A) and the reference stationary phase (S+A') must be measured. Data concerning the amount of the stationary phase, inlet and outlet pressures, room temperature, and gas velocity are not required. Further, any effects of adsorption of the solutes on the stationary phase or on the support of non-zero sample size and of non-ideality of the column on the retention volume, will largely cancel out.

Equation (66) has been used to determine the association constants of complexes of substituted alkenes with silver ions in ethylene glycol with a precision of about 5%.[108]

Vitt et al.[109] have studied the dependence of the partition coefficients of acetone, tetrahydrofuran, dimethoxyethane, ethyl acetate, ether, and hexane at 50°C on the mercury(II) bromide concentration in different stationary phases (polyethylene glycol 400, dinonylphthalate, and octadecyl bromide). They concluded that eqn. (63) is valid only when using poorly solvating solvents and at a concentration of complexing agent smaller than its solubility at the working temperature.

The influence of the solvent on the formation constants of electron donor–acceptor complexes as measured by gas–liquid chromatography has also been studied.[110–112] Purnell et al.[110] have examined the influence of an inert solvent on the experimentally determined stoichiometric formation constants of a series of simple aromatic hydrocarbons complexing with 2,4,7-trinitro-9-fluorene. They found a substantial change in the formation constants with a change of solvent. Eon et al.,[111, 112] when discussing the same problem, find that it is necessary to introduce an entropy correction to account for the different molecular sizes of additive, solute, and solvent. They show that K_s [from eqn. (63)] is equal to the stoichiometric formation constant only when the molar volumes of the additive and the solvent are equal. In no other instance can K_s equal the ratio [AB]/[A] [B], and in certain examples the difference can be as much as 50%. Other data concerning the stability constants of different complexes obtained from gas chromatographic determinations of partition coefficients may be found in the references.[110–120]

The advantages of using gas chromatography for the study of chemical equilibria are classified by Purnell[105] as follows:

(i) As gas chromatography is a sensitive separation technique, trivial amounts of relatively rare and/or impure compounds can be used.
(ii) As an extension of (i), by using mixtures, many measurements can be made in a single experiment.
(iii) Temperature can be varied at will, which is unusual with conventional methods.

(iv) Non-aqueous systems are ideal for gas chromatography, whereas in other methods they offer considerable difficulties, as is evidenced by the paucity of data for stability constants in non-aqueous media in the references.

(v) From (iv) it follows that hydrolysable or water-insoluble complexes may be studied, and so previously unexplored regions are opened up.

(vi) The method is technically simple.

2.2 ACTIVITY COEFFICIENTS

The equation that links the activity coefficient of the solute at infinite dilution γ_2^∞ to the specific retention volume V_g^0 is[121, 122]

$$\gamma_2^\infty = \frac{17.04 \times 10^6}{M_L f_2^0 V_g^0}, \tag{67}$$

where M_L is the molecular weight of the liquid stationary phase and f_2^0 is the fugacity of the pure saturated solute vapor in millimetres of mercury.

One of the hypotheses that facilitated the derivation of eqn. (67) is that no solute–solute, solute–carrier gas, or carrier gas–carrier gas interactions exist in the vapour phase. It has been shown that this assumption, that implies an ideal behaviour in the gas phase, is partly valid when the carrier gas is helium.[123]

But in order to obtain the true activity coefficient γ_2^∞ corrected for imperfections in the vapour phase, the following equation should be used:[92]

$$\ln \gamma_2^\infty = \ln \gamma_p^\infty + \frac{\bar{p}}{RT_c}(2B_{12} - V_2^0) - \frac{p_2^0}{RT_c}(B_{22} - V_2^0), \tag{68}$$

where

$$\gamma_p^\infty = \frac{17.04 \times 10^6}{M p_2^0 V_2^0}, \tag{69}$$

p_2^0 is the vapour pressure of the pure saturated solute in millimetres of mercury, $\bar{p}$ is the average column pressure, V_2^0 is the molar volume of the pure solute, B_{22} is the second virial coefficient of the solute vapor, B_{12} is the second virial coefficient for the interaction between the solute and the carrier gas, R is the gas constant, and T_c is the column temperature in °K. When using helium as carrier gas, eqn. (68) simplifies to

$$\ln \gamma_2^\infty = \ln \gamma_p^\infty - \frac{p_2^0 B_{22}}{RT_c} . \tag{70}$$

Equation (70) is identical to eqn. (67) because

$$\ln f_2^0 = \ln p_2^0 + \frac{p_2^0 B_{22}}{RT_c} . \tag{71}$$

Martire and Pollara[123] underline the fact that it is not permissible to replace f_0^2 by p_2^0 in eqn. (67) for the simple reason that sometimes the non-ideality of the pure solute can be significant.

If the values for B_{22} cannot be found in the references, they can be calculated from the Berthelot equation,[124] which represents the behaviour of non-polar vapours fairly accurately, as was demonstrated by Lambert et al.[125] Another means of calculating B_{22} is by using the equation developed from the principle of corresponding states[126-128] which is applicable to non-polar and certain slightly polar vapours. Thus the value of B_{22} may be obtained from[127, 128]

$$B_{22}/V_c = 0.430 - 0.886(T_c'/T_c) - 0.694(T_c'/T_c)^2 - 0.0375(n-1)(T_c'/T_c)^{4.5}, \tag{72}$$

where n is the effective carbon number, T_c' and V_c are the critical temperature (°K) and volume (cm³/mole) of the solute. If the critical constants are not accesible,[129] they may be calculated using the parachor equations. For example, for aromatic compounds[130]

$$T_c'/T_b = 2.640 - 0.4634 \log P \tag{73}$$

and

$$V_c = 2.92 P^{1.2}/T_c'^{0.3}, \tag{74}$$

where P is the molecular parachor and T_b the normal boiling point (°K) of the solute.

For polar compounds, virial coefficient data are rather scarce and the approximate methods for estimating them are not valid. When this data is not available, eqn. (69) can be used, which offers a reasonable approximation to the true activity coefficient. In such instances, the value of the activity coefficient is a measure of non-ideal behaviour in both the gas phase and in the solution rather than in the solution alone.[123]

Besides the corrections concerning non-ideal behaviour in the gas phase, for the very accurate calculation of activity coefficients by gas chromatography the possibility must be taken into account of adsorption of the solute at the gas–liquid interface in certain systems (e.g. non-polar solute–polar stationary phase). Such phenomena may introduce great errors in the determination of the specific retention volume, the value for which must be used in the equation for the activity coefficient.

The accurate knowledge of the activity coefficients for different systems is very useful for the analytical chemist as it allows the best stationary phase for a given application to be selected. This is a valuable advantage especially when the gas mixture contains compounds with very similar vapour pressures. At the same time, gas chromatography can be regarded as one of the most rapid and accurate methods for the determination of activity coefficients and other constants.

The values of the activity coefficients of some normal saturated hydrocarbons, cycloparaffins, alkenes, and aromatic hydrocarbons in solvents of various polarities are available in the literature.[131–146] The activity coefficients calculated using gas chromatography proved to be in agreement with the results obtained by static methods (McBain balance).[132, 147, 148]

The importance of the determination of the activity coefficients is underlined by the correlation that exists between their value and other thermodynamic quantities, a correlation that helps the understanding of interactions in solutions. Thus the activity coefficient at infinite dilution is related to the excess partial molar free energy by

$$\Delta G^E = RT \ln \gamma_0^\infty. \tag{75}$$

But at the same time,

$$\Delta G^E = \Delta H^E - T \Delta S^E, \tag{76}$$

where ΔH^E is the excess partial molar enthalpy and ΔS^E is the excess partial molar entropy. By assuming that ΔS^E and ΔH^E are constant for a small change in temperature, and differentiating eqns. (75) and (76) with respect to the reciprocal temperature,

$$\Delta H^E = R \frac{d \ln \gamma_0^\infty}{d(1/T)}. \tag{77}$$

Thus by studying the thermal variation of the activity coefficient at infinite dilution by gas–liquid chromatography, the excess partial molar enthalpy may be determined. When the excess partial molar enthalpy and free energy are known, the excess partial molar excess entropy may be calculated, using eqn. (76).

Although, in the light of the above statements the importance of activity coefficients for the analytical chemist is obvious, at the present time data are not available that would allow the best stationary phase for a given mixture of components to be chosen. Gas chromatography has been used until now as a rapid and accurate method for the determination of the activity coefficients that were necessary for the interpretation of different theories of solutions. The theory of solutions is one of the basic problems of physical chemistry. It is a theory that makes possible the interpretation and calculation of the thermodynamic properties of mixtures of non-electrolytes. Although a great number of papers has been devoted to this study, there is not yet any general theory of solutions. The more elaborate and sophisticated statistical mechanical theories require much information about the proporties of the pure component, and are usually strictly applicable only to systems involving simple non-polar molecules. At the other extreme are the purely empirical relationships which may describe the properties of various classes of solutions well enough, but require much experimental data to establish the particular class correlation. Furthermore, having no theoretical foundation, they cannot easily be extended to more generalized forms. There are, however, a number of solution theories which, although they are not highly exact, have a sound theoretical basis, and their parameters are open to physical interpretation.

Three such theories are the quasi-lattice theory, regular solution theory, and a recently developed "perturbation approach". The determination of activity coefficients by gas–liquid chromatography made possible the verification of these theories for several classes of mixtures of hydrocarbons.[144]

For solutions consisting of molecules of widely differing size, the deviation from Raoult's law is determined by two factors: (a) the difference in size between the solute and solvent molecules, and (b) disparities in the intermolecular forces.

In the limit of random mixing, this deviation can be expressed as the sum of two contributions to the logarithm of the solute activity coefficient ($\ln \gamma_2$):[149]

$$\ln \gamma_2 = \ln \gamma_2^{\text{ath}} + \ln \gamma_2^{\text{th}}, \tag{78}$$

where $\ln \gamma_2^{\text{ath}}$ (the athermal part) is associated with the statistical effect arising from the size difference between the molecules, and $\ln \gamma_2^{\text{th}}$ (the thermal part) is associated with the interchange energy arising from interactions between the molecules.

The magnitude of the athermal contribution to the activity coefficient is given[149, 150] by

$$\ln \gamma_2^{\text{ath}} = \ln 1/r + (1 - 1/r), \tag{79}$$

where r is taken as the ratio of the molar volume of the solvent to that of the solute (V_1^0/V_2^0).

It is found that the contribution of (a) to the excess entropy term is always positive. In the light of the statistical treatment of the solutions, a positive value implies that the number of possible configurations is greater than that corresponding to an ideal solution. This inequality in the volumes of molecules then makes a negative contribution to $\ln \gamma_2^{\infty}$, leading to a smaller solute partial pressure than would be predicted by Raoult's law.

The thermal contribution to the activity coefficient ($\ln \gamma_2^{\text{th}} = \chi$), the so-called interaction parameter, is strictly a free energy term and may be calculated by substracting the athermal contribution from the logarithm of the value of the experimental activity coefficient.

Contact point lattice treatement[151, 152] shows that for a binary mixture of molecules, 1 and 2, with two types of contact points a and b, the equation that gives the thermal contribution to the excess partial molar free energy at infinite dilution is

$$\ln \gamma_2^{th} = \chi = C_2(f_1^a f_2^b + f_2^a f_1^b - f_1^a f_1^b - f_2^a f_2^b)\frac{\omega^{ab}}{kT}, \tag{80}$$

where C_2 is the total number of contact points of the solute molecule, f_i^j is the fraction of contacts of molecule i which are of type j, and ω^{ab} is the interchange energy for the formation of an ab contact pair (ω^{ab} is a measure of the deviation of ab–ab interactions from the arithmetic mean of aa and bb interactions, i.e. $\omega^{ab} = \frac{1}{2}(\omega_{aa} + \omega_{bb}) - \omega_{ab}$).

For n-alkane solutes in n-alkane solvents, where there are only methyl and methylene types of contacts points, eqn. (80) reduces to

$$\ln \gamma_2^{th} = \chi = C_2(f_1^a - f_2^a)^2 \left(\frac{\omega^{ab}}{kT}\right), \tag{81}$$

where f_1^a and f_2^a refer to the fraction of hydrogen atoms belonging to methyl groups in the solvent and solute (one contact point is attributed to each hydrogen atom, and methyl and methylene hydrogens are considered as being different[152]) and ω^{ab} is the methyl-methylene interchange energy.

Using this approach, attempts to treat $\equiv$ C—H and —CH$_2$—hydrogens differently yielded results which were unrealistic and inconsistent. On the other hand, treating methylene and methyne hydrogens as equivalent, yielded interchange energies considerably lower than those obtained for n-alkane mixtures. Furthermore, these values decreased with increased branching.[144, 152]

Two contact points are attributed for every double bond in an n-alkene. Thus an ethylenic group (—CH=CH$_2$—) is assigned a total of five contact points, and all these contacts are treated equivalently without distinguishing π-electrons from ethylenic hydrogens atoms.

Benzene is considered to have twelve contact points, six for the hydrogen atoms and six for the π-electrons. These contact points are regarded simply as benzene contacts without distinguishing between π-electrons and hydrogen atoms.[153]

Toluene, ethylbenzene, and the xylenes thus have two types of contact points, — benzene type and aliphatic hydrogen types, which are treated as the corresponding aliphatic hydrogen atoms in alkanes. Comparisons of experimental (obtained through gas–liquid chromatography) and theoretical interaction parameters using lattice theory for alkane solutes and benzene in n-alkane solvents have been made.[152, 154, 155] Those for alkanes, alkenes, and aromatic compounds are given[144] in Table 3.

TABLE 3. COMPARISON OF EXPERIMENTAL AND THEORETICAL INTERACTION PARAMETERS USING LATTICE THEORY

Values of $\ln \gamma_2^{th}$ (theor)—$\ln \gamma_2^{th}$ (exptl)[144]

Solute	Solvent		
	$n\text{-}C_{24}H_{50}$	$n\text{-}C_{30}H_{62}$	$n\text{-}C_{36}H_{74}$
n-Hexane	+0.002	+0,019	+0.006
n-Heptane	−0.005	+0.019	+0.011
n-Octane	−0.024	+0.014	+0.000
n-Nonane	−0.081	+0.005	−0.027
1-Hexene	+0.007	+0.002	−0.010
1-Heptene	+0.005	+0.013	−0.009
1-Octene	−0.007	+0.016	−0.023
1-Nonene	−0.007	+0.017	−0.014
Benzene	−0.011	+0.009	−0.006
Toluene	+0.013	+0.017	−0.009
Ethylbenzene	−0.002	+0.001	−0.024
Cumene	−0.037	−0.011	−0.023
p-Xylene	+0.022	+0.035	+0.016
m-Xylene	−0.005	+0.007	−0.012
o-Xylene	−0.036	−0.019	−0.043

As can be seen from Table 3, for the majority of systems the values were in agreement to within twice the probable experimental uncertainty.

The theory of regular solutions[156] gives the following expression for the thermal contribution to the activity coefficient at infinite dilution:

$$\ln \gamma_2^{th} = \chi = \frac{V_2^0}{RT}(\delta_1 - \delta_2)^2, \tag{82}$$

where V_2^0 is the molar volume of the solute and δ_1 and δ_2 are the solubility parameters (for the stationary phase and for the solute, respectively).

The solubility parameter may be determined from

$$\delta_2 = \left(\frac{\Delta E_2^v}{V_2^0}\right)^{1/2} = \left(\frac{\Delta H_2^v - RT}{V_2^0}\right)^{1/2}, \tag{83}$$

where ΔE_2^v and ΔH_2^v are, respectively, the molar energy and enthalpy of vaporization of the pure solute.

Molar volumes of solutes can be calculated from the published densities given by the law of rectilinear diameters, and molar enthalpies of vaporization of solutes can be calculated from published Antoine constants[98] (B and C) using the equation

$$\Delta H_2^v = 2.303 \ RB\left(\frac{t + 273.2}{t + C}\right)^2, \tag{84}$$

where t is the temperature in °C.

It is well known that Hildebrand's regular solution theory[156] is applicable to solutions of non-polar or slightly polar molecules of equal size. These solutions possess a zero excess entropy term (ideal entropy of mixing) and contain a random distribution of molecules (no orientational or associative effects).

The consistency and utility of combining the Hildebrand and Miller–Guggenheim theories[156, 157] for mixtures of molecules of unequal size has been demonstrated by Martire[122, 136] for non-polar aliphatic solutes in the slightly polar aromatic solvent di-n-butyl phthalate. Martire shows that the combined theory in the form of eqn. (85) should provide a useful approach to the prediction of activity coefficients for this class of systems:

$$\ln \gamma_2^\infty = \left(\ln \frac{1}{r} + 6 \ln \frac{6r}{5r+1}\right) + \frac{V_2^0}{RT}(\delta_1 - \delta_2)^2. \tag{85}$$

Everett and Munn[134] similarly found that eqn. (85) represents the best available experimental data with an accuracy at least as good as that achieved with equations based on more sophisticated theories. This is strengthened by the work of Everett and Stoddart[132] and of Gainey and Pecsok[158] who determined the activity coefficients at infinite dilutions of C_5–C_8 hydrocarbons in n-alkyl benzenes.

Recently, Luckhurst and Martire[159] have proposed that the interactional contribution to the solute activity coefficient can be obtained by perturbation treatment based upon an earlier statistical approach by Longuet-Higgins.[160] Consider a reference solute and a "perturbated" solute in a common solvent. These two solutions are assumed to differ only in the strength of the intermolecular attractions between solute molecules and between such molecules and the common solvent. Under identical conditions of temperature T, pressure P, and mole fraction of solute χ, the interaction parameter is given by where the primed quantities

$$\ln \gamma^{th} = \chi = \chi' + \frac{V_2'^0}{RT}\left[\frac{E_1}{V_1^0} - \frac{E_2'}{V_2'^0}\right]\left[\frac{T_2^c}{T_2^{c'}} - 1\right], \tag{86}$$

refer to the reference solute and the unprimed quantities to the perturbed one; E_i, V_i, and T_2^c are the intermolecular cohesive energy, the molar volume and the critical temperature of component i, respectively, and 1 and 2 refer to the solvent and solute. For large molecules, T_2^c is replaced by $(T_2^c/V_2^*)^{1/3}$, where V_2^* is the molar volume of the component at temperature $T_2^* = 0.6T_2^c$.[161]

The final conclusion is that the interactional contribution for a series of solutes in a given solvent at the same temperature should bear a linear relationship to $(T_2^c/V_2)^{1/3}$:

$$\ln \gamma^{th} = \chi = \alpha + \beta[T_2^c/V_2^*]^{1/3}, \tag{86a}$$

where α = the intercept and β = the slope of the plots of $\ln \gamma^{th}$ against $(T_2^c/V_2^*)^{1/3}$. A negative slope means that the cohesive energy densities $(-E_i/V_i^0$ or $\Delta E_i^v/V_i^0)$ of these solutes are less then that of the solvent (n-alkanes, monobranched, and dibranched alkanes, 1-alkenes, n-$C_{30}H_{62}$) and positive slopes mean that the solute cohesive energy densities are larger than that of the solvent (monosubstituted and disubstituted benzenes, n-$C_{30}H_{62}$).[144]

It must be remembered that the possibility of calculation of activity coefficients with the help of the theories discussed above is only possible for systems formed of non-polar or slightly polar solutes and stationary phases. There is a series of systems for which the physical parameters featured in the appropriate equations are unknown.

Many of the stationary phases used in gas chromatography for the separation of different compounds do not have a definite chemical structure (they are polymers), but when a quantitative estimation of the activity coefficients *is* possible on the basis of molecular structure of the solute and stationary phase, this is always very useful for the analytical chemist.[162–165]

2.3 LIQUID SURFACE EFFECTS IN GAS–LIQUID CHROMATOGRAPHY

Martin,[166] when studying the retention of non-polar compounds as a function of the quantity of stationary phase deposited on the solid support, observed modifications in the retention of these compounds especially in the system composed of a non-polar solute and a polar stationary phase. He chose as representative stationary phases n-hexadecane (non-polar); 1-chloronaphthalene (slightly polar) and β, β'-thiodipropionitrile (TDPN) (polar). He showed that these modifications are very small for n-hexadecane, greater for 1-chloronaphthalene and very important for TDPN.

Eliminating the supposition that components were adsorbed on the solid support, the conclusion may be reached that solute retention depends upon adsorption on the liquid as well as upon dissolution in the bulk liquid. In such circumstances, in the absence of support adsorption, and following the notation of Martin,[166] the corrected net retention volume per gram of packing $(V_{R_g}^0)$ is given by

$$V_{R_g}^0 = K_L V_L + K_A A_L, \tag{86b}$$

where K_L is the bulk liquid partition coefficient, K_A is the adsorption coefficient for the gas–liquid interface, and V_L and A_L are the volume and surface area of the liquid per gram of packing. $V_{R_g}^0$ is related to the specific retention volume V_g by

$$V_{R_g}^0 = \frac{V_g W_L T}{273}, \tag{87}$$

where W_L is the weight percentage of the liquid phase. By rearranging eqn. (87) and then plotting $V_{R_g}^0/A_L$ against V_L/A_L, the values of K_L (from the slope) and K_A (from the intercept) may be determined:

$$\frac{V_{R_g}^0}{A_L} = \frac{K_L V_L}{A_L} + K_A. \tag{88}$$

Martin further explained that for high values of A_L/V_L, the polarity of the liquid phase should largely determine whether solute will adsorb preferentially on the liquid surface or on the solid support. He concluded that adsorption on the liquid should be more important with polar phases, especially for non-polar solutes, while adsorption on the solid should be more important with non-polar phases especially for polar solutes. Martin's data appear to be thermodynamically consistent. Surface tensions of solutions at low solute concentration (measured with a tensiometer) were used to calculate solute surface excesses from the Gibbs adsorption formula. These independent results agreed reasonably well with those determined from his gas chromatographic data.

Pecsok et al.[167] extended the determinations of Martin to a great number of components on columns which have n-hexadecane and TDPN as liquid phase deposited in proportions between 0.30 and 14.90% on a solid support (Chromosorb W, 60–80 mesh, $A_L = 1.4$–0.7 m²/g and Chromosorb P, 42–60 mesh, $A_L = 3.9$–2.0 m²/g). The results agreed well with those of Martin; any differences were attributed to the different purity of TDPN. The authors also mentioned that the liquid surface effects could not be completely disregarded for polar oxy compounds on TDPN.

Martire[168] studied systems of components formed by polar compounds (C_1–C_4 alcohols) on polar liquid phases (glycerol and β, β'-oxydipropionitrile). Adsorption on the solid support (Chromosorb W) is excluded by the author as this was completely treated, the liquid phases were polar, the elution peaks were symmetric, and the retention time was reproducible. For methanol, a systematic variation of the specific retention volume was observed with the variation of the percentage (by weight) of the liquid phase. This obviously implies the existence of surface effects. On the contrary, for ethanol, propanol, and butanol systems, only small random variations in specific retention volume were observed, denoting the absence of liquid surface effects.

Martire[168, 169] tried to find a common property for all systems that present adsorption phenomena at the gas–liquid interface, by correlating certain thermodynamic data with the magnitude of the adsorption effect. He found a partial solution by relating the ratio K_A/K_L (this ratio is actually a measure for the relative magnitude of the adsorption) to the activity coefficient of the component at infinite dilution (γ_2^∞) $(K_A/K_L$ increases with increasing non-ideality of the solution). The author distinguishes two types of adsorption at the gas–liquid interface. The first is characterized by large deviations from Raoult's law; the greater the activity coefficient the less ideal the solution. Large positive deviations indicate that solute molecules are being forced from the bulk liquid to the surface or into the vapour phase, primarily because solvent–solvent interactions are much greater than solute–solvent interactions. The solute cannot compete with a solvent molecule for interaction with a neighbouring solvent molecule. It therefore has minimal solubility in the bulk phase, and this results in a high activity coefficient and a large value of K_A/K_L. The second type of adsorption occurs when the solutes are of comparable polarity to the solvent molecules. This results in dissolution of the solute molecules in the bulk solvent and adsorption of solute caused by dipole–dipole interactions or hydrogen bonding between the solute and the liquid surface.

A qualitative analysis of experimental data which took into consideration, in addition to dissolution in the stationary liquid phase (SLP), the adsorption of the chromatographed substance on the SLP-solid support interface, was first carried out by Berezkin et al.[170, 171] For a solute undergoing partition, liquid–gas interfacial adsorption and liquid–solid adsorption as it passes down the column, Berezkin and Fateeva gave the following equation:[172–174]

$$V_{R_g}^0 = K_L V_L + K_A A_L + K_L K_S A_S, \tag{89}$$

where K_S is the liquid–solid adsorption coefficient and A_S is the surface area of the solid support per gram of packing. Equation (89) is theoretically correct if it is only these three mechanisms which occur. It has been given in a slightly different form by Conder *et al.*[175], and was used by them to determine accurate values of the limiting activity coefficients of C_3–C_5 alcohols in squalane over the range 50–70°C.[176]

For the determination of K_L and K_A, Berezkin and Fateeva[173] used the following expression:

$$\frac{V_{R_g}^0 - \bar{V}_{R_g}^0}{V_L - \bar{V}_L} = K_A \left[\frac{A_L - \bar{A}_L}{V_L - \bar{V}_L} \right] + K_L, \tag{90}$$

where $\bar{V}_{R_g}^0$ is the net retention volume obtained on a reference column of a given $\bar{A}_L$ and $\bar{V}_L$. Typically, this reference column would be lightly loaded. K_A and K_L result from a plot of the left-hand side of eqn. (90) vs. the term in parentheses on the right-hand side of this equation. K_S can then be calculated from eqn. (89) if A_S is known. Note that since A_S is constant and independent of V_L, eqn. (90) is valid whether liquid–solid adsorption is important or not. The authors mention that the experimental data agree satisfactorily with the equations obtained.

Eon *et al.*[177] determined K_A and K_L for n-heptane, CCl_4, $CHCl_3$ and CH_2Cl_2 chromatographed on SLP-water 3.8–40%; solid support— Spherosil XOB30. As the plots of $V_{R_g}^0/V_L$ vs. A_L/V_L are straight lines for all components, the authors conclude that liquid–solid adsorption is not an important contribution so that eqn. (89) may be rewritten

$$K_L = \frac{V_{R_g}^0}{V_L} - K_A \frac{A_L}{V_L}. \tag{91}$$

The values of K_L and K_A were obtained by three methods. In method 1 the coefficients were obtained from least squares analysis of the plot of $V_{R_g}^0/V_L$ vs. A_L/V_L. In method 2 the intercept and slope were reversed through the use of a plot of $V_{R_g}^0/A_L$ vs. V_L/A_L. In method 3 the Berezkin–Fateeva approach was used [eqn. (90)]. In the first two methods the authors found good agreement between the results, but when using the third method (i.e. taking the differences in $V_{R_g}^0$, V_L, and A_L from experimental data points) they obtained erronous results for K_L, resulting in incorrect estimation of liquid–solid adsorption. As eqn. (89) is fundamentally sound, Eon *et al.*[177] consider that the error must arise from the calculation. Writing the propagated error in K_L in the conventional manner, it was shown that most of the error in the Berezkin–Fateeva approach arises from the small values of $(\varDelta A_L)^2$ and $(\varDelta V_L)^2$, which contain a large percentage error.

In order to use eqn. (90) properly, the authors consider that the functions $V_{R_g}^0$ vs. V_L and V_L vs. A_L must be determined by regression analysis. This can be done easily using a polynomial equation. From this, $V_{R_g}^0$ and A_L can be obtained for various values of V_L. When this approach is used, eqn. (90) gives K_A and K_L values closely similar to those obtained from the two-term equation.

Suprynowicz *et al.*,[178] considering that the basis of the chromatographic process is the difference between the potential energies of solute molecules in the surface and in the free gas phase, proposed a chromatographic method for the investigation of bulk liquid solution, solid surface adsorption, and liquid–gas interface adsorption. At infinite dilutions of solutes, the interactions between solute molecules may be neglected, and there is a simple relationship between the energies E_m ($m = L, A, S$ corresponding to the contribution to bulk liquid partition, to liquid surface adsorption and to support surface adsorption, respectively) and the coefficients K_m:

$$K_m = \exp\left[\frac{-E_m}{kT}\right], \tag{92}$$

where k is the Boltzmann constant and T is the absolute temperature.

Correspondingly $V_{R_g}^0$ is given by:

$$V_{R_g}^0 = \sum_m V_m \exp\left[\frac{-E_m}{kT}\right]. \tag{93}$$

Finally, the three contributions to the retention volume, namely bulk liquid solution, solid surface adsorption and liquid–gas interface adsorption, are obtained in the form of pairs of values E_m and V_m.

The determination of K_L, K_A, K_S is conditioned by the accurate knowledge of the surface area.[179, 180] The BET method is undoubtedly the most widely used technique for indirect determination of area.

Surface areas estimated in this manner can generally be accepted as reliable but approximate (it contains many rough approximations, including the necessity to estimate the cross-sectional area of the adsorbate).

The presence of fine pores or capillaries may be such as to allow one gas to penetrate, whereas another gas with larger molecules finds the pores inaccessible. It has been shown that the discrepancy (close to a factor of two) between the K_A values determined chromatographically and those determined statically arises mainly from the use of erroneous surface areas (A_L) in the chromatographic investigations.[179, 180]

Martire[169, 180] has proposed a two-stage method for the correct evaluation of A_L:

(1) Determination of the retention volume dependence upon weight per cent liquid phase for a solute eluted from a series of otherwise identical columns would allow evaluation of $K_A A_L$ for the solute.
(2) Then a study of the solubility and surface tension characteristics of the solute in the chosen liquid (by means of the combined McBain balance and du Nouy tensiometer apparatus) in turn would allow unequivocal evaluation of K_A and hence A_L.

On the other hand, discussing the nature of the partitioning process for low-loaded columns ($< 2\%$), Urone and Parcher[181] have shown that a more complicated mechanism than that of simple liquid phase or solid support adsorption probably exists. Their feelings are that adsorption partition coefficients are very complex functions, and that the assumption of additivity of surface and bulk liquid contributions are, in general, an oversimplification.

Intuitively, one would expect the structure of the liquid layers nearest to the solid-liquid interface to be modified because of their partial orientation on the support (this is especially true for polar liquid phases). In fact, at very low loading it is conceivable that the entire liquid phase could be quite unlike a normal bulk liquid. They propose a more realistic model which considers a modified layer whose partitioning properties are affected by the support to varying degrees depending on the surface activity of the support and the amount of the liquid phase.

It is clear that the mechanism of solute adsorption at the gas–liquid interface is complicated. However, the existence of this difficulty led analytical chemists to some most interesting experiments concerning the the increased selectivity of stationary phases.

Rogozinski and Kaufman[182] have demonstrated that reduction in the surface area–volume ratio for a highly polar liquid phase improves aromatic–aliphatic separations. Since aliphatics are retained primarily by adsorption on the liquid surface and aromatics primarily by solubility in the bulk liquid, then aliphatic retention can be reduced, while aromatic retention is increased, simply by decreasing A_L/V_L (increasing the weight per cent of liquid). Maximum selectivity was achieved with a column packing containing 38% TDPN. When studying the selectivity of capillary columns, they found no marked advantage over the highly loaded packed columns for their particular separation. Thus, when liquid surface effects are significant, a marked variation in the specific retention volume with weight per cent of liquid phase occurs. One must keep in mind the very existence of these adsorption phenomena when calculating thermodynamic parameters from gas chromatographic data.[176]

Even if the analytical chemist is not interested in the study of these adsorption phenomena, the presentation of the retention data should always contain information concerning the type of support used, the particle diameter, and the percentage of liquid stationary phase used.

2.4 THERMODYNAMICS OF GAS–SOLID CHROMATOGRAPHY

Most of the investigations on the thermodynamics of gas-solid chromatography have been concerned with the linear part of the adsorption isotherm.[183–188] As for gas–liquid chromatography, the determination of thermodynamic values are of importance for accurate knowledge of the specific retention volume.

For elution gas–solid chromatography, the corrected retention volume V_R can be converted to the specific retention volume V_s^T by dividing it by the surface area of the adsorbent A. The corrected specific retention volume at the column temperature for a given solute (sorbate)–sorbent (stationary phase) pair is equal to the distribution constant K:

$$V_R/A = V_s^T = K = C_s/C_g \quad (\text{ml/m}^2) \qquad (94)$$

where K is the ratio of the equilibrium surface concentration to the gas phase concentration of the solute when the surface concentration is expressed as moles per square metre and the gas phase concentration as moles per millilitre. The free energy of adsorption is related to the distribution constant by the relation:

$$-\Delta G'_{ads} = RT \ln K = RT \ln V_s = -\Delta H'_{ads} + T \Delta S'_{ads}, \qquad (95)$$

which can be rearranged to give the specific retention volume as a function of the enthalpy ($\Delta H'$) and entropy ($\Delta S'$) of adsorption:

$$\log V_s^T = -\Delta G'_{ads}/2.3RT = -\Delta H'_{ads}/2.3RT + \Delta S'_{ads}/2.3R. \qquad (96)$$

By assuming idealized standard states, the standard state free energies and entropies of adsorption (ΔG^0_{ads} and ΔS^0_{ads}) can be determined from the gas chromatographic retention data.

For the gas phase, the standard state of the adsorbate is defined as a partial pressure of one atmosphere with the adsorbate vapour behaving as a perfect gas. The standard state for the adsorbed phase is that suggested by de Boer and Kruyer[189]—namely, a two-dimensional perfect gas at one atmosphere. Hence, the mean distance between adsorbed molecules is defined as being the same as in the three-dimensional gas phase standard state. Thus, by solving for the intermolecular distance, the area per molecule can be evaluated at standard conditions. This leads to a standard state surface $C^0_{(s)}$ which is given by the expression

$$C^0_{(s)} = 4.07 \times 10^{-9}/T \quad (\text{moles/cm}^2) \qquad (97)$$

at the column temperature T.[190]

Combining eqns. (94) and (97) (and expressing the surface concentration as moles per square centimetre so that the distribution constant is given as K' (ml/cm^2) $= 10^{-4}K$ (ml/m^2)) the gas phase sorbate concentration $C_{(g)}$ is given by

$$C_{(g)} = 4.07 \times 10^{-9}/TK', \qquad (98)$$

which, when combined with the perfect gas equation, gives

$$P_{equil} = C_{(g)}RT = 4.07\times10^{-9}\,R/K'. \tag{99}$$

Knowing the equilibrium partial pressure P_{equil} of the sorbate vapour equilibrium with the sorbent permits the evaluation of the differential molar free energy ΔG^0 for the transfer of one mole of vapour at one atmosphere to its equilibrium vapour pressure P_{equil},

$$\Delta G^0 = RT \ln[P_{equil}/1] = RT \ln[4.07\times10^{-9}\,R/K'] \tag{100}$$

or

$$\Delta G^0 = RT \ln[4.07\times10^{-5}\,R/K] = RT \ln[4.07\times10^{-5}\,R/V_s^T]. \tag{101}$$

Substitution of constants and combining terms gives

$$\begin{aligned}
\Delta G^0 &= RT \ln(4.07\times10^{-5}\times82.05) - RT \ln V_s^T \\
&= 1.99T\ \ln(40.7\times10^{-5}\times82.05) + \Delta G'_{ads} = \Delta G'_{ads} - 11.33T.
\end{aligned} \tag{102}$$

Further rearrangement gives a relation between the specific retention volume and the free energy of adsorption as a function of temperature:

$$\log V_s^T = [-\Delta G^0_{ads}/4.58T] - 2.48 = -\Delta G'_{ads}/4.58T, \tag{103}$$

with the specific retention volume V_s^T expressed in millimetres per square metre, and the free energy $\Delta G'_{ads}$ in calories.

Sawyer and Brookman,[190] studying the behaviour of some aliphatic and aromatic hydrocarbons on salt-modified F-1 alumina and porous silica bead columns, found that the corrected specific retention volume varies linearly with the number of carbon atoms in the molecule.

The linear slope for the normal alkane series indicates that the adsorptive interaction is linearly related to the number of carbons. A similar additive relationship is indicated for the methyl-substituted benzenes. Thus the slope of the normal alkane curve represents the incremental contribution per methyl group to the logarithm of the specific retention volume. The intercept of the normal alkane line $\Delta(\log V_s^T)_0$ gives the constant which must be added to the incremental contributions for a given hydrocarbon to give the correct value for the specific retention volume of the compound.

In a similar manner the contributions of aromatic substituents may be evaluated by measuring on the ordinate the distance between the point corresponding to benzene and the substance in question.

The contribution brought about by the "double bonds" in the benzene molecule may be obtained by measuring the distance between the n-hexane point (situated on the characteristic curve of the aliphatic hydrocarbons) and the benzene point.[190-192]

For a given molecule the evaluated functional contributions to the logarithm of the specific retention volume can be used to predict the retention volume of the compound:

$$\log V_s^T = (\log V_s^T)_0 + n(\log V_s^T)_c + m(\log V_s^T)_\pi + \Sigma(\log V_s^T)\phi_{\text{subst}}. \quad (104)$$

In this particular instance the contribution in calories to the free energy of a functional entity at the column temperature may also be calculated:

$$\Delta(-\Delta G')_x = RT(\ln V_s^T)_x = 4.58T(\log V_s^T)_x. \quad (105)$$

Functional contributions to the adsorption enthalpy and entropy may also be calculated:

$$(\log V_s^T)_x = \Delta(-\Delta H')_x/4.58T - \Delta(-\Delta S')_x/4.58. \quad (106)$$

So it may be concluded that the slope of the straight line plot of the functional group $(\log V_s^T)_x$ versus $1/T$ will be $\Delta(-\Delta H')_x/4.58$. Other information on determinations of thermodynamic quantities by gas–solid chromatography may be found elsewhere.[193-198]

Table 4 gives the adsorption heats for a series of hydrocarbons, chlorinated derivatives, alcohols, etc., on sodium, potassium, rubidium, caesium, and thallium tetraphenylborates. For the aliphatic hydrocarbons, ΔH does not noticeably differ from one column to another. Its value increases with increase of the molecular weight in the homologous series. For alcohols, the adsorption heat values diminished generally in the order primary > secondary > tertiary alcohol. The highest values were obtained on a column of caesium tetraphenylborate. For toluene, the highest value was reached on thallium tetraphenylborate.

TABLE 4. ADSORPTION HEATS, ΔH (kcal/mol)[197]

Compound	NaB(C$_6$H$_5$)$_4$	KB(C$_6$H$_5$)$_4$	RbB(C$_6$H$_5$)$_4$	CsB(C$_5$H$_5$)$_4$	TlB(C$_6$H$_5$)$_4$
n-Hexane	8.64±0.32	8.82±0.55	8.77±0.82	9.28±0.37	8.82±0.23
n-Heptane	9.32±0.41	9.51±0.69	9.00±0.55	9.69±0.69	9.05±0.37
Octane	9.96±0.64	10.56±0.46	10.33±0.32	10.83±0.64	10.65±0.50
Toluene	9.00±0.64	10.05±0.46	9.78±0.18	9.96±0.46	13.89±0.87
Chlorobenzene	8.59±0.41	9.83±0.59	10.56±0.82	10.10±0.69	12.02±0.69
n-Propanol	...	12.70±0.73	...	15.86±0.87	10.22±0.73
2-Propanol	11.56±0.55	11.61±1.23	10.83±0.78	13.89±0.78	10.05±0.41
t-Butyl alcohol	...	9.19±0.69	10.05±0.37	12.70±0.69	10.42±0.59
2-Pentanone	13.30±0.41	...	15.26±0.73	17.14±0.82	12.80±0.59
Methyl acetate	7.36±0.37	7.91±0.69	10.19±0.82	11.61±0.37	9.09±0.78
Methylene chloride	9.32±0.09	12.75±0.69	9.28±0.37	11.33±0.69	8.23±0.46
Chloroform	7.86±0.32	13.44±0.64	9.37±0.37	11.12±0.69	9.05±0.32
Carbon tetrachloride	8.77±0.55	13.62±0.78	9.37±0.37	17.96±0.91	9.32±0.69

Pentan-2-one and methyl acetate have ΔH values that increase on going from sodium to caesium tetraphenylborate. Except for methylene chloride on a column of sodium tetraphenylborate, the adsorption heats for chloromethanes increase from methylene chloride to carbon tetrachloride, the highest values being obtained on a column of caesium tetraphenylborate.

In Table 5 the modifications of the relative adsorption entropies may be seen, calculated on the assumption that the capacity factors are directly proportional to the thermodynamic adsorption constant in columns packed with sodium and caesium tetraphenylborates.[198]

TABLE 5. ENTROPY CHANGES (CAL/DEG. MOLE) FOR SELECTED COMPOUNDS, CALCULATED USING LOG K VALUES AT 129°C[198]

Compound	ΔS NaB(C$_6$H$_5$)$_4$	ΔS TlB(C$_6$H$_5$)$_4$	$\Delta S/\Delta H$ Na	$\Delta S/\Delta H$ Tl
1-Methyl-4-isopropylbenzene	−27.2	−38.5	2.65	2.45
Naphthalene	−29.9	−33.3	2.54	2.37
Phenyl-t-butylacetylene	−29.9	−42.8	2.56	2.38
Benzo-t-butylcyclobutene	−27.2	−38.0	2.60	2.41
p-Di-t-butylbenzene	−35.3	−45.8	2.49	2.36
1-Methylnaphthalene	−32.1	−38.1	2.51	2.36
n-Decane	−29.2	−32.3	2.64	2.48
n-Pentadecane	−41.8	−43.9	2.43	2.33

These values are greater for the thallium tetraphenylborate column than that of sodium tetraphenylborate, these differences becoming significant for aromatic compounds. The fact that the $\Delta S/\Delta H$ ratios decrease for the higher homologues demonstrates that the adsorption heats increase more rapidly than the corresponding entropy values.

CHAPTER 3

THE SELECTIVITY OF THE STATIONARY PHASES USED IN GAS CHROMATOGRAPHY

THE use of selective stationary phases is most important in gas chromatography because:

(i) it makes possible the separation of solutes with very similar properties;

(ii) on the basis of their behaviour with a solute, conclusions may be drawn concerning the nature and the general characterization of the particular solute;

(iii) they contribute useful data to knowledge of molecular interactions.

Although at the very beginning of gas chromatography it was clear that the solute–stationary phase chemical interactions play a most important role in the separation processes,[199–201] the choice of stationary phase has been mainly empirical. The greater part of the discussions concerning selectivity were in terms of polar and non-polar stationary phases and taking into account certain experimental data and phenomenological rules. It can easily be understood why the general discussions took this path if it is remembered that:[202]

(a) gas chromatography is such an efficient separation technique that knowledge of the individual interactions concerned with the separation process is not a prime necessity;

(b) the selective interaction in question may be unknown or it may be the result of a number of simpler interactions that cannot be individually evaluated;

(c) the exact composition of a great number of stationary phases is not known (this is so for those polymers that show stability at high temperatures).

The relatively recent necessity to separate isomers with very similar physical and chemical properties has stimulated increased attention to the principles that are fundamental to the selectivity of stationary phases. The basic physicochemical process involved in the chromatographic column and upon which the high selectivity of this method depends is the interaction of the molecules of the different components to be separated with the stationary phase, the solid surface of the adsorbent, or a stationary liquid. This interaction must not be of a strong chemical nature, but, instead, a weaker molecular one, so that the molecules introduced into the column do not lose their individuality, and elute in a sufficiently short period of time.

3.1 SOLUTE–STATIONARY PHASE INTERACTION

The interaction of a solute with a liquid stationary phase is very complex. It is influenced mainly by the different forces that act within the solution, arising from the chemical composition of the molecules. The shape and size of the molecules appears to have less influence, but this can sometimes become significant.

Among the forces that arise when the distance between molecules is great compared to their size (attractive forces) are:

(i) *London or dispersion forces.* Dispersion forces arise from the electric field, produced by the very rapidly varying dipoles formed between nuclei and electrons at zero-point motion of the molecules, acting upon the polarizability of other molecules to produce induced dipoles in phase with the instantaneous dipoles producing them. These are best treated quantum mechanically.[203, 204] For practical purposes, the energy of dispersion interaction is given by the London formula[205]

$$E_L = -\left(\frac{3}{2}\right) \frac{\alpha_1 \alpha_2 I_1 I_2}{r^6 (I_1 - I_2)}, \tag{107}$$

where indices 1 and 2 refer to two cohering atoms of polarizability α and ionization potential I, with centres distant r.

Though the London equation strictly applies only to atoms, it can be applied to molecules by summing the terms for each individual interaction between pairs of atoms, one atom from each molecule. It is seen from the r^{-6} term that dispersion forces (like all intermolecular forces) decrease very rapidly with increase in distance between the interacting centres. With complex organic molecules such as are encountered in gas chromatography, therefore, it may be considered that the coherence occurs mainly at the surface of the molecules.[206, 207] Dispersion forces are always present in any solute–stationary phase system. They are the only source of attraction between two non-polar substances and are not temperature dependent.

(ii) *Dipole–dipole interactions or orientation forces.* Orientation forces result from the interaction between two permanent dipoles. The average energy resulting from this interaction is[208]

$$E_K = -\left(\frac{2}{3}\right) \frac{\mu_1^2 \mu_2^2}{r^6 kT}, \tag{108}$$

where μ_1 and μ_2 are the permanent dipoles of solute and solvent respectively, r is the distance between these dipoles, k is the Boltzmann constant, and T is the absolute temperature. The orientation forces therefore decrease with increasing temperature, and approach zero at very high temperatures, when all orientations are equally probable. Liquid phases which depend upon orientation forces for selectivity at low temperatures become ineffective at high temperatures.

(iii) *Induction of dipole-induced dipole interactions.* Induction forces result from an interaction between a permanent dipole in either the solute or stationary phase and an induced dipole in the other. The mean energy of this interaction may be expressed by the Debye equation[209]

$$E_D = -(1/r^6)(\alpha_2\mu_1^2 + \alpha_1\mu_2^2), \qquad (109)$$

where α is the polarizability.

Dipole-induced dipole interactions are not the same in all directions and are dependent on relative molecular orientation.

The ability of a polar solute to induce a dipole in a non-polar solvent has been investigated using gas–liquid chromatography by Meyer and Ross[210] for acetone, chloropentane, and capronitrile (solutes) and Apiezon M (stationary phase).

The energy of solution at infinite dilution for a polar solute is the sum of dispersion and induction energies between itself and the solvent. Orientation cannot contribute because there is but a single dipole in a sea of non-polar material. Thus a reliable estimate of the dispersion contribution to the energy of solution for the polar solute allows evaluation of its induction energy by difference. It has been shown[210] that induction energies contribute (of the order of kcal) to the total energy of attraction. They are certainly not negligible, as has commonly been thought.[211–213]

Van der Waals forces consist of the sum of these three attractive forces.

It must not be forgetten, however, that besides the forces arising from the interaction of permanent or temporary electrical multipoles of the molecules, there are also other long-distance forces. Such a force is the one due to the interaction of magnetic multipoles.[214] This can be explained by the fact that only a limited number of molecules are paramagnetic, and even in this instance, the magnetic forces are relatively weak, so that the magnetic interactions play only a secondary role in most instances. It has been established that for certain molecular species the interaction forces remain attractive even when the molecules are very distant from one another compared with their size. But as soon as the molecules approach each other, the attractive forces switch to highly repulsive forces. These attractive forces are called medium distance forces and they lead to the formation of molecular complexes. The nature of the bonds in the molecular complexes is not entirely clear, but it is believed that they are more or less polarized, weak covalent bonds. But still there is a possibility that an important part of the stabilization energy of these complexes is due to the van der Waals forces.

Molecular complexes can be divided into two categories:

(a) Donor–acceptor complexes, where the bonding is a result of the partial transfer of electrons from a filled orbital on the donor molecule to a vacant orbital on the acceptor molecule. In this class one may include charge-transfer complexes and most likely the hydrogen bond complexes as well.

(b) Yielding–retro-yielding complexes, where the bonding is a result of double charge-transfer between the partners.

The most popular model for the time being seems the one that presumes that every molecule undergoes a partial electron transfer from one of its filled orbitals to a vacant orbital on the partner, which in its turn acts in the same manner with the original donor molecule.

From the first category defined above mention must be made of the gas–liquid chromatographic studies of electron donor–acceptor systems formed by 2,4,7-trinitrofluorenone acting as an electron acceptor stationary phase and electron donor compounds (amines, aromatic hydrocarbons, heterocycles).[215–217, 110] The results indicate that complex formation plays a prominent role in the separation of electron donors on this electron acceptor stationary phase, and have been used to predict orders of complexing ability for the donors.[110, 217]

The formation of hydrogen bonds between two molecules, one of which has a strong polar group AH (FH, OH, NH) and the other molecule which has a strongly electronegative atom B (F, O, N), is of great importance for a wide range of separations by gas chromatography. The exact nature of the forces that lead to hydrogen bond formation is far from being known exactly. The theory of the hydrogen bond is at a less-advanced stage than the experimental studies on hydrogen bond complexes.

As it is obvious from the literature[218–221] there is an analogy between the hydrogen bond complexes and the charge-transfer complexes. In fact, a hydrogen bond is a result of three types of interactions:

(a) an electrostatic interaction between AH and the B donor group;

(b) charge-transfer interactions arising from the partial transfer of the electron from donor B to acceptor AH;

(c) interelectronic repulsion interactions arising from Pauli-type exclusion forces.

Table 6 classifies hydrogen bonding compounds into four classes on the basis of their structures.

TABLE 6. CLASSIFICATION OF HYDROGEN BONDING COMPOUNDS

Class	Structural characteristic	Classes of compounds
I	Has multiple active H atoms Has multiple donor hetero-atoms	H_2O, polyols, amino alcohols, hydroxy acids, polyphenols, dibasic acids
II	Has active H atom Has donor hetero-atom	Alcohols, acids, phenols, primary and secondary amines, oximes, nitriles, and nitro compounds with α-H atoms
III	Has no active H atom Has donor hetero-atom	Ethers, ketones, aldehydes, esters, tertiary amines, nitriles, and nitro compounds with no α-H atoms
IV	Has active H atom Has no hetero-atom	Aliphatic halogenated compounds, unsaturated hydrocarbons

A classification of hydrogen bond strengths as a function of the nature of the ligand is given in Table 7.

It is obvious that this classification is only approximate as the strength of the bonds depend on the very nature of the interacting molecules. Nevertheless, the above classification can be of use to the chromatographer in the discussion of solute-liquid stationary phase interactions in the sense of deviations from Raoult's law and of retention.[222-224] Thus, in systems formed by the I–IV and II–IV classes, there is always a positive deviation from Raoult's law; the solute is generally not well retained by

TABLE 7. CLASSIFICATION OF HYDROGEN BOND
STRENGTHS FOR THE CLASSES DEFINED IN TABLE 6

	I	II	III	IV
I	vs	s	m	m
II	s	s–m	m–w	m–w
III	m	m–w	–	w
IV	m	m–w	w	–

vs = very strong; s = strong; m = medium;
w = weak.

the liquid phase. In systems formed by classes III–IV a negative deviation is always found; the solute is well retained by the liquid phase. In systems formed by classes I–I, I–II, I–III, II–II, and II–III there may be both positive and negative deviations that must be predicted on an individual basis. In gas–liquid chromatography, solutes are generally retained by solvents. Systems formed by classes III–III and IV–IV are quasi-ideal; the solutes separate according to their boiling points.

Evaluation of hydrogen bond energies by means of gas–liquid chromatography is discussed by Jogansen *et al.*[225] The method developed is based on the chromatographic heats of solution of volatile "acid" AH in two different stationary liquids, a "base" and an inert solvent, respectively.

The quantity taken as an estimate of the desired one is:

$$\Delta H_{ib} \equiv \Delta H_v(\text{liquid b}) - \Delta H_v(\text{liquid i}), \tag{110}$$

which is the difference of molar heats of solution at infinite dilution of the proton donating "acid" AH in two solvents, the first solvent (liquid b) being the "base" BR with the electron-donating site B, and the second one (liquid i) being an inert solvent that forms no hydrogen bonded complexes with AH. It is implicit that all other interactions (to which dispersion forces contribute most) have the same strengths in both solvents.

The potentialities of this method are of particular interest for hydrogen bonding studies of simple molecules. Aromatic ketones and a nitroso compound (N-nitroso-methyl-aniline) providing the oxygen donor atom for hydrogen bonding have been studied recently by Vernon.[226] The energies of the spectroscopic solvent shifts of the donor molecules have been correlated with gas–liquid chromatographic retention index data when the donor molecules are used as the stationary phases and chloroparaffins, alcohols, and amines as solutes.

The formation of complexes by yielding–retro-yielding, in which the bond is formed by a double charge-transfer between the partners, may be illustrated in chromatography using silver nitrate as a stationary phase for the selective retention of olefins from a mixture with saturated hydrocarbons. According to Dewar,[227] the bond between the olefinic ligand and the metal ion is formed by donation of electrons from the π-ethylenic bonds to the vacant s-orbital of the silver ion and back donation of d-electrons from the metal to the π-antibonding orbitals of the unsaturated compound. Some reviews of the use of charge-transfer complexes of metals in the gas chromatographic separation of organic compounds have been published.[228, 229]

Pearson[230, 231] has proposed a general principle which may be of special assistance in utilizing and classifying molecular interactions, and was discussed for the particular case of gas–liquid chromatography by Langer and Sheehan.[202] The principle is that of "hard and soft acids and bases". Essentially all interactions are viewed in the Lewis sense, where acids are electron acceptors and bases are electron donors. Thus in the hydrogen bonded —OH : $O(R)_2$, hydrogen is acting as a Lewis acid and the oxygen in the ether as a Lewis base. Similarly, in a silver–olefin complex, the silver ion is viewed as an acid and the electron-rich double bond as a donor.

The "hard and soft acids and bases" principle states that hard acids prefer to co-ordinate to hard bases and soft acids prefer to co-ordinate to soft bases. The terms are defined as follows.

A soft base is a donor centre which is highly polarizable, easily oxidized, and associated with empty, low-lying orbitals; it has low electronegativity (R_2S; RSH; RS^-; I^-; Br^-; R_3P; R_3As; $(RO)_3P$; RNC; C_2H_4; C_6H_6).

A hard base is a donor centre which has low polarizability, high electronegativity, and is hard to oxidize; it is associated with empty orbitals of high energy, which hence are inaccesible (H_2O, F^-, CH_3COO^-, ClO_4^-, NO_3^-, ROH, RO^-, R_2O, RNH_2, N_2H_4).

A soft acid is the acceptor centre which has a small positive charge, large size and easily excited outer electrons (tetracyanoethylene, quinones, trinitrobenzene, Cu^+, Ag^+, Au^+, Pd^{2+}, Cd^{2+}, Pt^{2+}, $Tl(CH_3)_3$, BH_3, Br_2).

A hard acid is the acceptor centre which has a large positive charge, small size, and does not have easily excited outer electrons (H^+, K^+, Cr^{3+}, Co^{3+}, Fe^{3+}, UO_2^{2+}, BF_3, $B(OR)_3$, $Al(CH_3)_3$, HX, hydrogen bonding molecules, RSO_2^+, SO_3).

Despite the simplicity of the hard and soft classifications of interactions, it must be remembered that the most commonplace gas–liquid separation can involve a combination of interactions.

For instance, with a polyglycol, $HO-(CH_2-CH_2O)_n CH_2CH_2OH$, stationary phase, and amine solutes, there are several interactions; in addition to dispersion and dipole-induced dipole interactions, there are the hard interactions between the amine hydrogens and ether oxygens and between terminal hydroxyl groups and the nitrogen electron pair of the solutes. The latter becomes less important as the molecular weight of the polyglycol increases.

The adsorption of molecules on a surface has a simpler mechanism than the solvation of molecules in a solvent. When the effect of the geometric heterogeneity of surfaces and pores can be eliminated by modifying their surfaces chemically or by adsorption, the adsorption properties of gas–solid columns will be influenced principally by the chemistry of the adsorbent surface. The chemical structure of the solid surface determines the nature and energy of the molecular interactions that occur between the molecules of the substances being separated and the solid. The interaction of the molecules of a gas mixture with a homogeneous solid surface and the state of the adsorbed molecules on a sufficiently homogeneous surface lend themselves more easily to theoretical treatment than when they are dissolved in the bulk of a liquid film. In solution, all the molecules are mobile, and the molecules of any given component are surrounded on all sides by other molecules, whereas on adsorption on a sufficiently smooth solid surface, the molecules interact primarily only with the nearest force centres of the solid, and these centres are fixed. During adsorption, depending on the chemical structure of the molecules and surfaces, various interactions may come into play, ranging from non-specific to specific, during which the chemical individuality of the interacting partners is retained and increasing as far as chemical interactions, in which this individuality is lost and a new surface–absorbate compound is formed.[232]

TABLE 8. CLASSIFICATION OF MOLECULES AND ADSORBENTS ACCORDING TO THEIR CAPACITY FOR NON-SPECIFIC AND SPECIFIC MOLECULAR INTERACTIONS[233]

Groups of adsorbate molecules	Types of adsorbents		
	I. Carrying neither ions nor active groups (graphitized carbon blacks, BN, surfaces carrying only saturated group)	II. Carrying concentrated positive charges (acid hydroxyls, exchange cations of small radius)	III. Carrying concentrated negative charges (ether, nitrile, carbonyl gropus, exchange anions of small radius)
(a) With spherical symmetrical shells or σ-bonds (noble gases, saturated hydrocarbons)	Non-specific interactions depending mainly on dispersion forces		
(b) With electron density locally concentrated on the peripheries of bonds: π-bonds (N_2, unsaturated, and aromatic hydrocarbons) and lone electron pairs (ethers, ketones, tertiary amines, nitriles, pyridine)	Non-specific interactions	Non-specific and specific interactions	
(c) With positive changes locally concentrated on the peripheries of bonds (e.g. certain organometallic compounds)	Non-specific interactions	Non-specific and specific interactions	
(d) With functional groups with both electron density and a positive charge concentrated on the peripheries of their individual bonds (molecules with OH or NH groups)	Non-specific interactions	Non-specific and specific interactions	

The non-specific interactions manifest themselves in all instances. In general they are the universal dispersion interactions which are associated with the consistent movement of the electrons of the interacting systems. The specific interactions are caused by local peculiarities of electron density distribution, i.e. by the concentration of negative or positive charges on the peripheries of bonds and links of interacting systems.

The terms "non-specific" and "specific" interactions were utilized in gas–solid chromatography by Kiselev[233] and are merely convenient terms for the systematization of experimental data. Table 8 gives a schematic classification of molecules and surfaces based on their abilities to interact non-specifically or specifically.[233] According to Kiselev, the specific interactions are, in general, non-classical, and only in particular examples do they pass to the electrostatic Coulomb interaction.

The magnitude of the heats of adsorption of isotopes on alumina was one of the factors which led King and Benson[186] to postulate that electrostatic forces were responsible for the adsorption of isotopes on alumina and other adsorbents. According to this electrostatic model, gases are polarized by strong surface electric fields and by the uncompensated charges on the surface as they approach the surface. This electrostatic interaction causes the molecules to be attracted to the surface, the energy of attraction being

$$\phi_{att} = -\frac{\alpha}{2} \frac{C_{eff}^2}{z^4} \tag{111}$$

or

$$\phi_{att} = -\frac{\alpha}{2} E_z^2, \tag{112}$$

where α is the polarizability of the adsorbed molecule, C_{eff} is an effective surface charge, z is the distance of the molecule from the surface, and E_z — which is usually strongly dependent on z — is the electric field intensity normal to the surface. In eqn. (111), z can be replaced by $(r-r_0)$, where r is the distance between the adsorbed molecule and the surface and r_0 is the radius of the adsorbed molecule.

Thus, according to King and Benson,[186] in the simple example of the noble gases on an Al_2O_3 column, the gases elute from the column according to their relative polarizabilities. Also, as predicted by the theory, a plot of the logarithm of retention time vs. polarizability is a straight line.

For molecules whose polarizabilities and sizes are essentially the same (such as the isotopes) other factors, such as the vibrational and rotational energies on the surface, govern the separations. An interesting example of this effect is the chromatographic separability of oxygen and argon. It is well known that these species are difficult to separate on an adsorption column, especially at or above room temperature. Not only are the two species similar in size, but also the average polarizability of oxygen (O_2) is 1.60, and the polarizability of argon is 1.63. These values are so close that there is little difference in the interaction energies of the molecules with the surface. The vibrational energies on the surface are also very close since they are functions of the square root of the total mass. Therefore to separate the gases some other parameters must be changed. One choice is to change the rotation of the oxygen on the surface by lowering the temperature. At the lower temperature the rotation can become hindered and the effective polarizability becomes, not the average value, but the polarizability along the internuclear axis. It is not surprising that oxygen and argon can be separated on a number of adsorbents at low temperatures.[234, 235]

In spite of all the investigations that have been carried out there is still general confusion as to why molecules separate on molecular sieves and which molecules are adsorbed on a particular sieve.[236, 237] The prevailing belief is that the adsorption and separation of gases on molecular sieves is governed by the sizes of the "holes" or "channels" in the molecular sieve structure.[238, 239] However, the fact that *ortho-* and *para-* hydrogen are separated by a molecular sieve column[240] suggests that adsorption on molecular sieves is similar to adsorption on other adsorbents used in gas–solid chromatography in that electrostatic interactions play a dominant role.[186]

Additional support for the hypothesis that electrostatic forces govern adsorption on molecular sieves comes from the investigation of Barrer and Gibbons.[241] They found that the heat of adsorption of ammonia on Linde sieve X is a function of the substituted cation in the sieve, the heat value increasing as the polarizing power of the cation increases.

Also, heats of adsorption of saturated hydrocarbons on zeolites contain‑ing singly charged cations increase for heavier cations, and have the highest value on silver-containing specimens.[242] That is proved by the fact that the polarizability of cations increases with the increase in the number of electrons in an atom, and hence the dispersion interaction with adsorbed molecules of saturated hydrocarbons increases. Gant and Yang,[243] in separating the isotopic methanes on a charcoal column (-3.5 to $150°C$), found that the retention times decreased in accordance with the decrease in polarizabilities as deuterium or tritium was substituted for the hydrogen in methane.

The irreversible adsorption of polar molecules or molecules with large quadrupole moments on chromatographic columns may be explained by electrostatic theory. Dipole or quadrupole interactions with an electric field are much stronger than those of induced dipoles with the electric field, and the molecules are held more strongly to the surface. In order for these molecules to be eluted, the highly active sites on the surface must be blocked. Thus electrostatics offers an interpretation of much of the data of gas–solid chromatography.

The discussions, however, are carried on mainly in qualitative terms. The theoretical investigation of these interactions and quantitative calculations meet with great difficulties. Nevertheless, the current theory for adsorption from the gas phase on homogeneous crystal surfaces is being developed on the molecular level by considering the geometrical and electronic structure of the solid body surface and of the molecule being adsorbed.[244–248]

For analytical applications of this theory, especially for highly sensitive detectors together with very low concentrations of adsorbed molecules and fairly high temperatures, the adsorbate–adsorbate interactions are insignificant, and thus it is only necessary to consider the adsorbate–adsorbent interaction. At higher surface concentrations, the second virial coefficient, which is a function of the adsorbate–adsorbate interaction in the vicinity of the adsorbent must be considered, or the adsorbate–adsorbate interaction must be taken into account in some other way.

It is very important to calculate the potential energy of the adsorbate–adsorbent interaction. The potential energy of adsorption of the molecule in its most favourable orientation on the surface is equal to the heat of adsorption at absolute zero temperature. The magnitude of the potential energy of adsorption also approximately expresses the heat of adsorption at the temperature of the gas chromatographic column. At fairly low temperatures, the order of the retention volumes usually corresponds to that of the heats of adsorption.

Calculations of the potential energy of adsorption have been carried out for a relatively large number of compounds on graphitized carbon black and compared with experimental values obtained by gas chromatography.[248-252] Graphitized carbon black is one adsorbent with a typical flat surface. The surface of this adsorbent is almost entirely formed of the basal faces of large crystals of graphite.[253] The adsorption properties of such carbon blacks are principally determined by the properties of the adsorbate–basal face system. The adsorption of molecules of different electronic structures on the surface of carbon black depends on dispersion forces and can be defined by the geometry of the molecule and by the polarizability of its bonds; a special role is played by the number of points of contact with the plane surface of the adsorbent.[254]

For adsorption of simple molecules on graphite the potential functions of the paired molecule-carbon atom interaction, e.g. $\varphi_{C \ldots Ar}$, $\varphi_{C \ldots NH_3}$ can be used. For the adsorption of complex hydrocarbon molecules and their derivatives, the atom–atom potential functions $\varphi_{C \ldots C}$, $\varphi_{C \ldots H}$, $\varphi_{C \ldots O}$ or group–atom potential functions $\varphi_{C \ldots CH_3}$, $\varphi_{C \ldots CH_2}$, $\varphi_{C \ldots CH}$, etc., must be used.

The energy of paired interactions is summed with respect to all atoms or atomic layers on the semi-infinite lattice of a graphite crystal, and, for complex molecules, with respect to the atoms or bonds which form these compounds, so that the potential energy of the molecule may be written

$$\phi = \sum_{C_i} \sum \varphi_{C \ldots i}, \tag{113}$$

where $\varphi_{C \ldots i}$ is the potential energy of interaction of a lattice carbon atom with an atom or bond i of the molecule:

$$\varphi_{C \ldots i} = -C'_{C_i} r_{C_i}^{-6} - C''_{C_i} r_{C_i}^{-10} + B_{C_i} \exp(-r_{C_i}/\varrho_{C_i}). \tag{114}$$

The constants C'_{C_i} and C''_{C_i} may be calculated with the aid of the Kirk-wood–Müller equation or an analogous equation for the polarizability and diamagnetic susceptibility of C and i bonds.[255] The constant B_{C_i} is determined from the equilibrium conditions and is given in terms of the equilibrium distance $z^0_{C_i}$ of the atom or bond centres i from a plane passing through the centres of the carbon atoms of the outer basal plane for the most favourable orientation of the molecules relative to this plane (i.e. the maximum number of molecular linkages which are in contact with the basal plane).

The constant ϱ_{C_i} is considered to be equal to 0.28 Å as in the molecular crystals. For adsorption of a dipolar molecule a term is introduced into eqn. (114) for the energy of the inductive attraction of the dipole to the polarized carbon atom in the lattice or for the energy of the reflected force of the dipole in the carbon layer of the lattice.[256] In both instances, the contribution from this energy is small.

Detailed calculations of the constants, the summation of eqn. (114) in eqn. (113) and the calculation of $-\phi_0$ for the equilibrium distances $z^0_{C_i}$ have been made.[244, 248] The calculated values agree well with the experimental data.

A theoretical calculation of the specific interaction energy is more difficult.[249] At present, to reveal the specific nature of the adsorption of molecules with different bonds and polar functional groups on various types of adsorbent surfaces, it is convenient to compare the specific retention volumes or heats of adsorption for non-polar molecules (n-alkanes and their derivatives) with the corresponding polar molecules on different types of adsorbents.[257]

It is recognized that, besides the chemical nature of the surface, adsorption depends on the pore structure of the adsorbent. The mechanical structure of the porous substances can be characterized by the following parameters:

(a) the average pore diameter $\bar{d}_p$ (Å);

(b) the specific surface S (m²/g);

(c) the specific volume of the pores V_p (ml liquid adsorbate/g);

(d) the distribution function $dV_p/dd_p = f(\bar{d}_p) = $ distribution of the pore size.

The values of these quantities are usually obtained by gas adsorption measurements.[258]

The most common adsorbates are nitrogen, noble gases, and the hydrocarbons. The most important criteria in the adsorption of gases and vapour on porous solid substances are the average pore diameter $\bar{d}_p$ and its ratio to the size of the adsorbate molecule. Depending on this ratio there are two limiting cases:

(a) $\bar{d}_p$ is some orders of magnitude greater than the adsorbate molecule diameter;

(b) $\bar{d}_p$ is of the same order as the adsorbate molecule diameter.

In the first case there are wide pores with a small specific surface ($\bar{d}_p > 2000$ Å; $S < 10$ m^2/g). To a first approximation the surface of the pores can be considered as a plane surface. The adsorption equilibrium is established very rapidly. With wide pore adsorbents the diffusion activation energy is very small, but for micropores it increases considerably.[259] With micropores the adsorption velocity depends greatly on the molecular size of the adsorbate, indicating that there is no adsorption surface but an adsorption cavity. With small pore-size adsorbents one must always have in mind the importance of the shape of the pores.

The classification of porous adsorbents based on the $\bar{d}_p$ value must take into account 30 Å $< \bar{d}_p < 2000$ Å (intermediary pores). For intermediary pore adsorbents there is a characteristic isotherm that has a hysteresis cycle between the adsorption and desorption branch. This can be explained on the basis of capillary condensation.[260, 261] The classification of adsorbents as microporous, intermediary and macroporous, is by Dubinin.[262] Figure 4 shows the adsorption isotherms obtained for SiO$_2$ of different pore size, the adsorbate being nitrogen.[260]

The specific surface area of the adsorbents with intermediary pores is in the range $10 < S < 500$ m^2/g, and for microporous adsorbents, $S > 500$ m^2/g.

Depending on the pore diameter, different methods can be used in order to measure the specific surface area. Thus, for macroporous adsorbents, the BET[263] method is used, for intermediate pore adsorbents, the

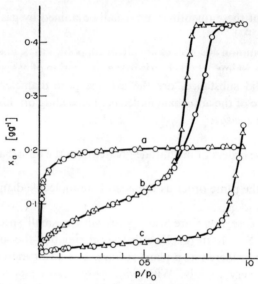

FIG. 4. Adsorption isotherm for an adsorbent with (a) micropores; (b) intermediary pores; (c) macropores[260], for nitrogen on porous silica, measured gravimetrically by an electronic microbalance. (○) adsorption; (△) desorption.

Kiselev[264] method is used, and for microporous adsorbents that give Langmuir type isotherms, the method of Harkins and Jura[265] is applied.

The determination of the distribution of the pore size and of the specific pore volume is made by adsorption measurements for micro and intermediary pores and by mercury penetration measurements for macropores[266–269]. A review of the methods used for the determination of the spatial structure of the adsorbents is given by Unger[260].

Static studies of adsorption of hydrocarbon vapours on silica gels have shown that the adsorption and heats of adsorption of hydrocarbons increase as the pores of the silica gel become narrower, and this increase occurs more rapidly the more carbon atoms there are in the molecule.[270] An investigation of the effect of geometric structure of fine-pore silica gels on the gas chromatographic separation of C_1 to C_4 hydrocarbons gases at high surface coverages showed that separation became more complete with increasing specific surface area and decreasing pore size.[271]

For samples with sufficiently wide pores ($d_p > 500$ Å), changes in pore size no longer affect the heat of adsorption or the specific retention volumes, irrespective of the specific surface area of the adsorbent. The heat of adsorption of n-alkanes on wide-pore silica gels increases linearly with the number of carbon atoms in the molecule.[272] When small samples are employed, the molecules are adsorbed mainly on the more favorable sites with respect to the total energy of the various kinds of adsorption interactions. If the average pore diameter of the adsorbent is greatly reduced and its specific surface area is increased, the retention volume increases[273] (thus it is advantageous to use wider pore silica gels for analysis because the time required for the analysis and the tendency of the peaks to become diffuse are greatly reduced). Therefore when silica gels are used to separate light gases, samples should be employed with an average pore size of not more than 20 Å; for the separation of light hydrocarbons (boiling point not more than 10°C) silica gels with average pore diameters between 50 and 200 Å should be used, whereas for rapid analysis of higher boiling hydrocarbons and some of their derivatives, wider-pore gels should be employed.

These facts show that, in general, the retention of solutes is determined, firstly, by the geometric pore structure of the adsorbent surface and its chemical nature, secondly, by the molecular weight and the geometric and electronic structure of the molecules of the substances being analysed and, thirdly, by the column temperature.

3.2 POLARITY OF THE STATIONARY PHASES

The large diversity of possible interactions, although valuable in practice for the chemist, impedes his attempts to understand the exact nature of the individual interaction terms because the measured quantity of the system, the retention index, is a composite function of many effects.

One of the most important problems in gas chromatography is to correlate retention data with the molecular structure of the solute–stationary phase system. Some results in this direction have been obtained for simple systems formed by non-polar solutes and non-polar stationary phases,[274-281] and for many olefins and alkyl benzenes on non-polar stationary phases.[282-285] Castello and D'Amato[286] attempted to correlate the retention indices of the alkyl iodides with their molecular structures and physical properties. They correlated retention indices with boiling points, substitution of an iodine atom in the chain, and molecular volume. The correlations were obtained for only one stationary phase, tricresyl phosphate.

In order to make these studies more general, it is desirable to correlate data for many solutes on a series of stationary phases ranging from non-polar to highly polar. This more general problem has not been solved because of the lack of an adequate procedure for isolating the controlling factors operative in determining retention indices. Present-day analytical chemists use an empirical notion (polarity) in order to characterize different stationary phases.

If gas chromatographic polarity is to be defined by any quantity, it should be possible to estimate from it the retention of as large a number of substances as possible with maximal accuracy. The quantitative characteristic of polarity which corresponds most closely to the term "polarity" as it has previously been interpreted, can be considered to be the retention ratio of a polar or polarizable solute RX to that of one which is non-polar RH, for a non-polar standard column to which a zero polarity is attributed:[287-289]

$$P_p = \log(V_p^{RX}/V_p^{RH}) - \log(V_n^{RX}/V_n^{RH}). \tag{115}$$

The retention ratio for another polar/non-polar pair of solutes in the liquid phase p is then expressed by:

$$\log(V_p^{RX'}/V_p^{RH}) = A^{RX'}P_p + \log(V_n^{RX'}/V_n^{RH}) \tag{116}$$

where $A^{RX'}$ is a constant characteristic of the solute RX'. This method of characterizing liquid phases has been applied by Maier and Karpathy[287] to commercially produced columns.

The method is, of course, only an approximation and is subject to considerable error. Chovin and Lebbe[290] expanded this polarity scale, which is based on a unit of the retention ratio of butadiene to n-butane and which is valid only at 30°C, to a wide temperature range, and used the retention ratio of two consecutive paraffins; the more polar the phase the smaller the value of α:

$$\alpha = V_{g(z+1)}/V_{g(z)}. \tag{117}$$

For practical reasons, the logarithm of the relative retention was chosen as the characteristic value, and the polarity was expressed on an empirical scale, taking zero as the value for squalene and unity as the value for β, β'-oxidipropionitrile. Equation (117) is based on the assumption that $V_{g(z+1)}/V_{g(z)}$ is independent of the number of carbon atoms, which is not strictly true.

As has been observed by Littlewood,[291] the relative retention of two neighbouring paraffins is, as their vapour-pressure ratio, dependent upon the number of carbon atoms, at least for less than nine carbon atoms. According to Littlewood, the relative retention of two n-alkanes is only characteristic for those phases in which the solutes are not very soluble, i.e. for polar and very polar stationary phases. On the other hand, for phases in which the solutes are more soluble, the variation of the relative retention with the phases is too small, and thus does not permit an accurate classification.

Lazarre and Roumazeilles[292] consider the following equation to be a measure of the polarity of the liquid phase:

$$P = T \log V_{g(z+1)}/V_{g(z)} = \alpha T. \tag{118}$$

Based on experiment, it is concluded[293] that the product αT (T being the temperature of analysis in °C) is not constant for a given stationary phase and thus cannot be taken into consideration as a convenient method of phase classification.

Bonastre and Grenier[294] consider as a measure of the polarity the quantity

$$P = 10^3 \times \gamma_{z+1}^\infty/\gamma_z^\infty, \tag{119}$$

where γ_{z+1}^∞ and γ_z^∞ are the activity coefficients at infinite dilution of two neighbouring n-alkanes.

It must not be forgotten that the retention of a solute in a given phase is dependent on the interactions between an isolated molecule of the solute and the surrounding molecules of the stationary phase; and as the solute and the stationary phase can have quite different structural configurations, it should be expected that the classification of stationary phases will be closely dependent on the criterion which is chosen for establishing it. In other words, if the criterion is changed, the order of classification would almost certainly change also. The system of retention indices developed by Kováts[19, 295-297] permits the prediction of the increase in the retention index of a polar solute when it is chromatographed on the non-polar or polar stationary phase used as a reference, the calculation being based essentially on knowledge of the chemical structure of the solute. Kováts suggested that liquid phases should be characterized by means of eleven index differences of different homologous series.

Schomburg[298] considers that the polarity of liquid phases should be defined by the expression:

$$P = \Delta I_{\text{benzene}} - \Delta I_{\text{cyclohexane}}. \tag{120}$$

He determined a number of column polarities and found that the index difference of olefins increases linearly with increasing polarity.

Rohrschneider[298] holds that a more simply and accurately measured quantity is the index difference of benzene itself as a simple measure for the characterization of columns, and which was given for twenty-two liquid phases.

Very generally it can be stated that by chromatographing a solute (other than an alkane) on two stationary phases a and b and expressing the result in Kováts' retention indices, if the index on phase b (I^b) is greater than the index on phase a (I^a), b may be considered to be more polar than a.

If the less polar stationary phase is chosen as the reference phase (the phase that gives the smallest index for the chosen solute), the polarity is measured by the difference

$$\Delta I = I^b - I^a. \tag{121}$$

It is obvious that the classification of stationary phases on the basis of eqn. (121) is very much dependent on the choice of the polar component. Any change in the polar component will have a drastic effect on this classification. Nevertheless eqn. (121) is of great practical importance as it offers a starting point for certain generalizations. One of these is that if ΔI is greater for a given solute, the polarity of the phase is greater, and, for a given stationary phase, a component is more polar.[299]

Thus

$$\Delta I = ax, \tag{122}$$

where a and x are the "polarity factors" of the solute and stationary phase, respectively. But eqn. (122) is much too simple and thus does not give a- and x-values that allow the evaluation of retention indices of any solute on any stationary phase.

The system was subsequently developed by Rohrschneider[300] by increasing the number of polarity factors for each solute and stationary phase up to five. In this instance, for a given solute–stationary phase couple:

$$\Delta I = ax + by + cz + du + ev, \tag{123}$$

where x, y, z, u, v and a, b, c, d, e are the polarity factors characterizing the stationary phase and the solute, respectively.

By chromatographing each of m solutes on each of n stationary phases, mn results are obtained that should allow the determination of $5m$ polarity factors of the components and $5n$ polarity factors of the stationary phases. A system of mn equations with $5(m+n)$ unknowns is obtained:

$$\left. \begin{aligned}
\Delta I_1^1 &= a_1 x_1 + b_1 y_1 + c_1 z_1 + d_1 u_1 + e_1 v_1, \\
\Delta I_2^1 &= a_1 x_2 + b_1 y_2 + c_1 z_2 + d_1 u_2 + e_1 v_2, \\
&\cdots \\
\Delta I_n^1 &= a_1 x_n + b_1 y_n + c_1 z_n + d_1 u_n + e_1 v_n, \\
\Delta I_1^2 &= a_2 x_1 + b_2 y_1 + c_2 z_1 + d_2 u_1 + e_2 v_1, \\
&\cdots \\
\Delta I_n^m &= a_m x_n + b_m y_n + c_m z_n + d_m u_n + e_m v_n.
\end{aligned} \right\} \tag{142}$$

The system is undetermined. In order to solve this system, Rohrschneider chose arbitrary values for $x, y, z, u,$ and v:

$$x = \Delta I_{\text{benzene}}/100; \quad y = \Delta I_{\text{ethanol}}/100; \quad z = \Delta I_{\text{methylethylketone}}/100;$$
$$u = \Delta I_{\text{nitromethane}}/100; \quad v = \Delta I_{\text{pyridine}}/100.$$

6*

In this instance, for a stationary phase j and for a solute i,

$$\Delta I_i^j = a_i \frac{\Delta I_{C_6H_6}^j}{100} + b_i \frac{\Delta I_{C_2H_5OH}^j}{100} + c_i \frac{\Delta I_{C_4H_8O}^j}{100} + d_i \frac{\Delta I_{CH_3NO_2}^j}{100} + e_i \frac{\Delta I_{C_2H_5N}^j}{100}. \quad (125)$$

The calculation of all the factors is laborious. But if the a, b, c, d, and e factors of a solute chromatographed on five different stationary phases (on these phases five reference solutes were chromatographed in order to obtain the polarity factors x, y, z, u, and v) are determined, then the behaviour of the solute can be predicted on any stationary phase for which the x, y, z, u, and v factors have been determined.

The method has been applied to twenty-five components and a stationary phase that was not included in the system formed (Carbowax-dioleate), and led to a calculation of the retention indices with an error of 4.1 index units (with respect to the experimentally determined value).

The five polarities of the substances are interpreted by Rohrschneider as a measure of the orientation forces (factor e), charge-transfer forces (donor and acceptor forces a and d), and hydrogen bonding (H-donor b, H-acceptor c). Bearing in mind that no quantitative relations have been established between the polarity factors and molecular physical parameters, this interpretation may be considered somewhat empirical. The polarity of the stationary phase and of the solute has no real physical significance as these quantities are dependent on arbitrarily chosen polarity factors. From experimental data the conclusion has been reached that substances that may form hydrogen bonds have a high b-value, and substances containing a carbonyl group have a positive c-value. It has also been concluded that fluoroalkylsilicone columns which retain selectively the carbonyl combinations have a very big z-value, and that the superior members of a homologous series have very similar polarity factors.

Rohrschneider's method is reminiscent of the work of Pierotti et al.[162, 297] who have developed an empirical approach for the prediction of activity coefficients in gas–liquid chromatography. Essentially they assign a term for each type of interaction between solute and liquid phase and sum all the terms to obtain the total interaction.

On the basis of Rohrschneider's concept, Takács et al.[301] supposed that the gas chromatographic interaction between the solute and the stationary phase could be characterized not only by the difference in the retention indices, but also by the quotients of their indices, which is very useful especially when a computer programme is used.

Recent discussions concerning the Rohrschneider method of characterizing liquid stationary phases usually consider the number and the chemical nature of the standard substances used and the most exact possible explanation of the physical significance of the polarity factors.[302-305]

McReynolds[306] recommends for the characterization of the liquid phases the use of the retention index differences of ten standard substances: benzene, butanol, 2-pentanone, nitropropane, pyridine, 2-methyl-2-pentanol, 1-iodo-butane, 2-octyne, 1,4-dioxan, and cis-hydrindane. Without discussing his proposal in detail, the following remarks must be made. There is a basic difference between Rohrschneider's method and that of McReynolds, which was developed after the Rohrschneider concept. While Rohrschneider's method is suitable not only for the characterization of the stationary phase polarity but also for the pre-calculation of the corresponding retention, the method of McReynolds, although it may possibly be used for this pre-calculation, can be used in its present form only for the characterization of the liquid phase. Certainly the optimum selection of the compounds for standard substances as well as the prediction of the physical signification of polarity factors can be the result only of widespread theoretical and practical co-operation.

Other means of stationary phase characterization may be found.[307, 308] Thus by comparing the frequency shifts in the infrared spectra of the chromatographed compounds, resolved by the polar groups of the stationary phase, with the logarithms of the retention relative to n-hexane, Ecking et al.[307] derived a relation that allows a polarity echelle of the stationary phases to be set up for every functional group by simple infrared spectroscopic measurements. Squalane was decided upon as the reference stationary phase, so that the frequency shift Δv_i^{ass} is defined as being the modification of the wave number of the valence vibration v_i of the polar group of the chromatographed compound under the influence of the stationary phase minus the wave number corresponding to the absorption in a non-polar stationary phase (squalane):

$$\Delta v_i^{\mathrm{ass}} = v_{i(\mathrm{squalane})} - v_{i(\mathrm{stationary\ phase})} \quad [\mathrm{cm}^{-1}], \qquad (126)$$

where $v_{i(\mathrm{squalane})}$ is the wave number corresponding to the absorption of a polar group i of the compound dissolved in squalane, and $v_{i(\mathrm{stationary\ phase})}$ the wave number of the same polar group i of the component dissolved in a polar stationary phase.

The frequency shifts are interpreted as being the product of the contributions between the interaction of the proton donors and acceptors.[309] If P_i is the term corresponding to the polar group of the compound and T_j is the term corresponding to stationary phase j, the frequency shift can be written as

$$\Delta v_i^{\mathrm{ass}} = P_i T_j. \qquad (127)$$

If $\Delta v_i^{\mathrm{ass}}$ is measured for the polar group i of the same compound in different stationary phases j, then v_i is constant, the only variable being T_j whose variation is a function of the properties of the stationary phase. Thus the variations of $\Delta v_i^{\mathrm{ass}}$ are a direct measure of the variation of the polar properties of the stationary phase. The classification of stationary phases on the basis of $\Delta v_i^{\mathrm{ass}}$ for a polar group i will lead to the establishment of a polarity scale with a defined physical value because $\Delta v_i^{\mathrm{ass}}$ is a measure of energy. A polarity scale corresponds to each different polar group of the component; thus the polarity scale is valid only for a well-defined class of substances. It must not be forgotten that the alkyl group has an important influence on the magnitude of a frequency shift.

Difficulties arise when the stationary phase in question (which is considered as the solvent) has the same functional groups as the molecule of the component. It is obvious that in such instances an exact measure of the frequency shift is impossible owing to the absorption band interference.

Robinson and Odell[308] proposed a system of "standard" retention indices for the characterization of stationary phases. They defined the standard retention index as

$$I_{\mathrm{std}} = 100n + 100\,\frac{\log BP_x - \log BP_n}{\log BP_{n+1} - \log BP_n}, \qquad (128)$$

where BP_x, BP_n, and BP_{n+1} refer to the boiling points of the compound and the n-alkanes with n and $n+1$ carbon atoms, respectively, similar to the method of Kováts.[296-297] A standard retention index difference (ΔI^*) is obtained by comparison of the standard and experimental retention indices:

$$\Delta I^* = I_{std} - I_x^T, \tag{129}$$

where I_x^T is the retention index of the compound at temperature T on stationary phase x.

The hydrocarbons studies were divided into seven classes according to their structure, namely, I, n-alkanes; II, branched alkanes; III, n-alkenes; IV, branched alkenes; V, cycloalkanes; VI, cycloalkenes; VII, aromatics. It has been found that ΔI^* values vary little within a group but there are marked differences between some groups. The stationary phases squalane and SE-30 thus have an enhanced retaining effect on cyclic, non-aromatic compounds, the retention indices of these compounds being higher than standard values. They have called this the "cyclic" effect. Because of this effect the authors consider squalane to be unsatisfactory as a reference stationary phase and suggest that standard retention indices be used in future to define departures from ideality (as measured by ΔI^*) which may be used to characterize stationary phases.

The average magnitude of ΔI^* for the different classes on a given stationary phase gives an indication of how the phase behaves with respect to these compounds. For instance, the large negative value for aromatics on benzyldiphenyl shows that aromatic compounds are retarded in their passage through the column in comparison with alkanes or alkenes.

The most widely used method of classification and characterization of stationary phases remains that of Rohrschneider. The method is used in order to characterize the polarity of the liquid stationary phases[310, 311] A wide range of gas chromatographic liquid phases produced in the USSR[312] and in the USA[313] are characterized by the manufacturers by the Rohrschneider constants. Table 9 gives a number of commonly used stationary phases together with their Rohrschneider constants, selected from the literature. Tabulation of the stationary phases with these constants has proved to be useful for the prediction of separations, from

TABLE 9. SOME STATIONARY PHASES WITH THEIR ROHRSCHNEIDER CONSTANTS

Liquid phase	Rohrschneider constants				
	x	y	z	u	v
Apiezon L	0.32	0.39	0.25	0.48	0.55
Diethylene glycol succinate	4.93	7.58	6.14	9.50	8.37
Methyl-phenylsilicone oil (DC-710)	1.05	1.50	1.61	2.51	1.90
Tricresyl phosphate	1.74	3.22	2.58	4.14	2.95
Fluorosilicone QF-1	1.41	2.13	3.55	4.73	3.04
Neopentylglycol succinate	2.68	4.88	3.87	6.13	5.21
Methylsilicone (OV-1)	0.16	0.20	0.50	0.85	0.48
Methylphenylsilicone (OV-17)	1.30	1.66	1.79	2.83	2.47
Liquid methylsilicone (OV-101)	0.16	0.20	0.50	0.85	0.48
Trifluoropropyl-methylsilicone (OV-210)	1.41	2.13	3.55	4.73	3.04
Cyanopropylmethylphenylmethyl-silicone (OV-225)	2.17	3.20	3.33	5.16	3.69
Silicone gum rubber SE 30 (methyl)	0.16	0.20	0.50	0.85	0.48
GE Nitrile silicone gum XE-60	2.08	3.85	3.62	5.33	3.45
Polyoxyethylene-sorbitan-mono-oleate (Tween 80)	2.14	4.20	2.78	5.20	3.65
Polyphenylether (5-ring)	1.75	2.27	2.34	3.26	2.84
Carbowax 20 M	3.18	5.33	3.81	7.02	5.04

the known structures of the compounds of interest. For exemple, dieldrin differs from aldrin only by the addition of an epoxy (electron donor) group. From the high z-value for QF-1, one would predict correctly that dieldrin is retarded more (and thus eluted later) on this phase than, for example, on DC-200, OV-1, or OV-17. Dieldrin is eluted after and separated from aldrin on QF-1.

Stationary phases with five similar values are essentially identical chromatographically (e.g. OV-1, OV-101, SE-30, DC-200).

With the exception of certain porous polymers[314], the Rohrschneider system has not been used for the characterization of solid stationary phases. This is quite understandable if we take into account that not so long ago the only adsorbents used were silica gel, molecular sieves, and active charcoal. But remembering that in the last years a very large number of new solid stationary phases have been proposed, we believe that the Rohrschneider method could be used succesfully for homogenous adsorbents with a well-defined chemical structure.

3.3 SELECTIVITY

The selectivity of a given stationary phase is the ability of the stationary phase to separate a pair of compounds (solutes) 1 and 2 having very similar properties such as boiling points,[315] molecular weights,[316] number of carbon atoms in the molecule,[317] or vapour pressures.[318] It is obvious that if the selectivity is not an intrinsic property of a stationary phase, it depends of the choice of components 1 and 2.

Selectivity can be expressed in terms of the value of the separation factor $\alpha_{1,2}$, a quantity that depends only on the stationary phase–solutes system ($\alpha_{1,2} = p_2^0 \gamma_2^\infty / p_1^0 \gamma_1^\infty$), where p^0 are the vapor pressures of the pure solutes at the column temperature and γ^∞ are the activity coefficients at infinite dilution of the solutes in the stationary phase. It is considered that a separation can be performed for $\alpha_{1,2}$ values greater than 1.05.

It is obvious that the chromatographic separation process is not related to the ratio of the vapour pressures of the two solutes. In practice, only the ratio of the activity coefficients at infinite dilution is considered, and the selectivity is defined as the possibility of a stationary phase separating two solutes that have the same vapour pressure at the column temperature. For the selectivity coefficient of a stationary phase $S_{1,2}$:

$$S_{1,2} = \gamma_2^\infty / \gamma_1^\infty. \tag{130}$$

From the relation between the activity coefficient and the excess partial molar free energy of mixing at infinite dilution per mole of solute ($\Delta \bar{G}_e^\infty = RT \ln \gamma^\infty$), eqn. (130) can be rewritten

$$S_{1,2} = \frac{e^{\Delta \bar{G}_{e,2}^\infty / RT}}{e^{\Delta \bar{G}_{e,1}^\infty / RT}} = \frac{e^{\Delta \bar{H}_{e,2}^\infty / RT} e^{-\Delta \bar{S}_{e,2}^\infty / R}}{e^{\Delta \bar{H}_{e,1}^\infty / RT} e^{-\Delta \bar{S}_{e,1}^\infty / R}}. \tag{131}$$

Negative $\Delta \bar{G}_e^\infty$, negative $\Delta \bar{H}_e^\infty$, and positive $\Delta \bar{S}_e^\infty$ values all favour the dissolution of the solute from the gas phase. $\Delta \bar{G}_e^\infty$ may be considered to be the result of a number of interactions between solute and stationary phase involving all permutations of stationary phase and solute functional groups. Considering that the total interaction between each solute and the stationary phase is the sum of a number of interactions x, each interaction having an associated heat H and entropy S, eqn. (131) becomes[319]

$$S_{1,2} = \frac{[e^{H_{21}/RT}e^{H_{22}/RT} \ldots e^{H_{2x}/RT}][e^{-S_{21}/R}e^{-S_{22}/R} \ldots e^{-S_{2x}/R}]}{[e^{H_{11}/RT}e^{H_{12}/RT} \ldots e^{H_{1x}/RT}][e^{-S_{11}/R}e^{-S_{12}/R} \ldots e^{-S_{1x}/R}]}. \quad (132)$$

From eqn. (132) it can be seen that favourable selective interactions reinforce each other and are desirable.

The effect of selective interactions is enhanced by increasing the concentration of selective interacting groups in the stationary phase provided unfavourable entropy factors do not appear and adsorption on the stationary phase does not complicate the separation. Thus, if it is desired to utilize a specific interaction for selective separation, it is appropriate to minimize other solvating (attractive) interactions which may act counter to the desired separation.

One effect of specific group interaction may be illustred from the results of investigations of the behaviour of some salts of di-2-ethylhexyl-phosphoric acid (HDEHP) with lithium, sodium, potassium, thorium, zirconium, uranium(VI), and [Co(en)$_3$]$^{3+}$ as stationary phases in comparison with Apiezon L and the free acid.[320] Chemical analysis and infrared spectrophotometry show that these compounds have the following formulae: LiDEHP; NaDEHP; KDEHP; [ZrO$_{0.5}$(NO$_3$)(DEHP)$_2$]$_n$; Th(NO$_3$)(DEHP)$_3$; UO$_2$(DEPH)$_2$; Co(en)$_3$(DEHP)$_3$.[321] Table 10 gives the differences between the free energy of solution for the various salts and HDEHP itself and Apiezon L ($\Delta G_{\text{Me(DEHP)}_n} - \Delta G_{\text{Apiezon L}}$).

The interaction of the various homologous series and the metal compounds is not noticeable for the alkanes, aromatic hydrocarbons, and ketones. The higher values measured for these solutes may be attributed to their solubility in Apiezon L. The strong interaction of 2-pentanone and of the alcohols with the HDEHP column may be ascribed to their ability to form hydrogen bonds. For alcohols, the complexing action increases from tertiary alcohols to primary alcohols and is more pronounced for the zirconium, cobalt, and thorium compounds. Very likely

TABLE 10. $\Delta G_{Me(DEHP)_n} - \Delta G_{Apiezon\ L}$ AT $125°C^{(320)}$

Solute	Salt							
	Li	Na	K	Zr	Th	U	Co	Acid
Heptane	+1.95	+1.78	+1.56	+0.87	+1.71	+1.12	+0.67	+0.06
Benzene	+1.33	+1.20	+0.98	+0.64	+1.18	+0.45	+0.40	−0.16
Monochlorobenzene	+1.37	+1.33	+1.10	+0.59	+1.15	+0.33	+0.34	−0.26
Aniline	+0.90	+0.61	−0.07	−0.12	+0.61	—	−0.82	—
Nitrobenzene	+1.32	+1.23	+0.90	+0.29	+0.92	−0.14	−0.08	—
Ethanol	+0.07	−0.28	−0.95	−2.76	−1.24	−0.08	−1.19	−1.20
s-Butyl alcohol	+0.83	+0.58	−0.26	—	+0.57	+0.13	−0.81	−1.09
t-Amyl alcohol	+1.34	+1.12	+0.25	—	+1.11	+0.37	−0.53	−0.98
2-Pentanone	+1.35	+1.31	+0.88	+0.51	+1.05	+0.39	−0.04	−0.34
α-Picoline	+1.19	+1.19	+0.93	−0.28	−0.03	−0.62	+0.14	−0.52

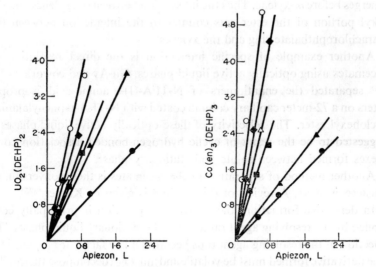

FIG. 5. Relative retention of a homologous series of organic compounds on $UO_2(DEHP)_2$, $Co(en)_3(DEHP)_3$ columns relative to an Apiezon L column.$^{(320)}$ (●) alkanes; (▲) aromatic hydrocarbons; (■) ketones; (△) tertiary alcohols; (◆) secondary alcohols; (○) primary alcohols.

the high value of interaction between these compounds and primary alcohols may be attributed to a structural change of the compounds. Figure 5 shows the behaviour of various homologous series on $UO_2(DEHP)_2$ and $Co(en)_3 (DEHP)_3$ columns vs. an Apiezon L column. Similar behaviour is observed for other metal compounds, the slope, however, being different for each homologous series. The slopes of these lines are a measure of the interaction which takes place between a set of homologous compounds and the stationary phase.

The effect of counterselective interactions is illustrated by the data for the m-/p-xylene separation on tetrachlorophthalates.[322] The interaction involved is a soft charge-transfer with the electron-deficient tetrachloro-phthalate ring. The p-isomer, which has the higher vapour pressure, is held selectively in the column because it has a lower ionization potential than the m-isomer, but the p-/m-xylene separation factor diminishes as the ester alkyl groups increase in length. The alkyl groups result in greater solubility, but from similar long-chain aliphatic substrates, p-xylene emerges before m-xylene. Thus the interaction between the xylenes and the alkyl portion of the ester acts counter to the interaction between the tetrachlorophthalate ring and the xylenes.

Another example of specific interaction is the direct resolution of racemates using optically active liquid phases. Gil-Av and coworkers[323, 324] separated the enantiomers of N-TFA-(I)-α-amino-acid isopropyl esters on a 72-meter capillary column coated with N-TFA-L-phenylalanine cyclohexyl ester. The selectivity of these optically active liquid phases is suggested to be the result of cyclic hydrogen-bonded association complexes formed between solute and stationary phase.

Another method of separation of the racemates is their conversion to diastereoisomers. A review of this method is given by Karger.[325]

In derivative formation, the degree of separation is principally controlled by the resolving agent rather than the stationary liquid phase. The selection of the resolving agent thus becomes a key factor for separation. The derivative formed must be volatile and must not decompose thermally. It should be readily synthesized, if possible, in very high yield. It should be simple to decompose chemically should it be necessary to isolate the pure solute after the separation. The particular problem will dictate the resolving agent selected. In order to provide a basis for prediction, the diastereoisomeric structural factors responsible for separation must be investigated.

Karger *et al.*[326, 327] have indicated that conformational immobility of groups attached to the asymmetric centres greatly affects resolution. From this finding, these authors have chromatographed N-TFA-L-prolyl-2-methylindoline (I) and found a $\Delta(\Delta G^0)$ value of -298 cal/mol, the largest value reported for diastereoisomeric separations.

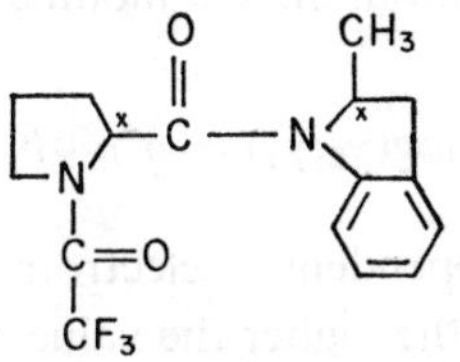

(I)

An examination of the structure of this diastereoisomer reveals that both asymmetric centres are part of ring systems, so that groups attached to these centres are to a large extent conformationally immobile. These systematic structural variation studies are thus of value in the prediction of resolving agents to produce good separations.

Derivative formation is not used only for the gas chromatography of racemates, this technique being used always when through "derivatization" an unstable or non-volatile material may be made volatile. From this point of view we consider as most important the use of gas chromatography in the separation of metal chelates.[328–330]

In the study of the selectivity of stationary phases it is often of interest to attempt to evaluate the effect of systematically varying substituents in the solute. For example, an attempt might be made to correlate the logarithm of the measured activity coefficient of a solute containing a benzene ring with the substituent on the ring. This could be a correlation with a Hammett substituent constant or one derived from a similar type of linear free energy relationship.[331–333]

The Hammett equation was originally written as follows:

$$\log(K_2/K_1) = \sigma^* \varrho, \tag{133}$$

where K_2 and K_1 are rate or equilibrium constants for reactions 1 and 2, respectively, σ^* is a parameter dependent only on the substituent and its position, and ϱ is a parameter dependent on the nature of the reaction but not on the substituents.[334] Equation (133) was found to be applicable to m- and p-substituted benzene derivatives but not to o-substituted derivatives. The equation was soon extended to o-substituted derivatives and also to aliphatic compounds, but different constants were defined in these instances.[335] For use in gas chromatography it is convenient to modify the Hammett equation. Such a modification was given by Karger et al.:[331, 332]

$$\log(\gamma_2/\gamma_1) = \sigma^*\varrho + b, \tag{134}$$

where b is a constant independent of electronic factors considered in the $\sigma\varrho$ term. It is evident that the higher the value of constant ϱ, the stronger is the influence of the stationary phase on the resolution. The constant ϱ is therefore a direct measure of the stationary phase selectivity.

Kremser et al.[333] have calculated from the plot of $\log(\gamma_2/\gamma_1)$ vs. σ^* the values of ϱ for m- and p-substituted phenols as solutes in triphenylmethanol and triphenylmethane as stationary phases at 171.5°C. The value ϱ for triphenylmethane (0.22) is much lower than for triphenylmethanol (0.625). This corresponds to the fact that ϱ measures the effect of the stationary phase on resolution, which is greater for triphenylmethanol because the hydrogen bonding between solute and stationary phase molecules is much stronger than for triphenylmethane. Almost all commercial polyester and other types of phases have much higher ϱ values, usually about unity.

Up to the present time, most workers have used stationary liquid phases which have low vapor pressures at the operating temperature of the column. These liquids have been employed in order to prevent the stationary phase from bleeding off the column with the subsequent change in column characteristics with time and subsequent decrease in the lifetime of the column. Thus the stationary phases used in gas–liquid chromatography have mainly been limited to poorly characterized polymers or other high molecular weight species. However, under such circumstances it would seem interesting to test simple, well-defined liquids such as water and formamide for use as stationary phases in gas–liquid chromatography. With simple polar solvents, the concentration of polar groups per unit volume will be much larger than with polymeric solvents. When solute–solvent orientation forces are selective for a given solute mixture, enhanced separation may result for such simple solvents relative to the usual stationary phases.

The use of volatile stationary phases in gas–liquid chromatography is a largely untapped approach to selectivity. Purnell and Spencer[336] wrote a preliminary report on the separation of chloromethanes with water as the stationary phase. Separation of n-C_1–C_5 alcohols with water as the stationary phase is made in reverse order to their molecular weights.[337] Separation is based almost totally on the polarity of the solute. Extremely rapid elution times of hydrocarbon are possible using water as a stationary phase.[338, 339]

Novak and Janak[340] have also shown this rapid elution of alkanes relative to polar molecules, using formamide as the stationary phase. This rapid removal of hydrocarbons can be highly useful if it is necessary to separate hydrocarbons from some polar solutes. However, at the operating temperatures used in gas chromatography the low molecular weight, simple solvents will probably be volatile and bleed off the column. The non-stationary character of the column is undesirable, but changes in the amount of solvent can be ascertained by weighing the column before and after a chromatographic run. If this weighing step is inconvenient, it should be possible to maintain a fairly constant liquid loading on the column by pre-saturating the carrier gas with the solvent. Pre-saturation can be accomplished either by using an earlier column in which the packing contains the volatile solvent or by bubbling the carrier gas through a reservoir of solvent before entering the column.[341]

The potential value of simple, low molecular weight stationary phases for gas–liquid chromatography would appear to be such that efforts should be made to develop the principles and techniques of the method. Highly selective and specific separations will probably result from this development.

Components that do not interact through their functional groups with the liquid stationary phase (for example aliphatic and aromatic compounds) may be treated by taking into account the Hildebrand–Scatchard equation for regular dilute solutions:

$$\gamma_{solute} = e^{V_2(\delta_s - \delta_2)^2/RT} \cong e^{\Sigma H/RT} \cong e^{\Delta H_m/RT}, \tag{135}$$

where δ_s is the solubility parameter of the stationary phase and δ_2 is the solubility parameter of the solute. V_2 is the molar volume of the solute. The term ΣH represents a combination of the general solution interactions which were indicated in eqn. (132) and is approximately the heat of mixing. The Hildebrand definition of a regular solution supposes that specific orienting and chemical effects are not present, and that the distribution and orientation of the solutes in the solvent are random.

Although eqn. (135) was derived with a number of restrictions for spherical non-polar molecules, it is surprisingly valid for a number of mixtures involving moderately polar molecules but without specific interactions of any considerable energy. Thus δ is a qualitative measure of the general solution forces present in pure materials, whether stationary phase or solute. From eqn. (135) it can be seen that the effect of any small differences between the Hildebrand parameters of two solutes is enhanced by using a stationary phase of very different δ-value so that the ratio of the heats of mixing of the solutes in the stationary phases is greatest and the selectivity coefficient is larger. Where selective interactions are desired, it still seems desirable to pick solutes or stationary phases with appending groups which tend to give the greatest difference in δ-values between the stationary phase and solutes. This approach gives the greatest difference in excess partial molar free energy.

The theory of regular solutions has been used by many authors for the prediction of activity coefficients and retention data.[122, 342-345]

For gas chromatographic behaviour in regular solutions, Rohrschneider[344] has proposed the following rules:

(i) Substances with identical boiling points can be separated only if they differ in their molar volume. The smaller molecule will be retained more strongly with increasing internal pressure (δ^2), i.e. increasing "polarity" of the stationary liquid. Two substances with identical boiling points differ in their "polarity" if their molar volumes are different. The smaller molecule will appear to be the more "polar".

(ii) Two substances with identical molar volumes differ in "polarity" if they have different boiling temperatures or heats of evaporation. The molecule having the higher evaporation heat or the higher boiling temperature will appear to be the more "polar".

(iii) For two solutes not separated on a liquid stationary phase one may write

$$\delta_s(V_1 - V_2) = 2(V_1 \delta_1 - V_2 \delta_2), \qquad (136)$$

where indices 1 and 2 refer to the components and s to the stationary phase.

In another separating column with higher internal pressure δ' ($\delta' > \delta$), the left-hand side of eqn. (136) will always be greater that the right-hand side:

$$\delta'_s(V_1 - V_2) > 2(V_1 \delta_1 - V_2 \delta_2). \qquad (137)$$

Thus, the size difference of the substances to be separated determines the retention sequence, according to the following equation:

$$\log(t'_2/t'_1) = \frac{(V_1 - V_2)\delta_s^2 - 2(V_1 \delta_1 - V_2 \delta_2)\delta'_s}{2.303RT}. \qquad (138)$$

If V_2 is smaller than V_1, the value of $\log(t'_2/t'_1)$ will have a positive value, and the smaller molecule will emerge after the larger one ($t'_2 > t'_1$). If two substances are not separated on a stationary liquid phase, the substance with the smaller molar volume will emerge after that with the greater volume, on another liquid phase with higher internal pressure.

(iv) The "polarity" of a stationary liquid phase is significantly dependent on the internal pressure. Of decisive influence on the retention at least in regular solutions are the internal pressure and boiling point of the stationary liquid phase and the molar volume of the dissolved substances.

As most systems in gas chromatography are composed of components with small molecules and non-volatile stationary phases with big molecules, the ability to neglect the entropy term is an important simplification.

The difference in size is expected to cause an appreciable deviation from ideality which may be expressed in terms of an athermal or entropy contribution to the activity coefficient.[123, 346, 347] With the Flory–Huggins treatment, this athermal contribution γ_a^∞ at infinite dilution may be expressed as[346]

$$\gamma_a^\infty = e^{-\Delta S/R} = (1/m)e^{(1-1/m)}, \tag{139}$$

where m is the size ratio of stationary liquid phase molecules to solute molecules, approximated as the ratio of molar volumes. Except for very polar stationary phases, the gas chromatographic situation will generally involve values of $m > 1$ if the stationary phase is relatively non-volatile. When $m > 1$ and the difference in molar volumes of solutes tends to reinforce separation, it is desirable to use higher molecular weight stationary phases where $m \simeq 5$–10. Where this effect does not reinforce the separation, lower molar volumes are desirable. About 90% of the size effect apparently occurs before $m > 10$, so that excessively high molecular weight stationary phases are not necessary to emphasize size differences.

Besides the entropy effects linked to the different size of the solute and stationary phase molecules, gas chromatography also uses the ability of some stationary phases to separate molecules partly according to their shape. The most obvious approach is to utilize an oriented liquid phase such as a liquid crystalline material as a stationary phase. Liquid crystals are unusual states of matter intermediate between crystalline solids and normal isotropic liquids.[348, 349] Mechanically they behave as liquids, but the molecules in them possess some order in that they cannot rotate freely. Generally, these compounds initially melt to give oriented mesophases which may be smectic, nematic, or cholesteric; these crystalline phases lose orientation at a transition temperature to give isotropic liquids. Smectic phases possess two-dimensional orientation in which rod-like molecules are arranged in parallel to give a structure with layers having a thickness approximately the length of the molecules comprising the layer. In the nematic phase, parallel orientation is maintained, but layered structure does not exist, and liquid molecules are free to move within the limits of parallel configuration. The smectic phase, occurring at a lower temperature than the nematic phase, is the more ordered mesophase because there is constraint not only in parallel but also a layer structure forms.

A liquid crystal that can exist in the smectic and nematic form is 4,4′-di-n-heptyloxyazobenzene:

$$n-C_7H_{15}O-\langle\bigcirc\rangle-N=NO-\langle\bigcirc\rangle-O-n-C_7H_{15}$$

This compound is in the smectic form between 75–95°C and in the nematic form between 95–127°C.

A third mesophase which sometimes occurs for other liquid crystal systems is the cholesteric phase. The cholesteric transition phase is one exhibited by cholesteryl derivatives and similar compounds in which some molecular rotation is permitted in the plane of each layer in a layered two-dimensional structure. Among the compounds that show cholesteric form are cholesteryl benzoate in the temperature range 149–180°C and cholesteryl nonanoate that has a smectic form at 75°C and a cholesteric form between 79–90°C. The cholesteryl esters can exhibit smectic and cholesteric mesophases but never a nematic phase.

When liquid crystals are used as stationary liquid phases, the solute retention mechanism must involve in part a fitting of the molecules in the lattice structure. The ability of the molecules to fit will depend on their shape. For example, a p-isomer, being more elongated than a m-isomer might be expected to fit between the molecular layers of a mesophase more readily, and thus be eluted later than the m-isomer. This result can markedly be seen in Fig. 6, which shows the separation of m- and p-chlorotoluenes on p-azoxyanisole (nematic temperature range 120–135°C).[350]

An examination of the figure reveals that the isomers are well separated at 127°C and very poorly separated at 141°C. Thus in passing from the nematic to the isotropic liquid state, the separation decreases because an isotropic liquid does not discriminate according to shape. It should also be noted that the p-isomer was eluted later than the m-isomer.

The solubility of solutes in liquid crystals also depends on factors other than shape; it is therefore neither surprising nor disappointing to find examples where liquid crystals form ineffective stationary phases.

Gas–liquid chromatography can also be used to determine the transition temperature for liquid crystals.[351–353] At the transition temperature there is a sharp change in retention volume for a given solute. A review

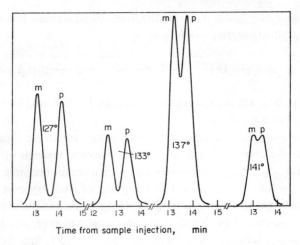

Time from sample injection,   min

FIG. 6. Chromatogram of the separation of *m*- and *p*-chlorotoluene at various temperatures. Liquid phase: *p*-azoxyanisole.[350]

on the use of liquid crystals in gas chromatography was written by Kelker and von Schivizhoffen.[354]

The simplest approach to a relation between spatial and energetic factors with respect to solubility is presented by the Hildebrand theory. From this point of view, Kelker and Verhelst[352] showed that it seemed feasible to start a discussion with a semi-quantitative picture and to consider that liquid–crystalline phases generally have a higher density (i.e. a smaller molar volume) and a larger heat of evaporation than the corresponding isotropic liquids. So the solubility parameters of a mesophase δ_s^A can be regarded quite generally as being larger than the δ-value of the isotropic phase δ_s^i under the same conditions. Hence, using the Hildebrand treatment once with respect to the isotropic phase and then regarding the anisotropic state of the same substance, it follows that δ_1 will increase during transition from the isotropic into the anisotropic state. From this a corresponding enlargement of the γ_{solute} value follows because the condition $\delta_s > \delta_2$ in eqn. (136) is fulfilled for both states of the stationary phase. More generally, the following rule may be stated: the excess free energy of a solute is more positive (or less negative) and the heat of mixing is correspondingly more positive (or less negative) than the corresponding values of these thermodynamic functions in the isotropic state. The reverse has never been observed.

The gas chromatography and thermodynamics of divinylbenzene separations on 4,4'-dihexoxyazoxybenzene liquid crystals (nematic range 81.0–128.2°C) was reported by Zielinski et al.[141] Thermodynamic treatment of the data has illustrated that the m-isomers have a lower solubility (resulting in shorter retention times) than the respective p-derivatives owing to a greater enthalpy requirement for solution in the rod-like ordered solvent. The partial molar enthalpies and entropies of solution are more negative for the p-isomers than for the m-isomers. More negative enthalpy values (more exothermic) are indicative of stronger solute–solvent interactions, whereas more negative entropy values reveal a more ordered solution state. This behaviour is consistent with the postulate that p-substituted solutes, being more rod-like spatially, "fit" better into, and thereby interact more strongly with, the rod-like ordered solvent. In other words, in going from a completely disordered vapour state (similar for both isomers), to an ordered liquid state, the p-isomer sacrifices more translational and rotational freedom, but in return, its favourable geometry allows it to interact more strongly with the aligned liquid crystal molecules. On balance, its entropy loss is overcome by its enthalpy gain, making it more soluble than its m-counterpart (which, although its motion is less restricted, has a greater enthalpy requirement for solution relative to an ideal solution); 4,4'-dihexoxyazoxybenzene liquid crystals as a stationary phase becomes far less selective in the isotropic temperature region (in which the long range disorientation of the solvent molecules prevents any appreciable alignment and, hence, selective retention of the p-isomers).

Chow and Martire[355] proposed a molecular interpretation of solubility in nematogenic solvents. Two nematogenic compounds, 4,4'-dimethoxyazoxybenzene (p-azoxyanisole, PAA), and 4,4'-dihexoxyazoxybenzene (DHAB) were studied:

$$PAA\left(solid \xrightarrow{117.5°} nematic \xrightleftharpoons{134.2°} isotropic\right),$$

$$DHAB\left(solid \xrightarrow{81.0°} nematic \xrightleftharpoons{128.2°} isotropic\right).$$

The following approximate expressions can be derived from basic statistical mechanics:

$$\gamma^{\infty}_{\text{solute}} \approx (1/\overline{Z}_s)_{\text{transl}} (\overline{Q}_g/\overline{Q}_s)_{\text{rot}} (\overline{Q}_g/\overline{Q}_s)_{\text{vib}} \, e^{-1}, \tag{140}$$

$$\Delta \overline{H}^{\text{soln}}_2 = \Delta E_{\text{pot}} + \Delta E_{r+v} - RT, \tag{141}$$

where $\gamma^{\infty}_{\text{solute}}$ is the activity coefficient of the solute at infinite dilution; $\overline{Z}_{\text{transl}}$ is the solute configurational molecular partition function (the potential energy contribution to the translational partition function); $\overline{Q}_{\text{rot}}$ and $\overline{Q}_{\text{vib}}$ are, respectively, the solute rotational and vibrational molecular partition functions (indices g and s are referred to the gas and the stationary phase respectively); $\Delta \overline{H}^{\text{soln}}_2$ is the heat absorbed or given out when an infinite dilution transfer of solute takes place from the ideal gaseous mixture to the solution; ΔE_{pot} is the solute internal energy loss owing to interaction with the solvent; and ΔE_{r+v} is the combined solute rotational and vibrational internal energy loss.

In the derivation of eqn. (140) it is assumed that:

(a) The translational, rotational, and vibrational modes act independently.

(b) In the gaseous phase (treated as ideal) the solute molecular energy states are unaffected by the environment; thus $(\overline{Z}_g)_{\text{trans}}$ is unity and $(\overline{Q}_g)_{\text{rot}}$ and $(\overline{Q}_g)_{\text{vib}}$ depend only on the solute molecular structure.

(c) In the liquid state (isotropic and anisotropic) much of the solute molecular motion may be restrained or lost because of solvent environmental effects, leading to changes in the configurational, rotational, and vibrational partition functions upon solution. Further, it is assumed that while the solute possibly affects the local structure of the solvent, the long-range structure and the molecular energy states (on the average) of the solvent are unaltered because of the infinite dilution condition of the solute.

According to eqn. (140), the magnitude of $\gamma^{\infty}_{\text{solute}}$ should be influenced by three effects — a potential energy effect, a rotational effect, and a vibrational effect.

1. *Potential energy effect.* Upon dissolution in an anisotropic or isotropic solvent, solute–solvent interactions take place. The stronger the potential energy of interaction, the greater the potential energy (or con-

figurational) partition function $(Z_s)_{transl}$ and the lower the value of γ^{∞}_{solute}. Thus solubility is favoured by strong solute–solvent interaction.

2. *Rotational effect*. The rotational partition function for a given free rotational mode is proportional to the square root of the moment of inertia corresponding to that mode. If, upon solution, certain rotational motion is restricted or lost, the ratio $(\bar{Q}_g/\bar{Q}_s)_{rot}$ becomes greater than unity. The greater the loss of rotational motion, the greater the value of $(\bar{Q}_g/\bar{Q}_s)_{rot}$ and the greater the value of γ^{∞}_{solute}. Hence, solubility is not favoured by a loss of rotational motion.

3. *Vibrational effect*. There are energy barriers of up to a few kilocalories, for internal rotation in flexible (non-rigid) molecules such as the n-alkanes, but many conformations are possible in the gaseous phase (although the zigzag conformation is most probable). However, if these molecules become confined to a solvent system where certain conformations become highly improbable (e.g. owing to space restrictions), then a conformational energy loss would occur. As this loss of motion would appear in the vibrational modes, it would lead to a value of $(\bar{Q}_g/\bar{Q}_s)_{vibr}$ greater than unity. Again, this situation does not favour solubility, and acts to increase the value of γ^{∞}_{solute}.

The corresponding free energy terms of the two additive contributory factors to $\Delta\bar{H}^{soln}_2$ act in opposite directions in influencing the magnitude of γ^{∞}_{solute}. Hence when examining a variety of solutes, no direct correlation between $\Delta\bar{H}^{soln}_2$ and γ^{∞}_{solute} (or $\Delta\bar{S}^{soln}_2$) values is to be expected. However, when considering a homologous series of solutes, Chow and Martire[355] find that there is a linear relationship between $\Delta\bar{H}^{soln}_2$ and $\Delta\bar{S}^{soln}_2$ in both phases of PAA and DHAB.

Utilizing eqns. (140) and (141), a molecular interpretation of solubility in nematogenic solvents can be made. In nematic PAA, the $\Delta\bar{H}^{soln}_2$ values for xylenes follow the trend p-xylene $>$ o-xylene $>$ m-xylene. As these isomers are fairly rigid, little vibrational energy loss is to be expected upon dissolution. Therefore, ΔE_{r+v} should follow the probable order of rotational energy loss, i.e. the order of solute molecular length-to-breadth ratio: p-xylene $>$ m-xylene $>$ o-xylene.

From the trends in $\Delta\bar{H}^{soln}_2$ and ΔE_{r+v}, m-xylene should have the smallest ΔE_{pot} value. o-Xylene (and, to a lesser extent, m-xylene) is capable of some dipolar interaction with the solvent. However, p-xylene should be in a spatially more favourable position for dispersion interaction with the solvent. Thus, as it is not certain which of the two isomers has the greater ΔE_{pot} value, it will be assumed that they are comparable. On this assumption, the trend in γ^{∞}_{solute} can be readily explained by the usual arguments based on eqn. (140). In the isotropic liquid phase, $\Delta\overline{H}^{soln}_2$ follows the trend of solute polarity, and γ^{∞}_{solute} follows the $\Delta\overline{H}^{soln}_2$ value (i.e. the larger is ΔH, the smaller is γ). The potential energy effect obviously governs these trends. The values of $\Delta\overline{H}^{soln}_2$ (kcal/mol) and $\Delta\overline{S}^{soln}_2$ (cal/mol deg.) are listed in Table 11.[355]

TABLE 11. PARTIAL MOLAR ENTHALPIES AND ENTROPIES OF SOLUTION IN PAA[355]

Solute	Nematic		Isotropic	
	$\Delta\overline{H}^{soln}_2$	$\Delta\overline{S}^{soln}_2$	$\Delta\overline{H}^{soln}_2$	$\Delta\overline{S}^{soln}_2$
o-Xylene	−4.87	−14.10	−8.54	−21.96
m-Xylene	−4.69	−13.74	−8.53	−22.03
p-Xylene	−5.27	−15.23	−8.24	−21.43

The proposed solution mechanism cannot be substantiated until definitive spectroscopic (e.g. infrared, microwave, and Raman) studies are made on similar systems. These studies are needed to help establish the extent of any rotational and vibrational changes that take place in the solute upon solution. Furthermore, with our present knowledge of liquid crystals and in the absence of any applicable statistical mechanical solution theories and additional thermodynamic data, a quantitative treatment is hardly possible at this time.

It should briefly be noted that in certain instances selectivity may be improved for a given separation when two or more stationary phases are employed in the gas chromatographic system.

Liquid phase combinations can be obtained in three ways.

(i) *Coupled columns.* Columns attached in series, with each column containing a different liquid phase.

(ii) *Mixed bed columns.* Each liquid phase is coated on the support particles, with the packings mixed in an appropriate manner in one column.

(iii) *Blended liquid phase columns.* The stationary phases are mixed together before coating onto the solid support.

A number of examples of the use of combinations of stationary liquid phases are given in the literature.[356–369, 287]

Mixed liquid phases find their greatest value in the separation of multi-component samples. For example, one liquid phase may resolve some solutes from a sample but not others, while another liquid may be selective only for the other solutes. Some combination of both solvents should then be able to separate the whole sample.

There has been a good deal of interest as to whether the three methods of producing liquid phase combinations give equivalent results.[370–376, 358] For coupled columns vs. mixed bed columns, similar retentions are expected for the same weights of liquid phases. Of course, this comparison must take into account the fact that because of a pressure drop the flow rates for the columns in series must be different in each column.

In the blended liquid phase method, similar results are obtained in comparison to the coupled column method if the solvent liquids mix ideally.[370] A convenient computer programme can be established if retention times or volumes are additive for multi-liquid phase columns.[377] Tabulation of retention data proves to be very useful in this instance.[378, 17]

If the stationary phases mix non-ideally, a new solvent may form whose characteristics differ from each single stationary phase. Such a new stationary phase may be able to resolve the given mixture. Purnell *et al.*[379] have studied retention on the binary stationary phase n-octadecanol + magnesium stearate in the range 40–91°C. The appearance of inflections on the curves of retention vs. composition of the binary stationary phase has been explained by the formation of complexes between the components of the binary stationary phase.

To ascertain the applicability of gas chromatography in revealing molecular compounds in solutions, Bogoslovsky *et al.*[375] investigated a number of binary stationary phases. Of particular interest as solutes are substances that may be both donors and acceptors of electrons and are thus able to interact with the two components of a binary stationary phase. Such substances include acetylenic hydrocarbons and allenes. Figure 7 shows the relationships between the composition of a binary

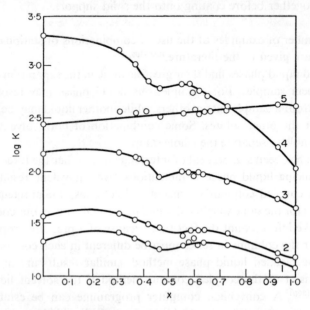

FIG. 7. Logarithm of partition coefficient of acetylenic hydrocarbons, allene, and n-hexane vs. mole fraction of tributyl phosphate in a binary stationary liquid phase (tributyl phosphate–nonyl alcohol, 20°); (1) allene; (2) methylacetylene; (3) vinylacetylene; (4) diacetylene; (5) n-hexane.[375]

stationary phase and the logarithm of the partition coefficient of such substances on the binary stationary phase, tributylphosphate + nonyl alcohol.[375] It shows that the relationships between retention and the composition of the binary stationary phase shows breaks (singularity points) that correspond to the equimolecular composition of a binary stationary phase (methods of physicochemical analysis show the formation of a molecular compound with a 1 : 1 molar ratio). Similar relationships for hexane are monotonic; this is typical of other non-polar compounds.

As the temperature is raised, the stability of molecular compounds decreases, and the relationship between the retention of the polar solute and the composition of the binary stationary phase becomes monotonic. A study of the temperature dependence of retention values on binary stationary phases will probably provide some interesting results for the determination of the stability constants of complexes in solutions. In some instances gas chromatography can furnish more reliable evidence of the formation of molecular compounds in a solution than the common physicochemical methods, provided that solutes are selected that are particularly sensitive to the interaction of the components of a binary system. The great advantage of gas–liquid chromatography is displayed in the study of systems that show weak interactions.

In contrast to gas–liquid chromatography, which has been used in the analysis of an extremely large and varied number of sample types, gas–solid chromatography has until now found only limited application as an analytical separation procedure.

For many years the adsorbents used in gas–solid chromatography were activated carbon, activated alumina, silica gel, and molecular sieves. These adsorbents have been used quite successfully for the separation of low molecular weight substances such as CH_4, C_2H_6, He, O_2, N_2, CO, and CO_2, but have not found much application in gas chromatography for the separation of larger compounds because strong adsorption and unsymmetrical peaks are usually obtained.

The neglect of gas–solid chromatography arose from a number of causes, the most important of which was the non-linear nature of the adsorption isotherms for the gas–solid interactions examined by the technique. Such non-linearity will always arise at relatively high vapour concentrations when the number of adsorbed molecules is a significant fraction of the adsorbing sites. The development of elution gas–solid chromatography has thus depended upon the development of high sensitivity ionization detectors. Non-linearity can also arise from the heterogeneity of the solid surface if the difference in activity and distribution of different sites is such that there is competition between vapour molecules for the most active sites. In the last few years, however, there has been a considerable renewal of interest in gas–solid chromatography because there are a number of ways in which surfaces may be prepared which do not suffer from such heterogeneity.

Thermal carbon black graphitized near 3000°C is a good example of a homogeneous and practically non-specific adsorbent of low surface area (6–30 m²/g). The surface of this adsorbent is almost entirely formed of the basal faces of large crystals of graphite.[380, 381] As a result, the surface of graphitized thermal carbon black is especially homogeneous. The graphitized carbon black surface has no points with electron densities locally concentrated on the periphery or with positive charges which can interact specifically with the functional groups of molecules. The low values of the heats of adsorption at low coverages of small polar molecules such as water, ammonia,[179] and methyl alcohol,[382] also prove that there are no strong electrostatic fields at the surface of the graphitized carbon black and that the adsorbent is non-specific.

Non-specific dispersion interaction during adsorption is determined by the polarizability and magnetic susceptibility of the individual linkages of the molecule as well as by its geometric configuration and orientation relative to the adsorbent surface. Therefore the flat surfaces of graphitized carbon black are well-qualified for the separation of molecules which differ only in their geometric structure, i.e. structural isomers and geometric stereoisomers. An example is the separation of the isomeric butylbenzenes. The number of linkages of the benzene ring side chain that can come in direct contact with the basal graphite plane is greatest for n-butylbenzene and least for t-butylbenzene.

Accordingly, the first to elute from a column with graphitized carbon black is the tertiary compound, and the last is the normal one. s-Butylbenzene and iso-butylbenzene occupy intermediate positions corresponding to their geometric structures and the possible orientations of their molecules.[383] Similarly, the cis and trans isomers of 1,2-, 1,3-, and 1,4-methylcyclohexanols can be separated on the basis of the differences in their geometric structure and possible orientation on a plane.[383] The problem of terpenes is another similar example. Terpenes separate on graphitized carbon black mainly in accordance with their geometric structure (the terpenes with the flattest molecules have the largest specific retention volume values). As the molecule becomes less flat and the number of carbon and oxygen atoms in it decreases, the value of V_g decreases.

The sum of the inverse distances from the centres of the carbon and oxygen atoms of the terpenes molecule to the plane surface when the molecule is most favourably oriented with respect to the surface is a convenient characteristic of the order of elution of terpenes from a column packed with graphitized carbon black.[384] When the number of carbon atoms contacting the flat surface is the same for different molecules, their separation on graphitized carbon black is difficult. For example, because the van der Waals radius of the CH_2 group (2.0 Å) is larger than that of the CH group of the benzene ring (1.8 Å), for m-xylene only one CH group comes in contact with the plane and m-xylene is more weakly adsorbed than o- and p-xylene, so it elutes from a graphitized carbon black column before the latter two. For o- and p-xylenes, two CH groups of each are in contact with the flat surface, and, therefore, their retention times on graphitized carbon black are nearly the same.

Although certain inorganic salts and metal complexes may interact selectively with certain components from the mixture to be analysed, their use in gas chromatography has been long neglected. These compounds have not been used as adsorbents because of their relatively small specific surface areas. However, if sufficiently small samples of volatile substances are injected into chromatographic columns packed with such compounds, selective retention and symmetrical peaks will often be observed. As a result, a large number of potentially useful adsorbents are available.

As an example, the study concerning the behaviour of the sodium and thallium (I) tetraphenylborates can be mentioned.[385] It has been established that a column packed with thallium tetraphenylborate retains selectively olefins and aromatic hydrocarbons compared to sodium tetraphenylborate. Table 12 gives the relative retentions referred to n-pentane at 58°C for some olefins on columns packed with sodium and thallium tetraphenylborates. This table also gives the relative retentions of some aliphatic hydrocarbons that have their boiling points very near to the corresponding olefins. The similar relative retentions of olefins and saturated hydrocarbons with similar boiling points on sodium and thal-

TABLE 12. RELATIVE RETENTIONS (REFERRED TO n-PENTANE AT 58°C) OF SOME OLEFINS ON COLUMNS WITH $NaB(C_6H_5)_4$ AND $TlB(C_6H_5)_4$ [385)(a)]

Compound	Boiling point (°C)	$r_{NaB(C_6H_5)_4}$	$r_{TlB(C_4H_5)_6}$
1,3-Butadiene	−2.6	0.35	0.65
n-Butane	−0.5	0.29	0.31
2-Methylbutadiene-1,3	35.0	1.10	1.76
n-Pentane	36.0	1.00	1.00
cis-2-Pentene	37.0	1.14	1.24
2-Methyl-2-butene	38.6	1.20	1.38
Cyclopentadiene	41–42	1.36	—
4-Methyl-trans-2-pentene	53.9	2.18	2.35
2,3-Dimethyl-1,3-butadiene	69–70	3.29	6.00
n-Hexane	68.9	3.15	3.00

(a) Aluminium columns were used (4 mm i.d. and 6 m long). The column packing was 25% sodium or thallium tetraphenylborate on Chromosorb W (silanized) 0.11–0.18 mm (Carlo Erba).

lium tetraphenylborates indicates the lack of any significant interaction between these components and the stationary phase.

Olefins and especially dienes are strongly retained on the thallium tetraphenylborate column. This interaction results from the thallium ions as they contain low-energy d-orbitals that may form a partly covalent bond with the π-orbitals of the adsorbed molecules. Cyclopentadiene does not elute from a column packed with thallium tetraphenylborate (column temperature 58°C, flow rate of carrier gas 20.1 ml/min).

Table 13 gives the values of the heats of adsorption ΔH of some compounds on sodium and thallium tetraphenylborates.

The adsorption heats were calculated from the slope of the plot of $\log K$ vs. $1/T \times 10^4$ ($T = 106$, 129, or 146°C). The slope was determined by the method of least squares and the standard error of the slope evaluated by regression analysis.

The values of the adsorption heats for the different compounds are greater on thallium tetraphenylborate than on sodium tetraphenylborate. The differences are especially greater for aromatic compounds.

TABLE 13. HEATS OF ADSORPTION OF SOME COMPOUNDS ON $NaB(C_6H_5)_4$ AND $TlB(C_6H_5)_4$[385]

Compound	$-\Delta H$ $NaB(C_6H_5)_4$	$-\Delta H$ $TlB(C_6H_5)_4$
1-Methyl-4-isopropylbenzene	10.28 ± 0.87	15.68 ± 0.37
Naphthalene	11.79 ± 0.41	14.03 ± 1.55
Phenyl-t-butylacetylene	11.65 ± 0.37	17.96 ± 1.32
Benzo-t-butylcyclobutene	10.46 ± 0.41	15.77 ± 0.87
p-Di-t-butylbenzene	14.17 ± 0.14	19.42 ± 0.46
1-Methylnaphthalene	12.80 ± 1.05	16.13 ± 0.18
n-Decane	11.06 ± 0.59	13.02 ± 0.46
n-Pentadecane	17.18 ± 0.82	18.87 ± 1.07

This is explained by the interaction of the thallium ions with these compounds.

Of great utility for gas chromatography has been the introduction of the porous polymers as adsorbents by Hollis.[386] The polymers used by Hollis were synthesized from styrene, t-butylstyrene and ethylvinylbenzene with divinylbenzene as the cross-linking agent. The properties of these polyaromatic beads vary with the chemical nature, pore structure, surface area and particle size of the material. Very good results were obtained with an ethylvinylbenzene–divinylbenzene copolymer having a specific surface area of 660 m²/g and a relatively fine pore structure. The excellent symmetry of peaks obtained with highly polar solutes such as water, ammonia, amines, alcohols, and glycols demonstrates that the polymer surface is very homogeneous. Solute retention can be attributed to adsorption on the large surface of the solid aromatic polymer, similar to adsorption observed on the aromatic matrix of some ion-exchange resins. On the other hand, the surface of the polymer can also be considered as a highly extended liquid surface and the retention explained by partition. At the present stage of research it is rather difficult to distinguish conveniently between the two phenomena, dissolution and adsorption taking place simultaneously.

Pore dimensions and pore structure seem to affect the separation.[387, 388] By introducing monomers containing various functional groups into the polymerization mixture it is possible to control the chemical properties of the surface of the sorbents obtained and thus to influence the nature of the intermolecular interactions of the surface of the sorbent with the components of the mixture being separated. This leads to the porous polymers that are commercially available. As non-polar polymers may be mentioned Porapaks Q and P and Chromosorb 101–102, and as polar polymers, Porapaks R, N, S, and T and Chromosorb 103–105.

Sakodynsky and Panina[314] made comparative studies of the retention of various classes of molecules on non-polar porous polymers (styrene and p-divinylbenzene copolymer gives Porapak P or Chromosorb 101–102) and on polar sorbents containing ether groups (methylacrylate and p-divinyl-benzene gives Polysorbate 2). The polymer sorbents containing ether groups are called polysorbates by the authors.

The local electronic structure of unsaturated compound does not appear to provide any features which would contribute to their ability to specifically interact with non-polar polymeric sorbents. The elution of unsaturated compounds occurs in accordance with their boiling points and does not depend on the presence of double bonds in the molecules. This applies to hydrocarbons as well as alcohols and acids. The molecular retention is independent of the value of the dipole moment, increasing with the general growth of the polarizability of the adsorbate molecules. The elution of the solutes on a polar polymeric sorbent (a methylacrylate and p-divinylbenzene copolymer) is determined not only by the value of general molecular polarizability but also by the value of molecular dipole moment. It is also influenced by the ability of the compounds to form hydrogen bonds with the polar polymer surface.

The differences in the retention behaviour of solutes on these sorbents result from the changes in the chemical nature of the surface; the transition from an aromatic polyhydrocarbon (a styrene and p-divinylbenzene copolymer) to a sorbent containing ether functional groups (methylacrylate and p-divinylbenzene copolymers).

Studies on the retention of different compounds on porous polymers (polar or non-polar) have been published.[389–393]

As a general characteristic it may be said that the contribution of non-specific dispersion interactions to the total energy of the molecules of different compounds on the non-polar polymer surfaces predominates, whereas in the system formed by porous polar polymers and polar solutes only specific interactions are observed.

One of the most polar porous polymers available is Porapak T. It exhibits a characteristically strong interaction with the oxygen atom in alcohols, ethers, and ketones. The structure of Porapak T has not yet been disclosed by the manufacturers (Waters Associates Inc.). Sakodynsky[393] indicated the presence of a positive charge on the surface of Porapak T, thus giving an excess specific interaction energy term and making this porous polymer particularly suitable for the retention of oxygenated materials and compounds having a free electron pair (e.g. ethers and ketones).

Standard deviations of the gas chromatographic peaks of different types of organic compounds were determined on different varieties of Porapak by Dressler et al.[390] An abnormality in the peaks of the branched alkanes and cycloalkanes was observed on Porapak P and PS columns. The higher the degree of branching of the compound, the lower was its retention time and the greater was the peak broadening. Olefins, aromatic compounds, alcohols, and mercaptans did not show such behaviour. However, sulphides seemed to show a similar effect. The broadening of the peak of a less-retained compound being greater than that of compounds having larger retention volumes is not known in gas chromatography for any adsorbent or stationary phase other than Porapak P. The other types of Porapak do not show this behaviour. The silylation of Porapak P (Porapak PS) has no effect on the anomalous behaviour.

The authors considered that the behaviour is a result of phenomena occurring within the pores of the polymer (as the pores are gradually filled with a stationary liquid, the effect decreases) and to elasticity of the structure, which is dependent on temperature (the effect disappears on increasing the temperature). It is not clear why the anomalous behaviour occurs only with alkanes, cycloalkanes, and sulphides, and not with olefins, aromatic compounds, alcohols, and mercaptans.

The anomalous retention of branched-chain compounds suggests, to some extent, that sieve effects occur as in gel permeation chromatography, where the peaks of the compounds with shorter retention times (compounds with higher molecular weights) display greater broadening.

B : SPGC 8

TABLE 14. PHYSICAL PROPERTIES OF SOME POROUS POLYMER ADSORBENTS[394]

Porous polymer	Copolymer type	Specific surface (m²/g)	Average particle diameter (Å)	Specific weight (g/cm³)	Temperature limit (isotherm) (°C)
Porapak P	Styrene-DVB[a]	—	—	0.28	250
Porapak Q	Ethylstyrene-DVB	600–650	75	0.35	250
Porapak Q=S	Silanized polymer	—	—	—	250
Porapak N	Styrene-DVB with added polar monomers	437	—	0.39	200
Porapak R		547	76	0.33	250
Porapak S		536	76	0.35	
Porapak T		306	91	0.44	200
Chromosorb 101	Styrene-DVB	30–40	3500	0.30	275
Chromosorb 102	Styrene-DVB	300–400	85	0.29	250
Chromosorb 103	Polystyrene	15–25	3500	0.32	275
Chromosorb 104	Acrylonitrile-DVB	100–200	600–800	0.32	250
Chromosorb 105	Polyaromatic polymer	600–700	400–600	0.34	250
PAR-1	Styrene-DVB	100	200	0.69	250
PAR-2	Styrene-DVB	300	90	0.67	250
Synachrom	Styrene-DVB	620	45	—	250
Polysorb	Styrene-DVB	300–350	130	0.29	250

[a] DVB = divinylbenzene.

Table 14 gives physical properties of some porous polymers used as adsorbents.[394]

A very general class of suitable surfaces may be prepared by modification of alumina, silica, or silica–alumina surfaces. Suitable modifiers include a very wide range of inorganic salts, metal complexes, and organic molecules. The exact nature of these modified surfaces has not yet been established, but a plausible model would be one in which the original surface has been completely covered by one or more layers of the modifier so that an adsorbent surface is formed with an area somewhat less than that of the supporting surface and of activity considerably altered by the modifier. The resultant activity is characteristic of the combination of support and modifier because the support affects the activity of the modifier either by altering the proportions of different crystal faces at the surface or by altering the exact crystal structure adopted by the modifier. Thus it is possible for one modifier to exhibit quite different activities depending on the active solid use to support it.

The modifiers may be classified approximately into four classes.[395]

(i) *Chemically inactive salts* (such as the alkali halides).These presumably interact with all adsorbed molecules via London and Debye (dipole-induced dipole) forces and in addition via Keesom (dipole–dipole) forces with polar molecules such as alcohols, ketones, and amines.

(ii) *Salts capable of chemically bonding with some adsorbed molecules.* Examples include silver nitrate and copper (I) chloride which form π-complexes with olefins, acetylenes, and aromatic compounds, and many salts which form complexes with amines.

(iii) *Metal complexes.* Examples include copper–alanine, silver nitrate-dodecene, and metal stearates. Interaction of the adsorbed molecules with the metal is reduced by the presence of the complexing species in the modifier, but there is an additional retardation resulting from solubility in the rest of the complex. This latter may be markedly affected by the ordered nature of the organic part of the modifier.

(iv) *Organic modifiers.* The degree of selectivity of the molecular inter-actions with the adsorbent surface may be effectively controlled by covering it with monomolecular layers of some organic compounds with functional groups such as —COOH, —OH, —NH$_2$, —CN, $\diagdown$O, CO and —C$_6$H$_5$. The monomolecular layers are much less volatile that the corresponding volume of the stationary phase because they are localized in the strong adsorption field of the adsorbent surface.[396] Covering the stationary phase with the above monomolecular layers can be brought about either by adsorption from the solution or by direct adsorption from the carrier gas on to the surface of the adsorbent. Covering a silica gel surface with monomolecular layers of glycerine, sorbitol, or triethanolamine leads to the formation of strong hydrogen bonds with the —OH groups from the silica gel surface. Because of this, the adsorbent interactions as a result of hydrogen bond formation with the solute molecules from the gas phase are reduced by the covering with the mono-molecular layer.[249, 397] Thus C_1–C_4 alcohols are weakly retained on such columns; they are easily eluted, even at low temperatures.

An extremely wide range of selectivity is available with gas–liquid–solid columns.[398] By proper choice of the relative quantity of the liquid phase with respect to the adsorbing material, a "tailor-made" column can easily be achieved to solve many analytical problems when they are difficult with only a gas–liquid or gas–solid column.

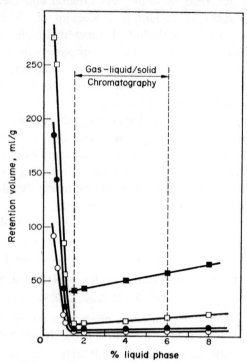

FIG. 8. Plot of the specific retention volumes for various compounds vs. percentage of liquid phase. (○) $C_6H_5N(CH_3)_2$; (●) $C_6H_5N(C_2H_5)_2O$; (□) $(C_6H_5)_2O$; (■) $(C_6H_5)_2NH$. Solid stationary phase: Sterling FT (hydrogen-treated carbon black). Liquid stationary phase: Dexsil 300 (poly-carborane siloxane).[399]

Figure 8 gives the behaviour of the corrected retention volumes per gram of adsorbent (hydrogen-treated carbon black) as a function of the percentage of liquid phase (Dexsil 300).[399]

As can be seen, in the first region, about up to 1.5% of liquid phase, there is a linear decrease of the retention volume as the surface coverage increases. This is the range in which further addition of liquid phase, if the liquid has no particular affinity with the adsorbates, has the effect of reducing the surface area of the adsorbent. As soon as a monolayer is obtained, no further decrease of retention volume is observed and a more or less pronounced increase takes place. This region, starting from about 1.5% of liquid phase is the range where the interactions of both the liquid phase and the adsorptive surface are acting synergistically with the molecules being eluted, giving a characteristic chromatographic process. The upper coverage limit of the region of gas–liquid–solid chromatography (GLSC) is, of course, indefinite, and it is difficult to identify the stage at which gas–liquid partition becomes the only process.

An example of the applicability of this technique to the solution of a particular problem is the separation of some heterocyclic compounds, obtained by Di Corcia and Bruner.[399] The second compound (Fig. 9) is unresolved from the third, when analysed with a gas–liquid column with Dexsil as liquid phase, while the separation on a gas–liquid–solid column is quite large.

A column with Porapak Q which was loaded with polyethyleneimine 5% w/w separates aliphatic amines; symmetric peaks are obtained.[400] The separation of methylamine and dimethylamine that are present as impurities in commercial trimethylamine has been carried out very acceptably on a Graphon column covered with 2% tetraethylenepentamine (TEPA).[401] Some polar isotopic pairs (methyl alcohol–methyl alcohol-D_3, methyl cyanide–methyl cyanide-D_3) were separated by gas–liquid–solid chromatography employing Graphon partially coated with 1.5% w/w of TEPA.[402]

It is obvious that gas–liquid–solid chromatography using packed column is useful for the investigation of the isotope effect (the heavier species is eluted first) and offers a very simple method of analysis of isotopic mixtures, at the same time allowing isotopic separation at more con-

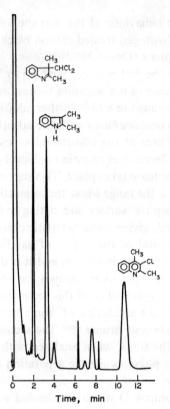

FIG. 9. Chromatogram showing the elution of some heterocyclic compounds. Column: 1.7 m×2 mm hydrogen-treated Sterling FT+2% Dexsil at 250°C; carrier gas, hydrogen.[399]

venient temperatures (separation of methyl alcohol from methyl alcohol-D₃ and of methyl cyanide from methyl cyanide-D₃ was performed in the following conditions. Column: 15 m×4 mm i.d. stainless steel column packed with Graphon, 40–60 mesh, coated with TEPA 1.5% w/w; 63°C; flow rate (hydrogen) 120 ml/min). It is to be remembered that the separation of methyl alcohol and methyl alcohol-D₃ in these conditions is superior to that obtained from capillary columns.[403] Gas–liquid–solid chromatography can also be brought about by mixing a solid and a liquid stationary phase and then depositing this mixture onto a solid, preferably inert, support.[404]

Pecsok and Vary[404] showed that a dispersion of finely divided metal phthalocyanine in silicone oil (on Chromosorb T as a solid support) is an effective substrate for gas chromatography. The technical difficulty of making a reproducible solid chromatographic column is diminished by this technique. For the phthalocyanine compounds (Cu, Zn, Fe(II), Co, Ni, Al, Cr(III)) the finely divided powder dispersed into a conventional liquid phase (less than 1 : 5 ratio) gives a reproducible surface area, which is totally available to the injected compound. The surface area exposed to the sample in this way is not only more reproducible but also is greater than if it were packed in a column or deposited on the wall of a capillary column or on the surface of glass beads.

An example of selective interaction using such columns is the separation of a mixture of n-hexane, cyclohexene, and 3-pentanone. A column with 8% silicone oil DC 200 on Chromosorb T at 60°C does not separate the last two compounds, which therefore elute together. Under the same conditions but with 0.1Miron(II) phthalocyanine dispersed in the silicone oil, the strong interaction between iron(II) phthalocyanine and 3-pentanone, and a weaker interaction between the phthalocyanine and cyclohexene, allows the mixture to be separated very easily. Iron(II) phthalocyanine does not affect the retention time of n-hexane.

It can be said that the experimental results show that gas–liquid and gas–solid chromatography can be superimposed for improvement of separation. The additional variable which is introduced by the solid phase in the conventional gas–liquid chromatographic column improves the selectivity of the column to such an extent that its use should not be overlooked for difficult separation problems.

3.4 CRITERIA FOR THE CHOICE OF PROPER STATIONARY PHASES IN GAS CHROMATOGRAPHY

The choice of the proper stationary phase is one of the most important decisions in gas chromatography. Several hundred stationary phases have been described and are commercial available. Several compilations are available from suppliers and may be used as a guide when selecting the stationary phase. For the time being, it can be said that the selection of the stationary phase for a new analytical problem is done on a trial-and-error basis, the success of a certain separation depending greatly on the experience accumulated previously by the analytical chemist. But qualitative knowledge of the molecular interactions, combined with the theory of solutions, helps in understanding the way in which the stationary phase works; it may be an important criterion in the selection of the proper stationary phase for a given application.

For mixtures of close-boiling solutes of different chemical classes, separation is relatively easy, and the general principle that "like dissolves like" is applicable.

Certainly this qualitative rule cannot be applied to the separation of isomers or of components with very similar properties. In such instances, in order to perform a given separation, one must remember the molecular characteristics of the components to be separated and of the stationary phases.

The selectivity of a stationary phase when non-polar isomers are separated increases with an increase in the concentration of heavy atoms (such as Cl, Br, and S) in the stationary phase and with a decrease in the concentration of hydrogen atoms.[405]

Perfluorinated stationary phases have a very low selectivity in isomer separation because fluorine atoms have only a weak ability for dispersion interaction. The number and type of chemical bonds within a molecule determine the polarity of the substance being analysed. This characteristic is important when separations of polar and non-polar compounds and isomers of the same formula but containing different functional groups are considered. It is desirable that the stationary phase possesses a high molar concentration of selectively interacting groups, a minimum concentration of groups which interact to oppose the desired separation, and a minimum concentration of solvating, non-selective interacting groups to allow low-temperature operation. A single strong selective interaction is preferred.

Under some conditions, the inclusion of groups which contribute positive heats of solution or negative entropies will make possible a practical separation. Thus a stationary phase of a different chemical nature from the solutes should seriously be considered.[202] The molar volume of the stationary phase may differentially affect the activity coefficient of isomers or closely similar compounds. Under these conditions, the molecular weight of the stationary phase should be selected to reinforce the desired interaction.

The general classification of stationary phases as polar or non-polar often can be eliminated in favour of more specific designations in terms of group interactions, stationary phase properties, and sometimes the thermodynamics of interaction.

The tabulation of stationary phases with their Rohrschneider constants has proved its utility for predicting separations from known structures of the compounds of interest. All of these assist in choosing and designing selective stationary phases.

Specific molecular interactions should also be considered. Complications may result from the relationship between excess enthalpies and entropies of interaction. Derivative formation may ease the separation of related compounds by altering relative vapour pressures, by eliminating dominant undesired interactions or by allowing the utilization of favourable solution interactions. However, available stationary phases should influence the choice of derivative.

The use of some mixtures of liquid stationary phases or a mixture formed by a solid and a liquid stationary phase offers to the analytical chemist a wider choice of a convenient stationary phase for a given separation.

Because the equilibrium between the gas and the solid phase is more rapidly established than between the gas and the liquid phase, the use of some solid stationary phases may give better chromatographic performance.

The use as adsorbents of porous polymers and inorganic salts, either porous or non-porous and modified oxides, as well as surfaces of dense monomolecular polymer layers adsorbed on a sufficiently developed and homogeneous surface of non-porous and wide-pore adsorbents, opens extensive possibilities for selecting and controlling the nature of the adsorbent surfaces and, therefore, for controlling the selectivity of gas adsorption columns. The use of some solid stationary phases is also recomended when programmed temperature chromatography, especially at high temperatures, is used. Solid stationary phases are preferred to liquid phases for determining micro-impurities in quantities near the limit of sensitivity of ionization detectors.

Despite the great advantages of gas–solid over gas–liquid chromatography in many instances, these methods should supplement one another rather than compete. The requirements which should be considered in selecting the appropriate stationary phase are the following:

(a) low vapour pressure at the desired operating temperature;
(b) chemical stability at the operating temperature;
(c) selectivity for the components to be separated;
(d) sufficient dissolving power for the components;
(e) low viscosity at the desired operating temperature;
(f) adequate wetting of the support surface or the column wall;
(g) reasonable solubility in some common volatile solvents.

CHAPTER 4

TYPES OF STATIONARY PHASES USED IN GAS CHROMATOGRAPHY

4.1 LIQUID STATIONARY PHASES

For the great majority of analytical problems, gas–liquid chromatography has proved to be a very efficient technique.

The use of liquid phases has the following advantages:[406]

(i) The adsorption isotherm is linear under usual operating conditions so that symmetrical peaks can be obtained.

(ii) Liquid phases are available in great variety; thus adequately selective phases can be found for a particular separation.

(iii) The amount of liquid phase in the column can be varied easily; therefore both preparative and high-efficiency columns can be made with the same liquid phase.

(iv) Liquid phases are available in great purity and in well-defined quality; thus retention values are reproducible.

(v) Both packed or open tubular columns can be fabricated with liquid phases in a simple manner.

The greatest disadvantage of liquid phases is their volatility.

The liquid stationary phase in the column always has a vapour pressure, the magnitude of which depends primarily on the nature of the phase and on the working temperature. It is exposed to the continuously streaming carrier gas which removes a certain amount of the liquid phase per unit gas volume. This process is known as bleeding. This bleeding is undesirable because it can disturb the proper function of the detector; it causes baseline shift in programmed temperature operation, it contaminates the trapped solutes, and because of the loss of the stationary phase, it gradually changes the separation character of the column until an unacceptable deterioration occurs.

Recommended maximum temperatures for different stationary phases can be found in tables. However, these values should always be considered only as guides and not as absolute values.

The sensitivity of the detector, the mode of operation (isothermal or temperature programmed), the column construction, etc., always affect the maximum temperature at which the column can practically be operated. The vapour pressure of the stationary phase in the column might be influenced by the support material and also by the loading. The maximum operating temperature for a liquid phase in an open tubular column is 25–50°C lower than in a packed column. Some liquid phases are not homogeneous materials. For example, commercial polyglycols or silicone greases are composed of homologues and may also contain some chemically different materials. Column conditioning, which results in the removal of the volatile constituents, can extend the upper temperature limit of these stationary phases.

The thermal stability of the liquid phase is also an important factor. Conformational changes, polymerization, cross-linking of silicone oil, condensation of polypropylene sebacate, etc., have been observed at high temperatures.[407, 408] An interesting approach to the investigation of the thermal behaviour of the stationary phase is the use of thermogravimetry. The information obtained by this technique is particularly valuable in practice if the effect of temperature on the column material (liquid phase plus support) is studied.[409, 410]

The lower temperature limit for operating a column with a given liquid phase is generally the melting point of the stationary phase. In many instances, however, increasing viscosity of the liquid by decreasing the temperature results in a loss of column efficiency.[411] On the other hand, the solidified stationary phase can also occasionally be effective as an adsorbent.[412] The lower temperature limit is, for example, 75°C for Apiezon L and 175°C for Versamid 900.

Taking into account the huge number of compounds suggested for use in gas chromatography as liquid stationary phases, and remembering that these are very different in structure and properties, it is easy to understand the difficulties encountered in any attempt at classification. There are two ways to discuss the separating properties of stationary phases:

(1) Those stationary phases are grouped together which are used for the separation of a whole class of solutes (alcohols, hydrocarbons, steroids, etc.).[413-415]

(2) Stationary phases are classified according to certain common properties, their possible applications being listed.[416]

In the following discussion the second classification, as suggested by Littlewood,[417] is used. The groups discussed are:

   I   non-polar stationary phases (paraffinic);

   II   dilute stationary phases;

   III   concentrated stationary phases;

   IV   specific stationary phases.

4.1.1 Non-polar stationary phases (paraffinic)

Non-polar stationary phases are those containing neither polar nor polarizable groups. Thus they include all alkanes and mixtures of alkanes and alkyl silicones. The intermolecular forces in non-polar stationary phases consist of dispersion and induction forces.

The paraffinic stationary phases are a very good medium for the solubilization of hydrocarbons that generally elute in the order of increasing boiling point. The hydrocarbons are much more soluble in these stationary phases than any polar solute with a similar boiling point. This property is used in the separation of these hydrocarbons from different mixtures of solutes.

Among a set of isomers, the linear alkane has the greatest retention, and the general rule among branched alkanes is that the greater the branching the smaller the retention. This rule, however, is, subject to systematic exceptions; for example, there are two doubly branched heptanes with retentions smaller than the triply-branched 2,2,3-tri-methylbutane.

Dyson and Littlewood[207] have proposed the idea that $\log V_g$ is proportional to molecular surface area measured as a suitable contour around the molecule, and have illustrated this by showing a linear relation between $\log V_g$ and the collisional cross-section in the gas phase, the latter being proportional to molecular surface area. According to this idea, branching decreases retention because it causes the molecule to become more compact with a smaller surface-to-volume ratio.

Polar solutes also separate according to size because their polarity is irrelevant. Though retentions in non-polar stationary phases are inde-pendent of solute polarity, the boiling points of pure liquid solutes are increased as a result of polar–polar interactions. For a mixture of a polar solute and a non-polar solute of about the same boiling point, the less polar solute will have the greater retention as it necessarily is a larger molecule than the polar substance in order to have a comparable vapour pressure. For polar molecules not containing halogen atoms, the size is very roughly proportional to molecular weight, so that for molecules of roughly the same shape, $\log V_g$ is proportional to molecular weight. Such a relation has been illustrated by Littlewood[418] for an alkane, alkene, nitroalkane, thiophene, aromatic compound, and ester as solutes and squalene as stationary phase,

$$\log V_g \cong [M(n^2 - 1)]/(n^2 + 2), \tag{142}$$

where M is the molecular weight of the solute and n is the refractive index of the solute. This is principally because the dispersion energy can be expressed as a function of the polarizability of solute and stationary phase (which is constant), and the polarizability of the solute is expressed to a large extent as a function of $[M(n^2-1)]/(n^2+2)$. Equation (142) is useful for the prediction of the magnitude of retentions but cannot be considered as quantitative.

Because non-polar stationary phases do not interact specifically with the solutes to be separated, they can be used in gas chromatography as reference stationary phases.

The most important members of this class of stationary phases are:

Squalane (2,6,10,15,19,23-hexamethyltetracosane)

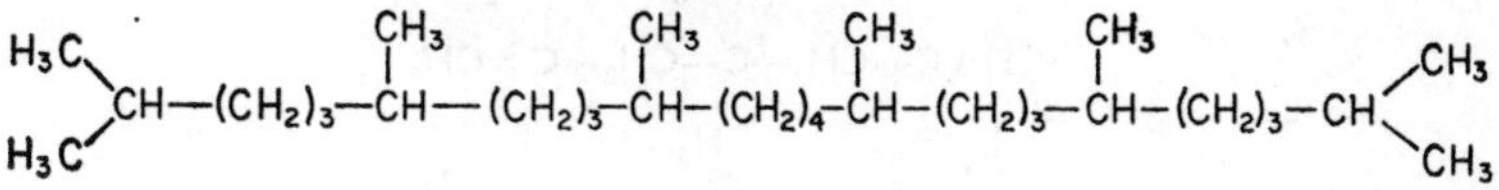

$$H_3C \diagdown CH-(CH_2)_3-\overset{CH_3}{\underset{|}{CH}}-(CH_2)_3-\overset{CH_3}{\underset{|}{CH}}-(CH_2)_4-\overset{CH_3}{\underset{|}{CH}}-(CH_2)_3-\overset{CH_3}{\underset{|}{CH}}-(CH_2)_3-CH \diagup^{CH_3}_{CH_3}$$

b.p. $\approx$ 375°C; density $d_4^{20} = 0.805$ g/cm³; minimum and maximum operating temperatures 20 and 150°C; commercial names: Embaphase, Squalane, Perkin Elmer U. Retention indices for 38 hydrocarbons on squalane at temperatures of 22, 30, 40, 55, 60, 70, 80, and 100°C have been published.[419] Squalane may be used also for the separation of C_5–C_8 hydrocarbons,[420, 421] some mercaptans,[422] and anaesthetics.[423]

Among the saturated hydrocarbons (with well-established structures) the following are used as stationary phases:

n-*Hexadecane*

$$CH_3-(CH_2)_{14}-CH_3$$

b.p. = 280°C; $d_4^{20} = 0.775$ g/cm³; minimum and maximum operating temperatures 20 and 50°C.

n-*Tetracosane*

$$CH_3-(CH_2)_{22}-CH_3$$

b.p. = 370°C; $d_4^{20} = 0.779$ g/cm³; minimum and maximum operating temperatures 52 and 140°C.

n-*Hexatriacontane*

$$CH_3-(CH_2)_{34}-CH_3$$

b.p. = 265°C/1 mm; $d_0^0 = 0.764$ g/cm³; minimum and maximum operating temperatures 76 and 150°C.

These compounds are used in the same way as squalane, mainly for the hydrocarbon separation.[424]

Tri-isobutylene

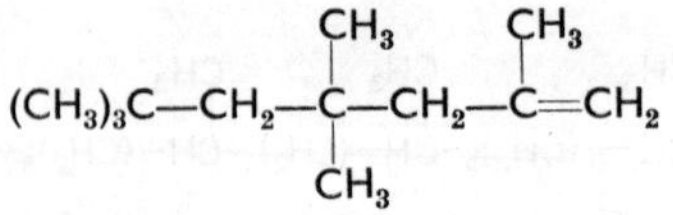

$$(CH_3)_3C-CH_2-\underset{\underset{CH_3}{|}}{\overset{\overset{CH_3}{|}}{C}}-CH_2-\overset{\overset{CH_3}{|}}{C}=CH_2$$

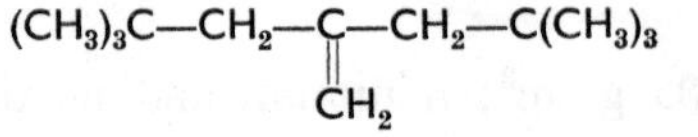

or $(CH_3)_3C-CH_2-\underset{\underset{CH_2}{||}}{C}-CH_2-C(CH_3)_3$ or $(CH_3)_3C-CH=\underset{\underset{CH_3}{|}}{C}-CH_2-C(CH_3)_3$

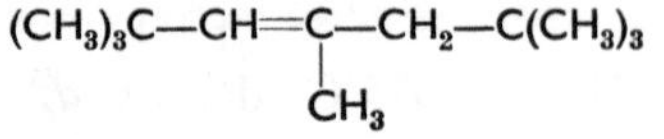

$d_4^{20} = 0.759$ g/cm^3; maximum operating temperature 20°C.

Paraffin oil (mixture of paraffins and cycloparaffins)

Maximum operating temperature 120°C; commercial name: Nujol.

Polyethylene

Structural unit: $-CH_2-CH_2-$

Small contribution of CH_3 groups (more in the high pressure type); molar ratio 10,000–50,000 (for high-pressure type); 50,000–300,000 (for low-pressure type); density 0.92 g/cm^3 (high-pressure type); 0.94–0.98 g/cm^3 (low-pressure type); minimum operating temperature 112°C (high-pressure type); 125–134°C (low-pressure type); maximum operating temperature 300°C.

Apiezon greases

Apiezon greases are widely used as liquid phases for the separation of many classes of compounds. Literature reviews report over 300 papers where Apiezon greases have been used as liquid phases.

Apiezons are obtained by subjecting selected lubricating oils to a high-temperature treatment. The most labile fraction is thus destroyed and the residue is purified by molecular distillation. According to the temperature of the heat treatment, the Apiezons are classified alphabetically: C, E, H, J, K, L, M, N, T.

The majority of published work that report separations on Apiezon greases confirm that Apiezon L is the one most widely used. It has the best thermal resistance (300°C). An evaluation of the properties of Apiezon greases (L, M, and N) as stationary phases in gas–liquid chromatography has been made by Castello.[425] Paraffins, n-alkyl iodides, n-alkyl chlorides, n-alcohols, methyl ketones, and acetals of normal alcohols were analysed at temperatures ranging from 50 to 300°C, at liquid phase concentrations ranging from 2.5 to 20%, and a carrier gas flow rate from 20 to 40 ml/min. Experimental results show that the three Apiezons are completely interchangeable from the point of view of their separatory properties. As Apiezon L has a lower vapour pressure and allows higher operating temperatures, it seems reasonable to employ mainly this type of grease.

For polar compounds, when Apiezon was used as received, with its characteristic yellow–brown colour and without purification, some effects (tailing) ascribable to a small degree of polarity were observed, but after purification by column chromatography on alumina, the colour disappeared and non-polar behaviour was fully restored. It seems probable that the slight polarity arises from the presence of some impurities, as infrared spectroscopy of the original Apiezon has shown the presence of carbonyl and carboxylic compounds whose characteristic bands were completely absent from the spectra of the purified fractions.[425]

Among the numerous gas chromatographic separations for which Apiezon L has been used as the stationary phase are the separation of the methyl esters of high molecular weight fatty acids,[426–428] fatty acid amides,[429] fatty amines,[430] phenothiazine and derivatives,[431] lead alkyls,[432] and pentaboranes and decaboranes.[433]

Methyl silicones

This group of compounds is among the most widely used non-polar stationary phases. Some papers dealing with the chemistry and the use of methyl silicones as stationary phases in gas chromatography have been published.[434–436] The methyl silicones are practically inert, being attacked only by halogens, halogenated acids, and a few strong bases.

The higher molecular weight methyl silicone fluids are stable at 150°C in air for extended periods. Above this temperature, the viscosity increases with time, and eventually the material gels. This viscosity change may be inhibited by the addition of an antioxidant.

Their general formula is:

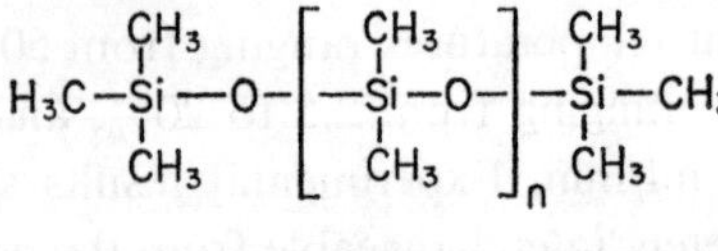

$$H_3C-\underset{\underset{CH_3}{|}}{\overset{\overset{CH_3}{|}}{Si}}-O-\left[-\underset{\underset{CH_3}{|}}{\overset{\overset{CH_3}{|}}{Si}}-O-\right]_n-\underset{\underset{CH_3}{|}}{\overset{\overset{CH_3}{|}}{Si}}-CH_3$$

Dimethly silicone oils can generally be regarded as liquids with low solubility parameters, high compressibility, low heats of fusion, and little tendency for specific interaction.

Gums are frequently high molecular weight linear polysiloxanes ($n > 4000$); dimethylsiloxanes may be modified or cross-linked by treatment with peroxides.

Their relative low viscosity and high thermal stability allows the use of silicone oils over wide temperature ranges (-50 to $300°C$).

These stationary phases are generally loaded in a low proportion on a solid support. In order to overcome undesired effects at the surface of the solid support, this is treated with hexamethyldisilazane or dimethyldichlorosilane.[437-439] The effect of the treatment is that it eliminates the hydrogen from the —OH group, thus removing any possibility of interactions of this group with alcohols, ketones, aromatic hydrocarbons, etc., which would give asymmetric peaks.

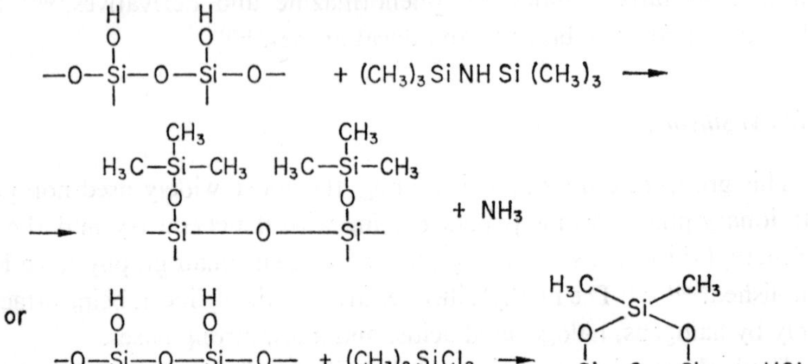

$$-O-\underset{|}{\overset{\overset{H}{\overset{|}{O}}}{Si}}-O-\underset{|}{\overset{\overset{H}{\overset{|}{O}}}{Si}}-O- \quad + \ (CH_3)_3\,Si\,NH\,Si\,(CH_3)_3 \longrightarrow$$

$$\longrightarrow \quad -\underset{|}{\overset{\overset{H_3C-\underset{|}{\overset{\overset{CH_3}{|}}{Si}}-CH_3}{\overset{|}{O}}}{Si}}-O-\underset{|}{\overset{\overset{H_3C-\underset{|}{\overset{\overset{CH_3}{|}}{Si}}-CH_3}{\overset{|}{O}}}{Si}}- \quad + \ NH_3$$

or

$$-O-\underset{|}{\overset{\overset{H}{\overset{|}{O}}}{Si}}-O-\underset{|}{\overset{\overset{H}{\overset{|}{O}}}{Si}}-O- \ + \ (CH_3)_2\,SiCl_2 \longrightarrow \ -\underset{|}{Si}-O-\underset{|}{Si}- \ + \ HCl$$

Methylsilicone stationary phases are marketed under various commercial names that are encountered quite frequently in scientific papers. Thus

General Electric supplies methyl silicones under the names: SE 30; SF 96; E 301; VISCASIL. Dow Corning under the names: DC 200; DC 11; DC 220; DC 500; DC 123; E 410. Union Carbide under the names: W 95; W 98; L 45; W 96; KS 1014. Applied Science under the names: JXR; OV-1; OV-101. VEB Chemiewerk Nünchritz under the names: MO 50; MO 200; NM 1-50; NM 1-200; NG 100.

Most of the references concerning the use of methyl silicones as stationary phases in gas chromatography are concerned with the analysis of steroids, pesticides, terpenes, and phenols.

Surface-bonded silicones

In 1966 Abel et al.[440] described the preparation of a new type of gas–liquid chromatographic phase where the stationary phase is chemically bonded and attached to the support.

The formula

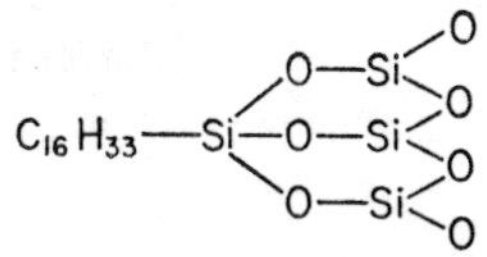

$$C_{16}H_{33}-Si\overset{O-Si}{\underset{O-Si}{\overset{|}{O-Si}}}$$

has been attributed to this stationary phase obtained by the polymerization of hexadecyltrichlorosilane on a celite surface.

Aue and Hastings,[441] studying the same problem, tried to initiate a reaction between the silanol groups of the support and dimethyldichlorosilane (DMCS) or a similar material (methyltrichlorosilane). They showed that if a great excess of DMCS is reacted with the silanol groups, it is likely that one chlorine atom will remain for a subsequent reaction with other silicone monomers during polymerization. Thus the silicone monomer (which is to form the particular liquid phase) is applied to the DMCS-treated support. Then, during polymerization, the silicone phase is chemically bonded via the DMCS to the solid surface. Once polymerization has been achieved, residual or newly formed hydroxyl groups, either on the surface or in the liquid phase, can be deactivated with a suitable silylating reagent (bis(trimethylsilyl)acetamide).

9*

The reactions may be written as follows:

Surface treatment

```
        O
        ‖
—Si⟋O⟍Si—·        H₂O, H⁺         OH     OH
   |       |        ――――――→          |      |
 Surface                           —Si—O—Si—
                                     |      |
                                   ·Surface
```

Initial reaction

```
                                         Cl              Cl
                                         |               |
                                   H₃C—Si—CH₃      H₃C—Si—CH₃
  OH      OH                             O               O
   |       |          DMCS               |               |
 —Si —O—Si—        ――――――→            —Si——O——Si—
   |       |                             |               |
 Surface                                   Surface
```

```
                                         Cl              Cl
                                         |               |
                            MTCS   H₃C—Si—Cl       CH₃—Si—Cl
                          ――――――→        O               O
                                         |               |
                                      —Si——O——Si—
                                         |    Surface    |
```

Polymerization

```
     R              R
     |              |
 Cl—Si—Cl      Cl—Si—Cl
     |              |
     Cl             Cl

     +                          H₂O        —O—Si——O——Si—O—
                             ――――――→          |          |
     Cl             Cl                        O          O
     |              |                         |          |
·H₃C—Si—CH₃   H₃C—Si—CH₃               H₃C—Si—CH₃  H₃C—Si—CH₃
     O              O                         O          O
     |              |                         |          |
   —Si——O——Si—                            —Si——O——Si—
     |    Surface   |                         |          |
                                               Surface
```

Deactivation

```
     R                                         R      CH₃
     |                                         |     ╱
 —O—Si—OH      CH₃CON[Si(CH₃)₃]₂         —O—Si—O—Si—CH₃
     ‖         ――――――――――――→               ‖     ╲
     O                                         O      CH₃


        R     CH₃                              R      CH₃
 DMCS   |      |          CH₃OH                 |      |
 ――――→ —O—Si—O—Si—Cl     ――――――→        —O—Si—O—Si—OCH₃
        ‖      |                               ‖      |
        O     CH₃                              O     CH₃
```

The most important step in the synthesis of surface-bonded silicones is fluidized bed polymerization. The way in which this polymerization occurs may lead to the formation of some surface-bonded silicones from low volatility monomer[440] as well as surface-bonded silicones from high volatility monomer.[441] This polymerization is a time-consuming step which requires suitable apparatus and is not easy to control. Consequently, further development in this area was sure to involve the conditions for polymerization, as they determine, to a great extent, success or failure in achieving support-bonding as well as the desired chromatographic performance.

Perhaps the easiest route to prepare silicones polymerized on adsorbent particles is to fill a chromatographic column with monomer-coated support, then saturate a suitable gas with water and blow it through the column until polymerization is complete. The feasibility of such a simplified *in situ* approach was investigated by Hastings *et al.*[442] An overall comparison of the two methods of on-support polymerization (*in situ* vs. fluidized bed) would seem to give preference to the latter in terms of flexibility and support-bonding. The *in situ* approach, however, is simpler, and definitely one to consider under favourable circumstances.

Per cent loading is calculated as 100 times the ratio of weight of silicone monomer (e.g. $C_{18}H_{37}SiCl_3$) to the total weight of silicone monomer plus solid support. For comparison with commercial phases, this loading is corrected to represent the polymer [e.g. $(C_{18}H_{37}SiO_{3/2})_n$] rather than the monomer. Among the surface-bonded silicones used by Aue and Hastings, the phases with best performance were: $(C_{18}H_{37}SiO_{3/2})_n$, $[(CH_3)_2SiO]_n$, $(CH_3HSiO)_n$, and the propyl and butyl derivatives of the trifunctional series.

These stationary phases were characterized by gas–liquid chromatography in terms of column bleed, solute retention, and resolution by liquid phase extractability, by thermogravimetric analysis, and carbon and hydrogen analysis. The phases showed virtually complete surface bonding and bleed comparable to chromatography-grade SE 30. Scanning electron microscopy showed that most of the liquid phase covered the periphery rather than the inside of the support grains.[443] This implies that the area covered by the visible liquid layer is much smaller than the

BET surface, and that this layer is about ten times thicker than would be assumed from an isotropic distribution throughout the internal and external regions of the particle. The average thickness and the total area of support-bonded silicone films would therefore be functions of the support particle diameter at a given polymer load. For example, for the same amount of polymer coated on a 120-mesh instead of a 60-mesh support, the thickness of the peripheral layer on the finer material would be half of that on the coarser material.

The advantages offered by the use of surface-bonded silicones as liquid phases in gas chromatography result from the following:[440]

(a) The active solvent group is chemically bonded to the —Si—O matrix with no apparent diminution of solute–solvent interaction, and thus there is no danger of deterioration of resolving efficiency owing to breakdown of uniformity of phase coating on the support particles. Such decrease in column efficiency is commonly observed when Apiezon greases and high molecular weight silicone oils are used at column temperatures up to 250°C.

(b) The greatly enhanced temperature stability enables the resolution which might be obtained from a n-octadecane solvent to be used up to at least 250°C. Preparative methods could be extended to include other active solvent groups in a suitable combination to give specificity for various classes of compounds.

(c) Resolution is notably superior to that obtained on a silicone oil column operated under similar conditions, both improved peak symmetry and better resolution being observed for a wide range of compounds.

(d) It should be possible to clean the phase, should it become contaminated, by the use of suitable solvents because there is little possibility of Si—O—C bond breakage. Once any low-boiling material is removed, there should be no bleed-off under any normal operating conditions.

These phases could well prove useful for employing with very sensitive detectors or for gas chromatography–mass spectrometry,[444] where high stability and minimum bleed-off is essential.

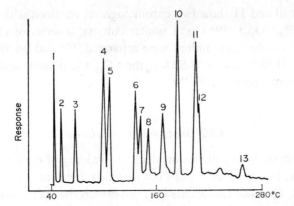

FIG. 10. Chromatogram of (1) acetone and ethanol, (2) benzene, (3) octane, (4) aminoethyl alcohol, (5) 2-octanone, (6) 3-(chloromethyl)-heptane, (7) benzyl chloride, (8) aniline, (9) tri-n-butylamine, (10) octanoic acid, (11) tri-n-butyl phosphate, (12) bromonaphthalene, (13) n-octadecane. Glass column: 50 cm × 4 mm i.d. containing 13.2 wt.% $(C_{18}H_{37} SiO_{3/2})_n$ on 60/80 Chromosorb G. Initial temperature: 40°C, increasing at 10°/min. Nitrogen flow rate: 40 ml/min.[441]

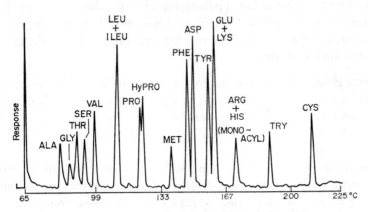

FIG. 11. Chromatogram of the N-trifluoroacetyl n-butyl esters of 19 amino acids. Column: as in Fig. 10 but 1 m long. Initial temperature: 65°C, increasing at 6°/min. Nitrogen flow rate: 45 ml/min. The injection mixture contained 5 μg of each amino acid.[441]

Figures 10 and 11 show two chromatograms obtained with a column with $(C_{18}H_{37}SiO_{3/2})_n$.[441] On a similar column, a series of chlorinated pesticides, alcohols, or amines were separated,[442] and on the column with $(CH_3(H)SiO)_n$ and $(CH_3SiO_{3/2})_n$ the n-C_{10}–C_{36} alkanes and n-C_2–C_{12} alcohols were separated.[442, 445]

4.1.2 Dilute stationary phases

These are stationary phases in which the majority of the molar volume is occupied by non-polar groups but in which there are also some polar groupings. The main characteristic of this class is that its members readily dissolve the polar solutes as well as those that are non-polar, thus being of universal use as stationary phases in gas chromatography.

Their retention characteristics are determined by the fact that they have polar groups with which polar solutes can interact, but not so many as to make a strongly cross-linked solvent lattice, the internal pressure of which is such as to exclude non-polar molecules.

Alkanes separate in dilute stationary phases in very much the same manner as in non-polar solvents except that specific retentions are somewhat smaller. The retention of polar solutes is determined in part by size, as in a non-polar solvent, but, in addition, there is a term in log V_g contributed by the specific interaction between solute and the polar group of the stationary phase.

Three classes of solvent polar groups can usefully be distinguished:
 (i) Those with purely polar groups, e.g. nitro groups or nitriles. These provide much extra retention for polar solutes but do not retain selectively compounds with hydroxyl groups or those which are hydrogen bond donors.
 (ii) Those with hydroxyl groups, which provide a very much extra retention for solutes with hydroxyl groups, and very considerable extra retention for solutes which are hydrogen bond donors.
 (iii) Those with groups which can act as hydrogen bond donors, such as ether groups. These have a moderate polarity and thus confer moderate extra retention on polar solutes, but interact very strongly

with solutes which have hydrogen atoms capable of forming hydrogen bonds. Included in this class of stationary phases are the silicone oils with different functional groups:

Phenyl silicones

Substitution of phenyl groups into methyl silicones would be expected to increase the solubility parameter to something approaching that of benzene. A value of 9.0 is reported for phenyl methyl silicones.[446] On the other hand, phenyl groups improve oxidation stability and lubricating properties. Phenyl silicone fluids are among the most heat resistant. Also phenyl methyl polysiloxanes would be expected to be more polarizable than the dimethyl compounds and to have some of the solvating properties of a benzene-like solvent without the accompanying volatility.

Furthermore, there is the possibility of weak hydrogen bonding to the π-electrons of the phenyl ring.[447, 448] Thus these stationary phases would be expected to be a good general purpose material for a variety of compounds, but not very specific or sensitive to functional groups.

Diphenyl silicones would be expected to be of a more aromatic nature, tending to reject aliphatic hydrocarbons and separating aromatic compounds by boiling point.

The diphenyl siloxanes might be moderately selective in terms of groups which might or might not interact with the benzene π-electrons. Among the phenyl silicones, there are:

Phenyl silicones with a small content of phenyl groups

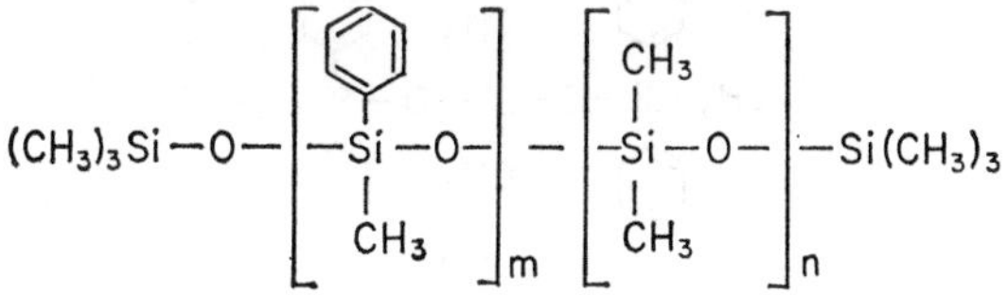

$$(CH_3)_3Si-O-\left[-\overset{\overset{\bigcirc}{|}}{\underset{\underset{CH_3}{|}}{Si}}-O-\right]_m - \left[-\overset{\overset{CH_3}{|}}{\underset{\underset{CH_3}{|}}{Si}}-O-\right]_n -Si(CH_3)_3$$

and

$$(CH_3)_3Si-O-\left[-\overset{\overset{\bigcirc}{|}}{\underset{\underset{\bigcirc}{|}}{Si}}-O-\right]_l - \left[-\overset{\overset{CH_3}{|}}{\underset{\underset{CH_3}{|}}{Si}}-O-\right]_n -Si(CH_3)_3$$

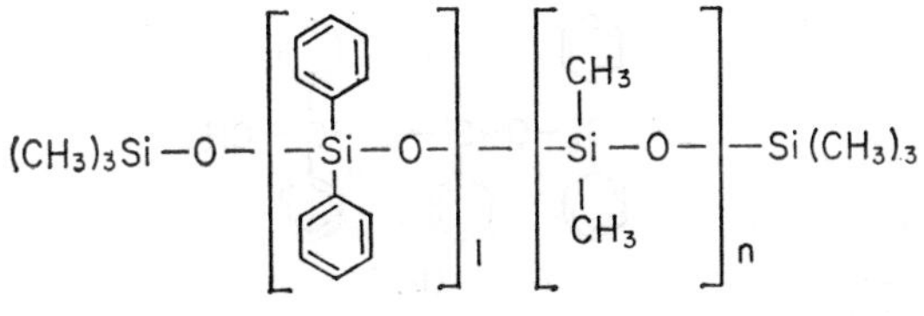

$d_4^{20} = 0.99$–1.04 g/cm^3; minimum operating temperature $-50°C$; maximum operating temperature $160°C$. Commercial names: DC 701; DC 702 (Dow Corning); MS 510 (Midland Silicones); NM 3-200 ($1 = 0$; $m : n \approx 1 : 14$) (VEB Chemiewerk Nünchritz); OV-3 (10% phenyl, 90% methyl, maximum temperature $300°C$); OV-7 (20% phenyl, 80% methyl, maximum temperature $300°C$) (Applied Science).

Phenyl silicones with a medium content of phenyl groups

$d_4^{20} = 1.04$–1.07 g/cm^3; minimum operating temperature $-30°C$; maximum operating temperature $300°C$ (OV-17; OV-61). Commercial names: DC 703 ($1 = 0$; $m = 3$; $n = 0$); DC 550 (Dow Corning); OV-17; OV-61 (35% phenyl; 65% methyl) (Applied Science); OE 4129; NM 4-500 ($1 = 0$; $m : n = 1 : 1$) (VEB Chemiewerk Nünchritz); MS 550 (Midland Silicones).

Phenyl silicones with a high content of phenyl groups

$d_4^{20} = 1.07$–1.11 g/cm^3; minimum operating temperature $-20°C$; maximum operating temperature $220°C$; $300°C$ (OV-25). Commercial names: DC 710 (Dow Corning); OV-25 (75% phenyl, 25% methyl) (Applied Science); OE 4011; NM 5-500 ($1 = 0$; $m : n = 7 : 3$) (VEB Chemiewerk Nünchritz).

Other phenyl silicones with a high phenyl group content but having a different structure are:

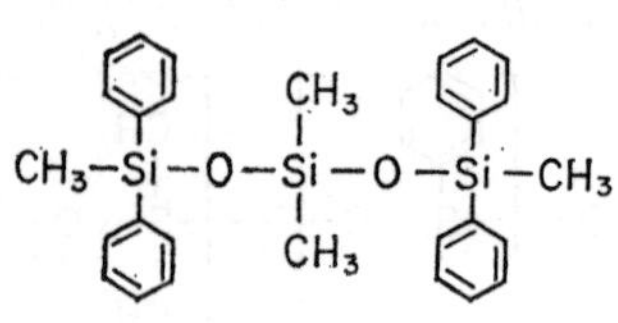

(a)

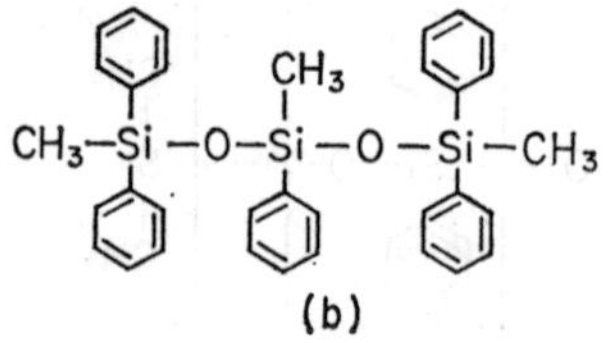

(b)

(a) $d_4^{25} = 1.066$ g/cm^3; minimum operating temperature $-34°C$; maximum operating temperature 180°C. Commercial names: DC 704 (Dow Corning); OE 4008 D (Institut für Silikon- und fluorocarbonchemie Radebeul).

(b) $d_4^{25} = 1.095$ g/cm^3; minimum operating temperature $-15°C$; maximum operating temperature 200°C. Commercial names: DC 705 (Dow Corning); OE 4007 D (Institut für Silikon- und fluorocarbonchemie Radebeul).

Methyl chlorophenyl silicones

The incorporation of a chlorophenyl group into silicone polymers would be expected to yield a high-temperature stationary phase with solvent properties resembling those of chlorobenzene.

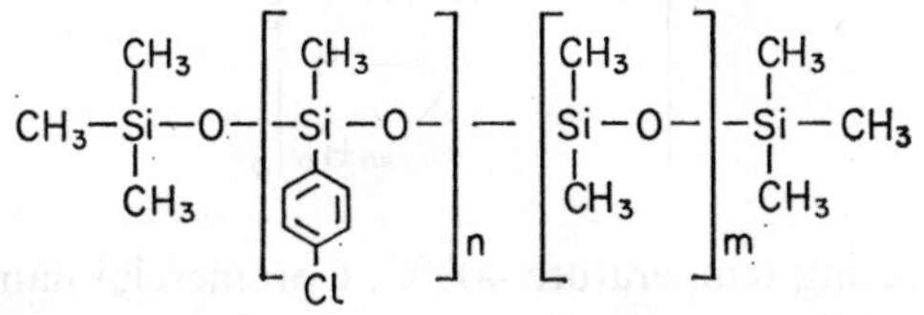

Chlorophenyl groups improve lubrication properties and also apparently reduce some dipole interactions between chains. Only moderate to slight selectivity to functional groups might be expected here. However, interaction between methylene groups adjacent to an ester group and chlorobenzene has been reported and studied.[449]

Maximum operating temperature 300°C. Commercial names: DC 560 (Dow Corning); F 50; F 60 (General Electric); SP-400 (Supelco).

Methyl phenyl vinyl silicones

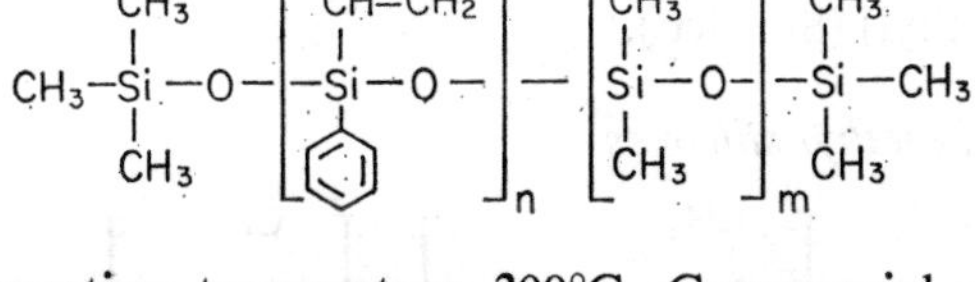

Maximum operating temperature 300°C. Commercial names: SE 54 (1% vinyl, 5% phenyl, 94% methyl) (General Electric).

Methyl cyanopropyl phenyl silicones

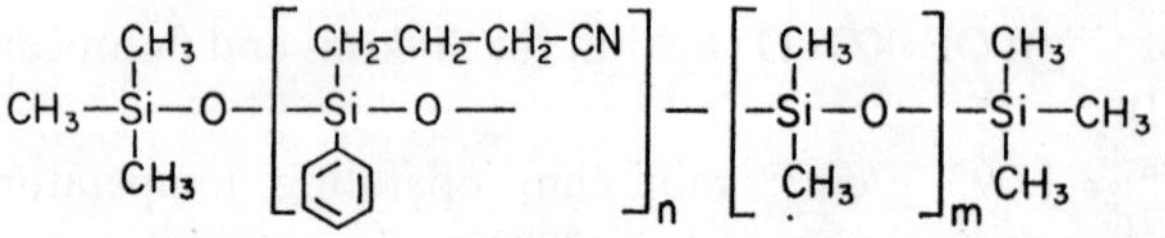

$$CH_3-\underset{\underset{CH_3}{|}}{\overset{\overset{CH_3}{|}}{Si}}-O-\left[\underset{\underset{\bigcirc}{|}}{\overset{\overset{CH_2\text{-}CH_2\text{-}CH_2\text{-}CN}{|}}{Si}}-O-\right]_n-\left[\underset{\underset{CH_3}{|}}{\overset{\overset{CH_3}{|}}{Si}}-O-\right]_m\underset{\underset{CH_3}{|}}{\overset{\overset{CH_3}{|}}{Si}}-CH_3$$

Maximum operating temperature 275°C. Commercial name: OV-225 (25% cyanopropyl; 25% phenyl, 50% methyl) (Applied Science).

By polymerization of carboranes with polysiloxanes, a thermally stable compound is obtained:[450]

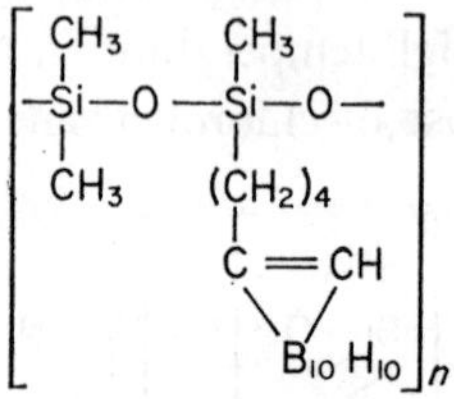

$$\left[\begin{array}{c} \underset{\underset{CH_3}{|}}{\overset{\overset{CH_3}{|}}{Si}}-O-\underset{\underset{(CH_2)_4}{|}}{\overset{\overset{CH_3}{|}}{Si}}-O- \\ \\ \underset{B_{10}H_{10}}{\overset{|}{\underset{\diagdown}{C}}}=CH \end{array}\right]_n$$

Maximum operating temperature 400°C. Commercial name: Dexsil 300 GC (Analabs). This compound is used as a stationary phase especially at temperatures over 300°C, where other siliconic stationary phases are unstable.[451, 452]

Methyl vinyl silicones

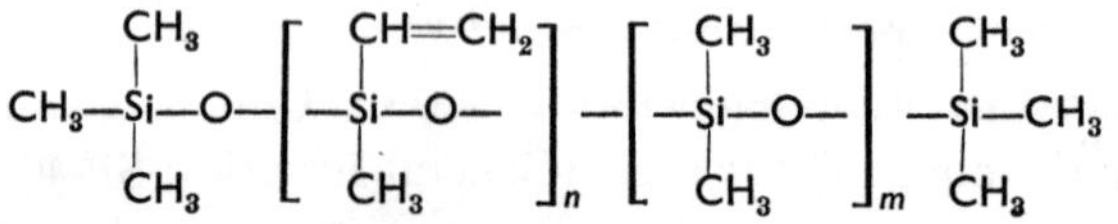

$$CH_3-\underset{\underset{CH_3}{|}}{\overset{\overset{CH_3}{|}}{Si}}-O-\left[\underset{\underset{CH_3}{|}}{\overset{\overset{CH=CH_2}{|}}{Si}}-O-\right]_n-\left[\underset{\underset{CH_3}{|}}{\overset{\overset{CH_3}{|}}{Si}}-O-\right]_m\underset{\underset{CH_3}{|}}{\overset{\overset{CH_3}{|}}{Si}}-CH_3$$

Maximum operating temperature 300°C. Commercial names: SE 31; SE 33 (General Electric); W 96; W 98 (Union Carbide); DC 430 (1% vinyl, 99% methyl) (Supelco).

Trifluoropropyl methyl silicones

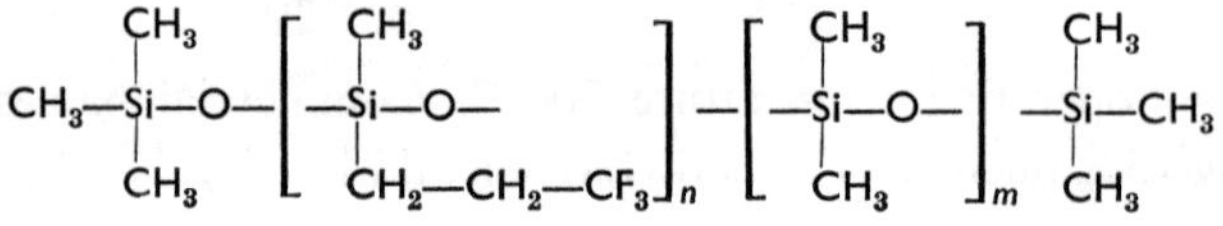

$$CH_3-\underset{\underset{CH_3}{|}}{\overset{\overset{CH_3}{|}}{Si}}-O-\left[\underset{\underset{CH_2-CH_2-CF_3}{|}}{\overset{\overset{CH_3}{|}}{Si}}-O-\right]_n-\left[\underset{\underset{CH_3}{|}}{\overset{\overset{CH_3}{|}}{Si}}-O-\right]_m\underset{\underset{CH_3}{|}}{\overset{\overset{CH_3}{|}}{Si}}-CH_3$$

The methylene group adjacent to the trifluoropropyl methyl silicone molecule might be polarized somewhat by the strongly electronegative trifluoromethyl group so that some positive charge is induced there. At high temperatures the polarity of this stationary phase decreases. When cooling, their polarity regenerates only after a long time. Thus, when frequently used, their retention data may undergo important modifications.

Trifluoropropyl silicones are particularly sensitive to strong bases, and depolymerize to form cyclotetrasiloxane. Thermal degradation produces primarily cyclosiloxanes, but does involve some silicon–carbon bond cleavage to produce 1,1-difluoropropene.

Trifluoropropyl methyl silicone has an unusual affinity for the carbonyl group.

Maximum operating temperature 250°C. Commercial names: QF-1; FS-1265 (Dow Corning); OV-210 (50% trifluoropropyl, 50% methyl) (Applied Science); SP-2401 (50% trifluoropropyl, 50% methyl) (Supelco).

Trifluoropropyl methyl and vinyl copolymer

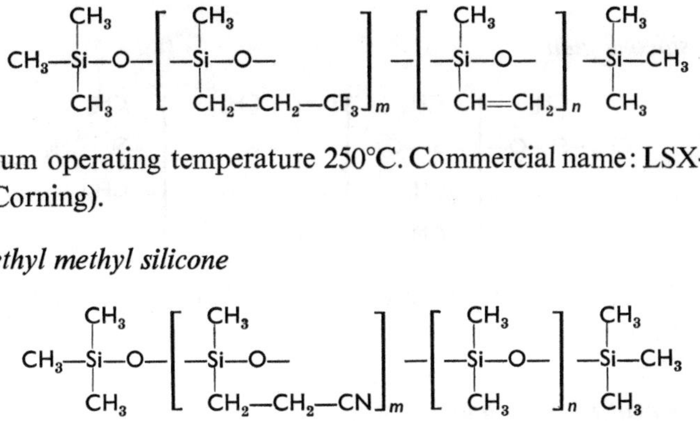

$$CH_3-\underset{\underset{CH_3}{|}}{\overset{\overset{CH_3}{|}}{Si}}-O-\left[\underset{\underset{CH_2-CH_2-CF_3}{|}}{\overset{\overset{CH_3}{|}}{Si}}-O-\right]_m-\left[\underset{\underset{CH=CH_2}{|}}{\overset{\overset{CH_3}{|}}{Si}}-O-\right]_n\underset{\underset{CH_3}{|}}{\overset{\overset{CH_3}{|}}{Si}}-CH_3$$

Maximum operating temperature 250°C. Commercial name: LSX-3-0295 (Dow Corning).

Cyanoethyl methyl silicone

$$CH_3-\underset{\underset{CH_3}{|}}{\overset{\overset{CH_3}{|}}{Si}}-O-\left[\underset{\underset{CH_2-CH_2-CN}{|}}{\overset{\overset{CH_3}{|}}{Si}}-O-\right]_m-\left[\underset{\underset{CH_3}{|}}{\overset{\overset{CH_3}{|}}{Si}}-O-\right]_n\underset{\underset{CH_3}{|}}{\overset{\overset{CH_3}{|}}{Si}}-CH_3$$

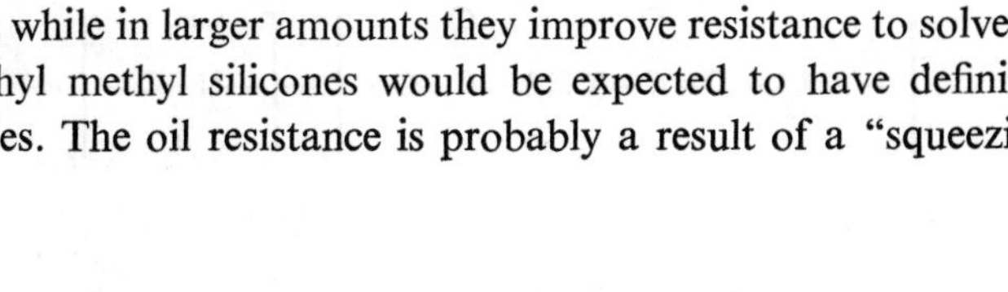

Cyanoalkyl groups, in the same way as fluorinated alkyl groups, in minor amounts impart better low-temperature properties to silicone rubbers, while in larger amounts they improve resistance to solvents. The cyanoethyl methyl silicones would be expected to have definite polar properties. The oil resistance is probably a result of a "squeezing out"

process owing to strong interactions between the cyano groups. This strong dipolar group would interact selectively, depending on the dipole moment and polarizability of materials being separated on the column. Aliphatic hydrocarbon solutes would have little solubility in the liquid phase. However, polarizable aromatics would be held selectively even at 250°C; the selectivity for aromatic hydrocarbons is very good. Thus naphthalene (b.p. 218°C) is retained twenty times more strongly than hexadecane (b.p. 287°C).[453]

In comparing the nitrile-substituted silicones with the trifluoropropyl silicones, the relatively high value of the cohesive energy density for aliphatic nitriles relative to the low value of δ for fluorocarbons should be noted. The properties of these materials as stationary phases are generally consistent with their cohesive energy density and specific polar interaction tendencies. Of course, in silicone polymers, cyano dipole-pair bonding would result in some cross-linking effects.

Maximum operating temperature for cyanoethylmethyl silicone is 250°C. Commercial names: SE 60; XF 1125; XF 1130; XF 1135; XF 1150 (General Electric).

Nitrile silicone gum

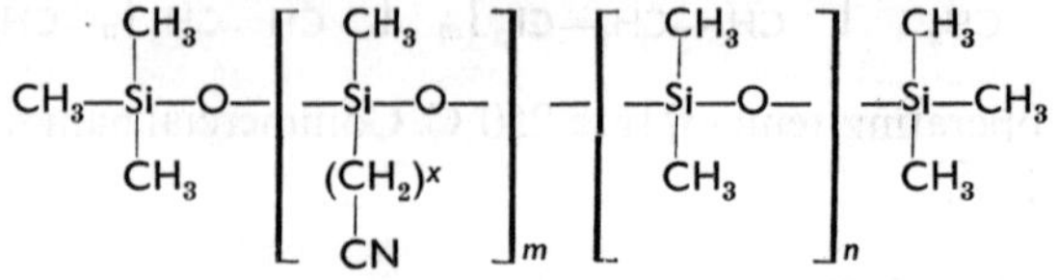

Maximum operating temperature 275°C. Commercial name: XE-60 (General Electric).

Copolymers of siloxane and esters

A copolymer of polyethylene succinate and methylsiloxane

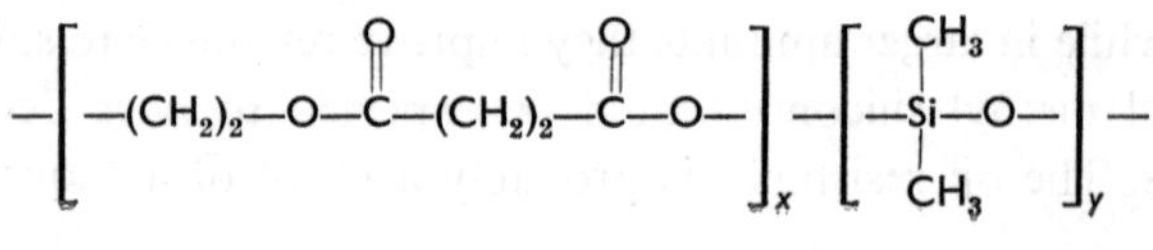

Maximum operating temperature 225°C. Commercial names: EGSS-X for $Y/X < 1$ and EGSS-Y for $Y/X > 1$.

A copolymer of polyethylene succinate and phenylmethylsiloxane

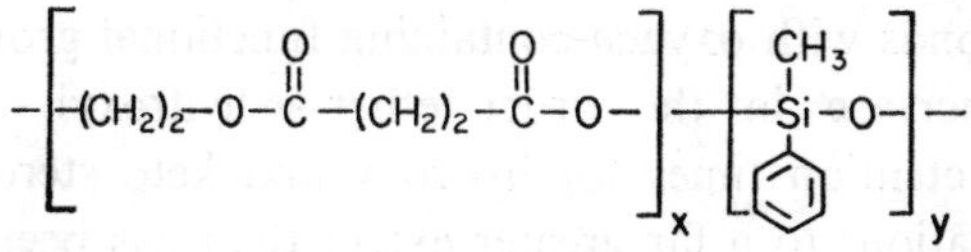

$$-\left[-(CH_2)_2-O-\overset{O}{\overset{\|}{C}}-(CH_2)_2-\overset{O}{\overset{\|}{C}}-O-\right]_x \left[-\overset{CH_3}{\underset{}{\overset{|}{Si}}}-O-\right]_y-$$

Maximum operating temperature 225°C. Commercial names: EGSP-A for $Y/X < 1$ and EGSP-Z for $Y/X > 1$.

A copolymer of polyethylene succinate and cyanoethylmethyl siloxane

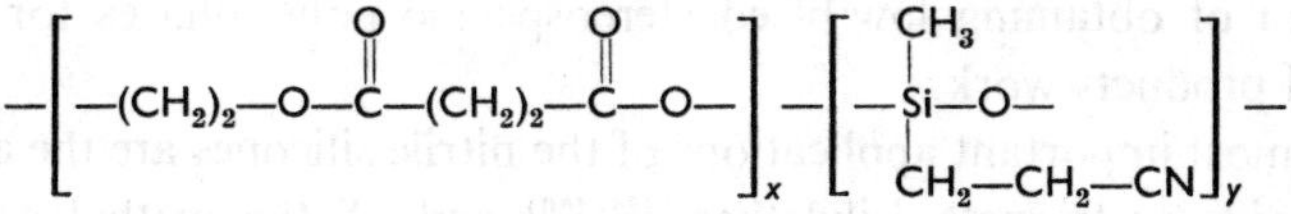

$$-\left[-(CH_2)_2-O-\overset{O}{\overset{\|}{C}}-(CH_2)_2-\overset{O}{\overset{\|}{C}}-O-\right]_x \left[-\overset{CH_3}{\underset{CH_2-CH_2-CN}{\overset{|}{\underset{|}{Si}}}}-O-\right]_y-$$

Maximum operating temperature 210°C. Commercial names: ECNSS-S for $Y/X < 1$ and ECNSS-M for $Y/X > 1$.

These stationary phases have a greater polarity owing to the presence of ester groups. These have donor oxygen atoms that may interact with the solutes to be separated by hydrogen bonding.

Silicones are widely used as stationary phases in gas chromatography for the most different of applications. A review of the use of silicones in gas chromatography would mean a huge number of literature references. Many times one can encounter for the very same separation the use of different silicone stationary phases, and certain stationary phases are being replaced gradually with other compounds from the same class but with better performances. Thus fluorosilicone QF-1 has been largely replaced by the more thermally stable, equivalent phase OV-210, and XE-60 has been replaced by OV-225, which is similar, but not identical, in composition and performance.

Phenyl silicones have been used for the analysis of aluminium, beryllium, and uranium trifluoroacetylacetonates,[454, 455] iodoperfluoroalkanes,[456] and some higher aromatic hydrocarbons,[457] and those of the OV series have been applied to the analysis of pesticides.[458, 459]

Fluoroalkyl silicones have been proposed for the analysis of hydroxy, acetoxy, oxo-stearic acid methyl esters,[460] and halogenated amine derivatives,[461] but the most important application is to the analysis of steroids.[462, 463]

For compounds with oxygen-containing functional groups, the retention times increase in the order ether < hydroxyl < ester < keto. Further, the retention times for hydroxy and keto steroids vary with structural variations to a far greater extent than has been observed for other phases. For example, QF-1 is useful for the separation of cholesterol and cholestanol as well as for the separation of 5α-pregnane-3β, 20α-diol, and 5α-pregnane-20β-ol-3-one.[462]

The use of fluoroalkyl silicones provides an alternate approach to the problem of obtaining low-bleed stereospecific polar phases for use in natural products work.

The most important applications of the nitrile silicones are the analysis of steroids as tri-methylsilylethers[464-466] and of the methyl esters of higher saturated fatty acids when their unsaturated counterparts are also present.[467]

Some aromatic compounds may also be classed as dilute stationary phases. On these, aromatic hydrocarbons are retained selectively. The presence of the alkyl groups in their molecules means a decrease of the selectivity.[468] As a consequence of the interaction with the π-electrons of the stationary phase, the electron acceptor solutes will be better retained than on non-polar stationary phases. The polar solutes will induce a dipole moment in the molecule of the stationary phase that finally leads to an increase of the retention.

Benzyldiphenyl

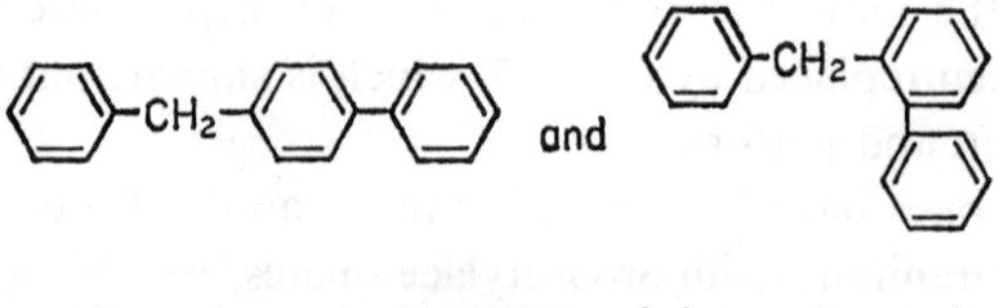

Maximum operating temperature 100°C; minimum operating temperature 50°C.

This stationary phase is recomended for the analysis of the C_4–C_{10} hydrocarbons.[423, 469]

Alkyl naphthalene

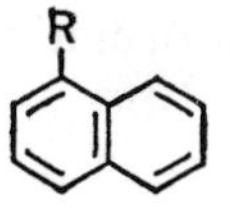

where $R > C_{20}H_{41}$.

Minimum operating temperature 50°C; maximum operating temperature 300°C. Commercial name: Fluhyzon.

This stationary phase has been used for the analysis of condensed aromatic hydrocarbons (phenanthrene, pyrene, benzopyrene, 1,2,5,6-dibenzanthracene[470]).

Radiation-induced polymerization of stilbene (C_6H_5—CH=CH—C_6H_5) in the presence of a heavy metal catalyst yields a mixture of phenyl substituted indane derivatives with a melting point of 50–60°C. As the mixture has a very low-vapour pressure, it was tested as a stationary phase in gas chromatography for the separation of pesticides.[471, 472]

Maximum operating temperature 300°C.

Novitskaya and Vigdergauz use as stationary phases colloidal solutions of *polystyrene*.[367]

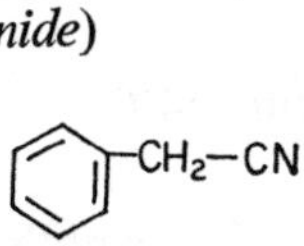

This stationary phase has been used for the analysis of C_1–C_5 alcohols.

Phenylacetonitrile (*benzyl cyanide*)

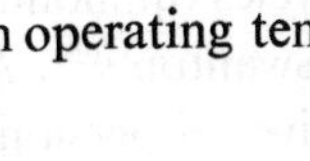

$d_4^{20} = 1.016$ g/cm³; minimum operating temperature -24°C; maximum operating temperature 20°C.

B : SPGC 10

Nitrobenzene

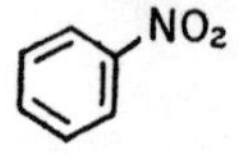

$d_4^{20} = 1.223$ g/cm³; minimum operating temperature 6°C; maximum operating temperature 20°C.

m-Nitrotoluene

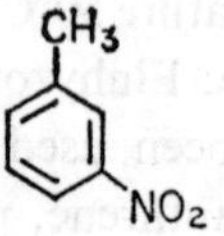

$d_4^{18} = 1.160$ g/cm³; minimum operating temperature 17°C; maximum operating temperature 30°C.

α-Naphthylamine

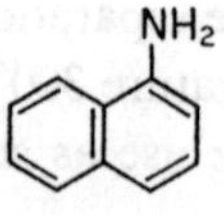

$d_4^{50} = 1.108$ g/cm³; minimum operating temperature 50°C; maximum operating temperature 85°C.

Diphenylamine

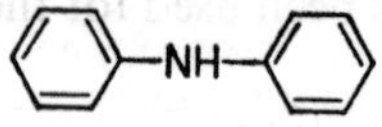

Minimum operating temperature 54°C; maximum operating temperature 85°C.

The separation properties of α-naphthylamine, diphenylamine, and some of the nitrogen heterocycles (quinoline (I); 7,8-benzoquinoline (II)) were studied by Desty and Swanton[133]. As with the aromatic amines, the nitrogen heterocycles give dispersion interactions with aromatic hydrocarbons, a fact that explains the separation of the isomeric xylenes.

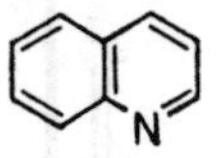

(I)

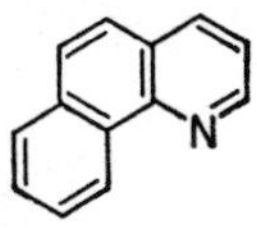

(II)

Minimum operating temperature −15°C; maximum operating temperature 50°C.

Minimum operating temperature 53°C; maximum operating temperature 150°C.

Generally the separation of *m*- and *p*-xylene is only slightly enhanced by replacing a carbon atom in the aromatic nucleus by a nitrogen atom; electron-donating groups such as amino and hydrogen groups slightly increase the separation whereas the electron attracting groups, choro-, bromo-, iodo-, and nitro-, markedly reduce the selectivity.

Also, methyl substitution appears to decrease the separation which is still further reduced by increasing the size of the alkyl group. In attempting to account for the separations achieved with these aromatic stationary phases it is clear that in addition to the dispersive forces between solvent and solute molecules the solvent exerts an inductive effect because of its own permanent dipole and the induced dipole in the solute molecule. Thus whether a hydrocarbon is selectively retained or not will depend on polarizability as well as solubility in the stationary phase.

Dibenzyl ether

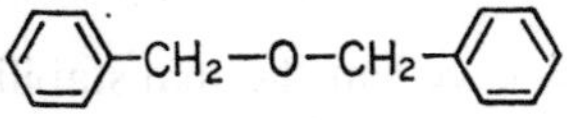

$-CH_2-O-CH_2-$

Maximum operating temperature 50°C. It is recommended for the analysis of chlorinated and aliphatic hydrocarbons.

Polymetaphenyl ethers

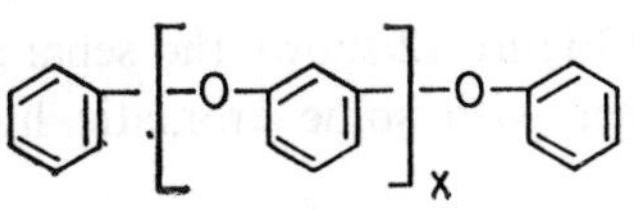

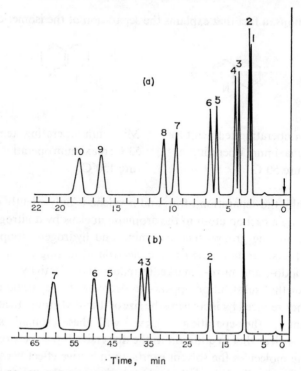

FIG. 12. Chromatogram of a mixture of hydrocarbons on polyphenyl ether. (a) Mixture of α-olefins and n-alkenes at 120°C; (1) hexane, (2) hexene-1, (3) heptane, (4) heptene-1, (5) octane, (6) octene-1, (7) nonane, (8) nonene-1, (9) decane, (10) decene-1. (b) Mixture of aromatic hydrocarbons at 100°: (1) benzene, (2) toluene, (3) m-xylene, (4) p-xylene, (5) o-xylene, (6) p-isopropyl benzene, (7) p-(n-propyl) benzene.[479]

with $x = 5$, 6, or 7 show excellent thermal stability and are used as high-temperature functional fluids, finding application as heat transfer agents. In addition, they have proved useful as the liquid phase in high-temperature chromatographic work.[473-477]

These stationary phases proved to be more useful for the separation of aromatic hydrocarbons[478, 479] and of a mixture of α-olefins and n-C_6-C_{10} alkanes.[479] Figure 12 shows the separation of some α-olefins and n-alkanes together with some aromatic hydrocarbons on 5-ring polyphenylether.[479]

The maximum operating temperature of these stationary phases is 400°C.

Recently, seven new polyphenyl ethers containing a biphenyl central core were synthesized by Hammann et al.[480] They included:

3,3'-bis(m-phenoxyphenoxy)biphenyl

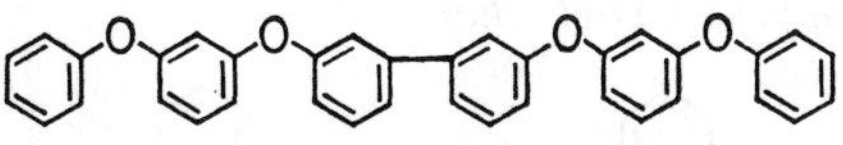

b.p. = 312°C/0.05 mm; m.p. = 85–88°C.

3-(m-phenoxyphenoxy)-3'-(p-phenoxyphenoxy)biphenyl

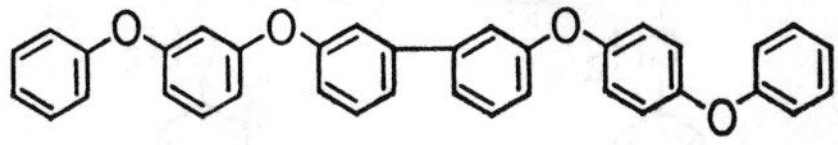

b.p. = 302–307°C/0.05 mm; d_4^{24} = 1.201 g/cm^3.

3,3'-bis[m(m-phenoxyphenoxy)phenoxy]biphenyl

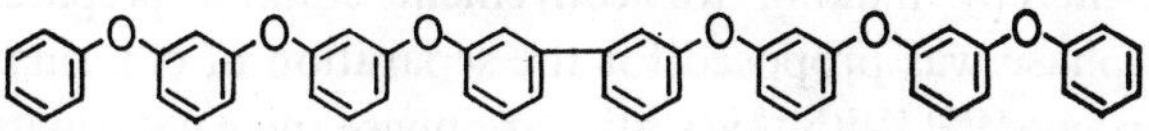

d_4^{24} = 1.231 g/cm^3.

4'-[m-(m-phenoxyphenoxy)phenoxy]-m-terphenyl

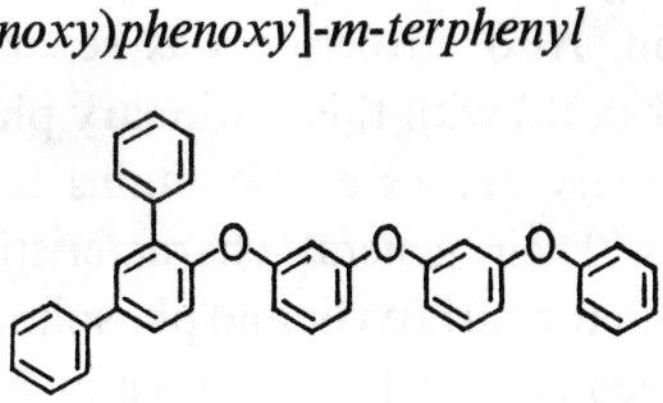

b.p. = 337°C/0.07 mm; d_4^{24} = 1.185 g/cm^3.

The toluene–formaldehyde and naphthalene–formaldehyde resins were also proposed as stationary phases.[481] These resins are obtained by condensation of the aromatic hydrocarbons with formaldehyde in an acidic medium.

$$(n+2)Ar—H + (2n+1)CH_2O \rightarrow Ar—(CH_2—O—CH_2—Ar)_n—CH_2O—Ar + nH_2O$$

Maximum operating temperature for the toluene–formaldehyde resin is 150°C while that for the naphthalene–formaldehyde resin is 200°C. The retention indices of a few polar solutes indicate the moderate polarity of these phases, which were used for the separation of some aromatic hydrocarbons and aliphatic alcohols.

4,4′-bis(tribenzyl-silyl)diphenyl ether

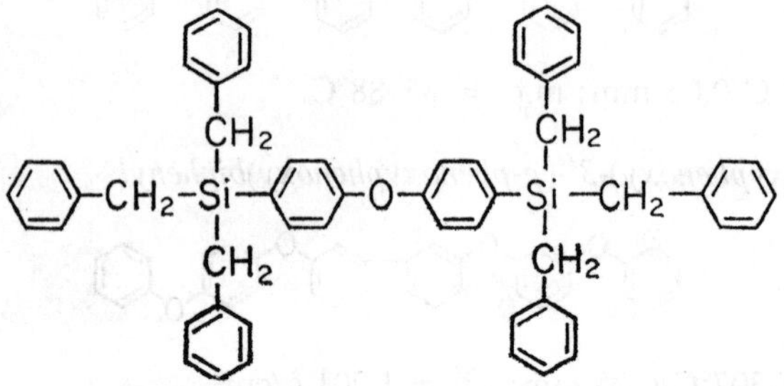

b.p. = 400°C/0.1 mm.

At slightly elevated temperatures (40°C), the viscosity approaches that of water, thereby making for convenient column preparation. This stationary phase was proposed for the separation of chlorinated derivatives of benzene.[482] With this stationary phase, nine chlorinated benzene derivatives (from monochloro to hexachlorobenzene) were separated using temperature programming between 75–260°C. The analysis lasted approximately 35 min. Also, glycerine was separated from triethylene glycol (b.p. 290°C for both) with this stationary phase.

This class of stationary phases also contains the esters of carboxylic and phosphoric acids. Their general characteristic is the presence of donor oxygen atoms (in the carboxylic and phosphoric groups respectively) that may form hydrogen bonds, thus resulting in high relative retentions for the solutes with protonated functional groups.

Dinonyl phthalate

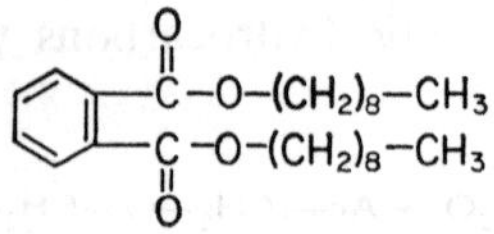

$d_4^{20} = 0.97$ g/cm^3; minimum operating temperature 20°C; maximum operating temperature 150°C.

Didecyl phthalate

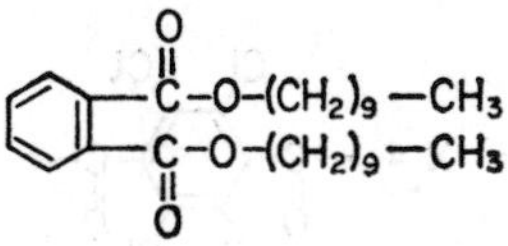

Minimum operating temperature 25°C; maximum operating temperature 150°C.

Diphenyl phthalate

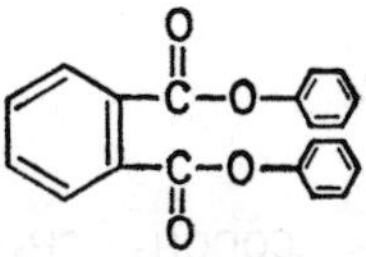

Maximum operating temperature 150°C.

These stationary phases present a medium solvation possibility for alkanes, aromatic hydrocarbons, ethers, esters, ketones, mercaptans, and thioethers.[483–490]

The tetrachloroterephthaloyl nucleus was chosen as the basic structural unit because of its potential participation in π-type charge transfer interaction with aromatic compounds and olefins, the tendency for the formation of linear, and therefore controlled, structures with *p*-isomers and the relative thermal and chemical stability of the tetrahalogenated aromatic nucleus.

Langer[491] prepared compounds of the following type:

$$R—O—X—O—(CH_2)_n—O—X—O—R$$

(I) R = butyl

    (a) $n = 2$; m.p. 183–184°C

    (b) $n = 3$; m.p. 108.3–109.7°C

    (c) $n = 4$; m.p. 184–185.5°C

(II) R = propyl

    $n = 3$; m.p. 127–128°C

(III)

$$R—O—X—O—(CH_2)_3—O—X—O—(CH_2)_3—O—X—O—R$$

R = propyl   m.p. 148–149.5°C

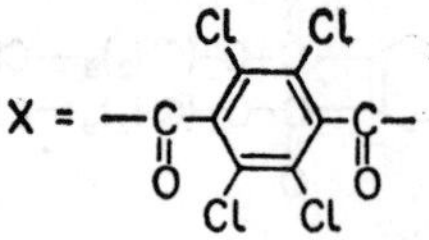

Tetrachloroterephthaloyl oligomers tend to stay liquid on supercooling especially in mixtures. These compounds were used for the separation of aromatic isomers.

Maximum operating temperature 200°C.

Dipropyl tetrachlorophthalate

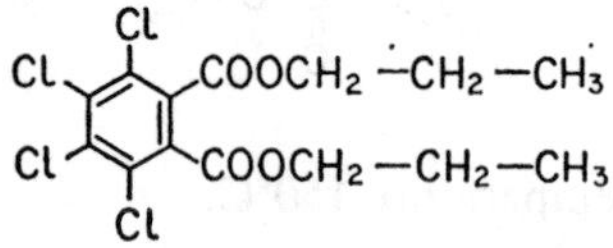

$d_4^{20} = 1.37$ g/cm³; minimum operating temperature 27°C; maximum operating temperature 150°C.

Dibutyl tetrachlorophthalate

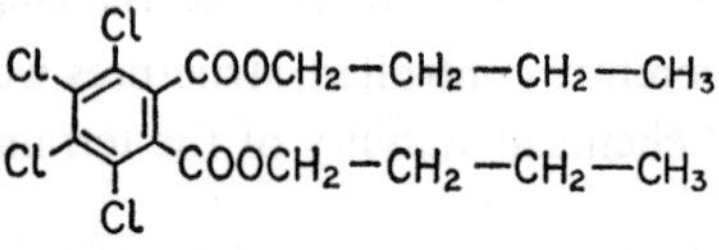

$d_4^{20} = 1.31$ g/cm³; minimum operating temperature 20°C; maximum operating temperature 150°C.

The affinity of these stationary phases for electron donor solutes is increased by the presence of the four chlorine atoms.

The propyl and butyl esters of tetrachlorophthalic acid proved to be useful for the separation of the xylenes[319, 322] and for a mixture of n-C_7–C_{10} mono-olefins.[492, 493]

Di-(2-ethylhexyl)sebacate

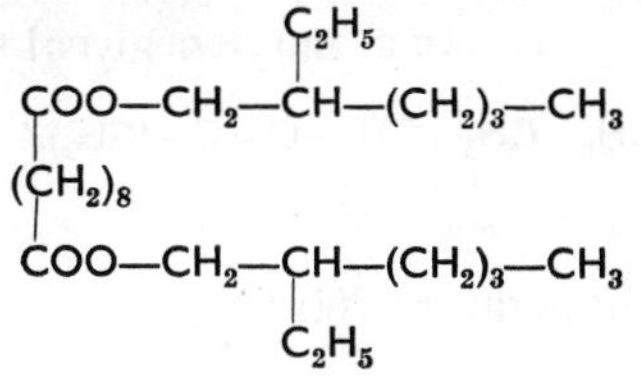

$$
\begin{array}{l}
\qquad\qquad\qquad C_2H_5 \\
\qquad\quad COO-CH_2-\overset{|}{C}H-(CH_2)_3-CH_3 \\
(\overset{|}{C}H_2)_8 \\
\qquad\quad COO-CH_2-\overset{|}{C}H-(CH_2)_3-CH_3 \\
\qquad\qquad\qquad C_2H_5
\end{array}
$$

Maximum operating temperature 130°C. Commercial names: Octoil S; Perkin Elmer B.

Di-(2-ethylhexyl) sebacate has good solubilization properties for hydrocarbons.[494-496]

Di-(2-ethylhexyl)adipate

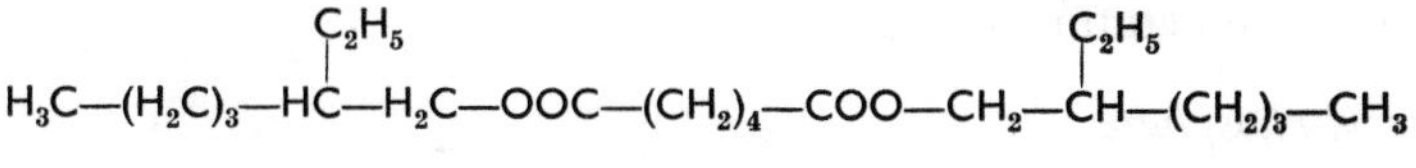

$$
\begin{array}{ll}
\qquad\quad C_2H_5 & \qquad\qquad\quad C_2H_5 \\
H_3C-(H_2C)_3-H\overset{|}{C}-H_2C-OOC-(CH_2)_4-COO-CH_2-\overset{|}{C}H-(CH_2)_3-CH_3
\end{array}
$$

Maximum operating temperature 125°C.

Didecyl adipate

$$H_3C-(H_2C)_9-OOC-(CH_2)_4-COO-(CH_2)_9-CH_3$$

Maximum operating temperature 125°C.

These last two stationary phases resemble Octoil S in their properties.

Trimethylolpropane tripelargonate

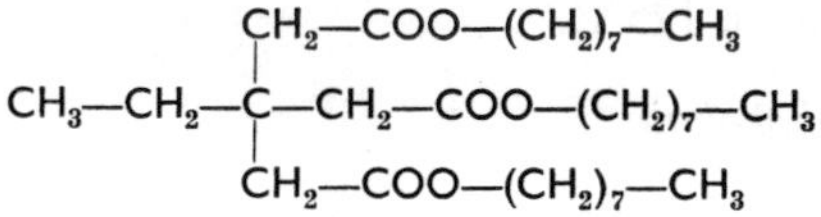

$$
\begin{array}{l}
\qquad\qquad\quad CH_2-COO-(CH_2)_7-CH_3 \\
CH_3-CH_2-\overset{|}{\underset{|}{C}}-CH_2-COO-(CH_2)_7-CH_3 \\
\qquad\qquad\quad CH_2-COO-(CH_2)_7-CH_3
\end{array}
$$

Maximum operating temperature 200°C. Commercial name: Celanese Ester no. 9. It is recommended for the separation of cresols and xylenols.

Ethylene glycol distearate

$$C_{17}H_{35}OOC-CH_2-CH_2-COOC_{17}H_{35}$$

Maximum operating temperature 250°C.

Polyesters with long methylene chains may be considered members of this class of stationary phases (e.g. butyleneglycol sebacates).

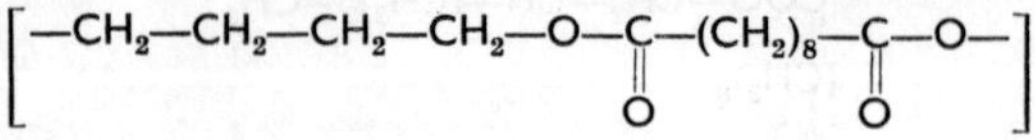

$$\left[-CH_2-CH_2-CH_2-CH_2-O-\underset{\underset{O}{\|}}{C}-(CH_2)_8-\underset{\underset{O}{\|}}{C}-O-\right]$$

Maximum operating temperature 200°C.

Triphenyl phosphate

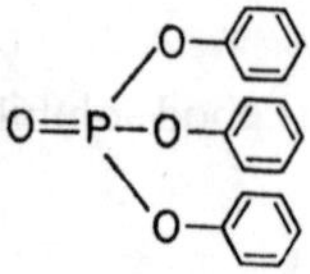

$d_4^{20} = 1.18$ g/cm³. Minimum operating temperature 50°C. Maximum operating temperature 100°C.

Tricresyl phosphate

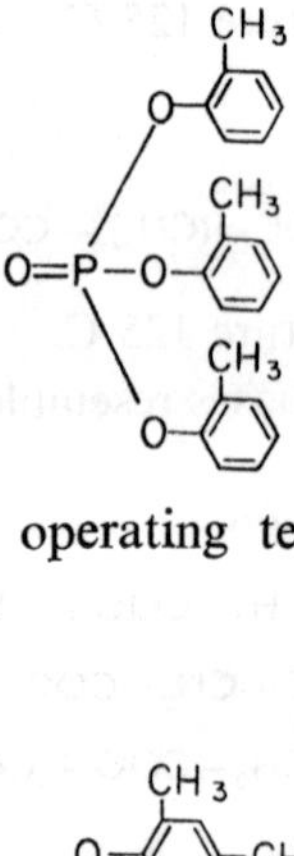

$d_4^{20} = 1.18$ g/cm³. Minimum operating temperature 20°C. Maximum operating temperature 125°C.

Tri-(2,4-xylenyl) phosphate

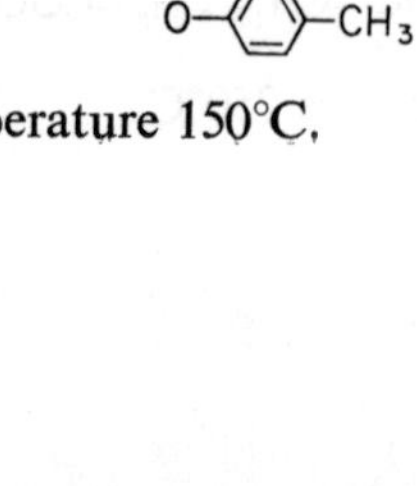

Maximum operating temperature 150°C.

These stationary phases are used for the separation of hydrocarbons, esters, ketones, alcohols,[497, 498] some chlorinated derivatives of hydrocarbons,[499] cresols,[500] xylenols, fluorophenols, and chlorophenols.[501]

Di-(2-ethylhexyl) phosphoric acid

$$O{=}P{-}OH \begin{array}{l} O{-}CH_2{-}CH{-}(CH_2)_3{-}CH_3 \\ \qquad\qquad | \\ \qquad\qquad C_2H_5 \\[4pt] O{-}CH_2{-}CH{-}(CH_2)_3{-}CH_3 \\ \qquad\qquad | \\ \qquad\qquad C_2H_5 \end{array}$$

Maximum operating temperature 150°C.

This stationary phase was used for the separation of some hydrocarbons, C_1–C_4 alcohols in the presence of C_3–C_4 ketones, and for the separation of some chlorinated derivatives of methane and ethane.[100]

Figure 13 shows the separation of C_1–C_4 alcohols in the presence of C_3–C_4 ketones on this stationary phase.[100]

The lithium, sodium, and potassium salts of di-(2-ethylhexyl) phosphoric acid have a better thermal stability, loss of material being noticeable only above 180°C.

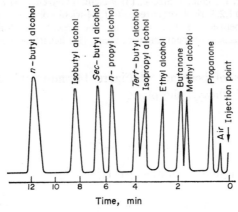

FIG. 13. Separation of C_1–C_4 alcohols and of C_3–C_4 ketones using 15% di-(2-ethylhexyl)phosphoric acid on silanized Chromosorb W (60–80 mesh), in a stainless steel column, 3 m×6 mm i.d. at 110°C. Hydrogen flow rate: 97 ml/min.[100]

The preparation of the lithium, sodium, and potassium salts is simple, their synthesis being an extraction of the respective metal ions from aqueous hydroxide solutions into an organic layer formed by the free acid and dibutyl ether followed by evaporation of the solvent.[321] These compounds were used for the isothermal separation of the chlorin-

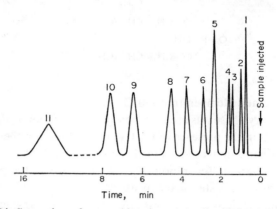

FIG. 14. Separation of some chloro-benzenes using 20% LiDEHP on silanized Chromosorb W (60–80 mesh) at 165°C.[502] Column: thermo-resistant glass, 1.8 m×3 mm i.d. Nitrogen flow rate: 46.2 ml/min.: (a) benzene, (2) chlorobenzene, (3) p-dichlorobenzene, (4) o-dichloro-benzene, (5) 1,2,4-trichlorobenzene, (6) 1,2,3-trichlorobenzene, (7) 1,2,4,5 tetrachlorobenzene, (8) 1,2,3,4-tetrachlorobenzene, (9) 1,2,4-trichloro-5-nitrobenzene, (10) pentachlorobenzene, (11) hexachlorobenzene.

ated derivatives of benzene.[502] Figure 14 shows this separation on lithium di-(2-ethylhexyl) phosphate.

Stearic acid

$$CH_3-(CH_2)_{16}-COOH$$

Maximum operating temperature 100°C.

Behenic acid

$$CH_3-(CH_2)_{20}-COOH$$

Maximum operating temperature 150°C,

These acids are usually added to non-polar or slightly polar stationary phases for the analysis of lower carboxylic acids, alcohols, or amino-acid esters.[503]

Information concerning the use of fractions of fatty acids (the C_{10}–C_{16} and C_{17}–C_{21} fractions) for the separation of C_6–C_{11} paraffins and C_6–C_8 alkylbenzenes has been published.[504]

Berezkin et al. used a long-chain carboxylic acid with about 200 carbon atoms as a stationary phase.[505] This compound can be used up to 270°C and was utilized for the separation of complex mixtures of polyphenols, alkyldiphenyl oxides, and C_7–C_{20} alcohols.

Hexadecylic alcohol

$$CH_3—(CH_2)_{14}—CH_2OH$$

$d_4^{49} = 0.798$ g/cm³; minimum operating temperature 50°C; maximum operating temperature 60°C.

Octadecylic alcohol

$$CH_3—(CH_2)_{16}—CH_2OH$$

$d_4^{59} = 0.812$ g/cm³; minimum operating temperature 59°C; maximum operating temperature 70°C.

Littlewood and Willmott[506] measured the specific retention volumes V_g of butanols and pentanols by 1-dodecanol (I) and 2-dodecanol (II).

$$CH_3—CH_2—(CH_2)_8—CH_2—CH_2OH \qquad CH_3—CH_2—(CH_2)_8—\underset{\underset{OH}{|}}{CH}—CH_3$$

(I)   (II)

It has been found that the ratio of V_g in 1-dodecanol to V_g in 2-dodecanol is approximately constant for alcohol solutes of a particular class, primary, secondary, or tertiary, but the value of the ratio for primary alcohol solutes differs significantly from that for secondary or tertiary alcohol solutes. Thus, if a primary alcohol happens to overlap a secondary or tertiary alcohol on a particular stationary liquid which is a primary alcohol, a reasonable generalization of the results leads to the conclusion that separation will be achieved on a stationary liquid which is a secondary alcohol, though otherwise of similar molecular structure. But in many applications these stationary phases are replaced with polyglycols, which are more thermal stable.

Bis(2-methoxyethyl)ether

$$H_3C—O—CH_2—CH_2—O—CH_2—CH_2—O—CH_3$$

Maximum operating temperature 50°C.

Bis[2-(2-methoxyethoxy)ethyl]ether

$$H_3C—O—CH_2—CH_2—O—CH_2—CH_2—O—CH_2—CH_2—O—CH_2—CH_2—O—CH_3$$

Maximum operating temperature 80°C.

2-(Benzyloxy) ethanol (benzyl cellosolve)

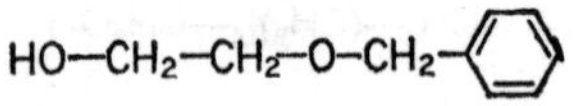

$$HO—CH_2—CH_2—O—CH_2—\langle\bigcirc\rangle$$

Maximum operating temperature 100°C.

These stationary phases were used for the separation of hydrocarbons up to C_6.[507]

Emulphor

This is a cetyl alcohol $(C_{16}H_{33}OH)$ or oleic alcohol $(C_{18}H_{35}OH)$ ethoxylated with 20–25 moles of ethylene oxide.

$$R—O—(CH_2—CH_2—O)_n—H \qquad R = C_{16}H_{35} \quad \text{or} \quad C_{18}H_{35}; \quad n = 20—25$$

Maximum operating temperature 175°C. Commercial name: Emulphor ON-870.

These stationary phases were used for the separation of iodoalkanes $(C_1–C_4)$, phenylalkanes, bicyclic hydrocarbons, terpenes, and phenols.[508, 509, 510]

Pentabenzoyl-α-glucose

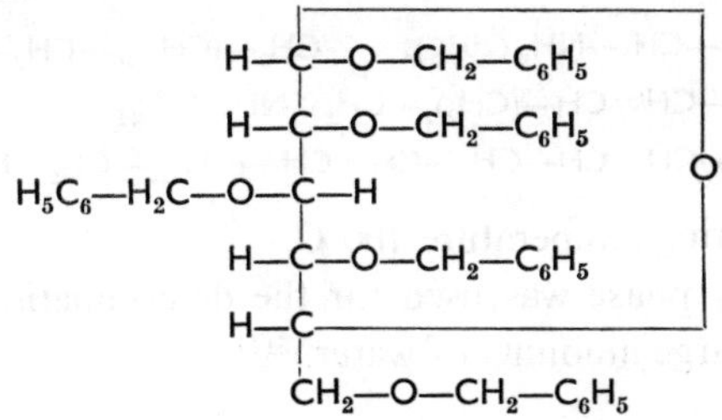

$$H-\underset{|}{C}-O-CH_2-C_6H_5$$
$$H-\underset{|}{C}-O-CH_2-C_6H_5$$
$$H_5C_6-H_2C-O-\underset{|}{C}-H$$
$$H-\underset{|}{C}-O-CH_2-C_6H_5$$
$$H-\underset{|}{C}$$
$$CH_2-O-CH_2-C_6H_5$$

Maximum operating temperature 200°C.

This stationary phase was used for the analysis of alcohols and aromatic hydrocarbons.[511]

Sodium dodecyl benzene sulphonate

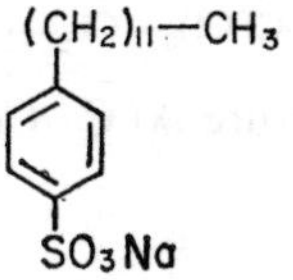

$$(CH_2)_{11}-CH_3$$

$$SO_3Na$$

Maximum operating temperature 210°C. Commercial name: Siponates DS-10.

Mixture of sulphonated lauryl alcohol and alkylaryl sulphonate

$$CH_3-(CH_2)_{10}-CH_2-O-SO_3Na$$

and

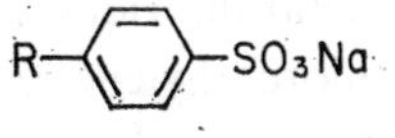

$$R-\langle\rangle-SO_3Na$$

where R = C_8–C_{18}. Maximum operating temperature 200°C. Commercial name: Tide.

Alkylaryl sulphonates are used as stationary phases for the separation of C_3–C_8 carboxylic acids, some aromatic amines and phenols.[512, 513]

Armeen SD is a product of the Armour Chemical Division and consists of normal hexadecyl, octadecyl, octadecenyl, and octadecadienyl primary amines:

$$CH_3—(CH_2)_{14}—CH_2—NH_2 \text{ (20\%)}; \quad CH_3—(CH_2)_{10}—CH_2—NH_2 \text{ (17\%)}$$
$$CH_3—(CH_2)_7—CH{=}CH—(CH_2)_7—CH_2—NH_2 \text{ (26\%)}$$
$$CH_3—(CH_2)_7—CH{=}CH—CH_2—CH{=}CH—(CH_2)_6—CH_2—NH_2 \text{ (37\%)}$$

Maximum operating temperature 100°C.

This stationary phase was used for the determination of alcohols in the presence of large amounts of water.[514]

2,2'-(2-Ethyl hexanamido)-diethyl-di-2-ethylhexoate

Maximum operating temperature 180°C. Commercial name: Flexol 8 N 8.

This is recommended for the separation of high acetylenic hydrocarbons.

Ethyl ester of tetrachloroperfluoro caprylic acid

$$Cl(CF_2—CFCl)_3—CF_2—COOC_2H_5$$

Maximum operating temperature 80°C. Commercial name: ethyl ester of the acid Kel-F 8114.

Polytrifluorochloroethylene

$$Cl(CF_2—CFCl)_n \, CF_3$$

The maximum operating temperature depends on the molecular weight. Commercial names: Kel-F oil 3 (50°C); Kel-F oil 10; Fluorolube HG 1200 (100°C); Kel-F wax 550 (200°C).

These compounds are very resistant to chemicals and are used as stationary phases for the separation of HCl, HF, HBr, Cl_2, Br_2, BCl_3, PCl_3, $POCl_3$, and volatile metal fluorides. As supports, polytetrafluoroethylene and polytrifluorochloroethylene were used.[515-517]

Other halogenated compounds used as stationary phases are: chlorinated biphenyl (Aroclor 1232) and chlorinated paraffin (Chlorowax 70°). These compounds are thermally stable up to 110°C.[518]

In this class we can also mention stationary phases chemically bonded to the support, obtained by the chemical modification of surface silanol groups. The reactivity of surface silanol groups offers three main possibilities for chemical modification. Such stationary phases appear more and more frequently in gas chromatography and in liquid chromatography.

Among the chemical modifications are the formation of Si–O–C bonds by esterification, of Si–C–C bonds by the reaction, for instance, of chlorinated surfaces with either organolithium or Grignard reagents, and Si–O–Si–C bonds, by silylation.[519] The esterification of siliceous surfaces with alcohols produced the so-called "brush" packings[520] (Waters Associates Durapak). These "Durapaks" were shown to obviate the necessity of long periods of column conditioning and to be satisfactory for the separation of the methyl esters of C_{18} fatty acids.[521] They suffer the disadvantage of relatively easy solvolysis of the Si–O–Si bond by polar compounds.

Locke et al.[522] reported a general method for preparing bonded stationary phase materials for gas and liquid chromatography. The fundamental reaction is between a Grignard reagent and a chlorinated siliceous surface to yield a Si–C bond. These bonds are especially thermally and solvolytically stable. Porasil C was halogenated and reacted repeatedly with naphthyl magnesium bromide to produce polynaphthyl Porasil, containing as much as 24% by weight of organic material. However, too few suitable reagents and the possible formation of magnesium occlusion salts during the bonding process make this approach much less realistic than esterification or silylation.

In situ polymerization of silane compounds possessing selective groups in the side chain produces some selective gas chromatographic packings.

Kirkland and de Stefano[523, 524] described and studied selective stationary phases with ether and cyanoethyl functions. These stationary phases show low vapour pressures at high temperatures, excellent column life, minimum bleed, homogeneous distribution on the support, and good column efficiencies for both gas and liquid chromatography.

Al-Taiar et al.[444] obtained chemically bonded stationary phases by silylation. These phases contain phenyl groups and are particularly suitable for gas chromatography–mass spectrometry. The relatively small number of such stationary phases proposed in the literature may be explained by the limited availability of commercial silane compounds possessing selective groups in the side chain.

In their recent work, Parr and Grohmann[525] synthetized some novel chlorosilanes for the derivatization of an inorganic silica carrier to form the following structure

$$
\begin{array}{c}
| \\
-Si- \\
| \\
O \\
| \\
-Si-O-Si-(CH_2)_n-\!\!\!\!\!\!\langle \bigcirc \rangle\!\!\!\!\!\!-CH_2-Cl \\
| \\
O \\
| \\
-Si- \\
|
\end{array}
$$

for further use in solid phase peptide synthesis. The structure suggests an obvious universal utility in the preparation of selective chemically

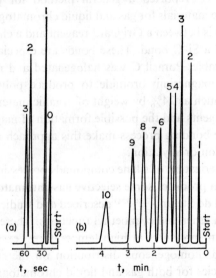

Fig. 15. Gas chromatographic separation on a packed column of Porasil C/3-hydroxypropionitrile (non-optimized) (15–100 μm)[520] on a 1 m× 2 mm i.d. column. (a) (0) methane, (1) ethane, (2) propane, (3) propene at 28°C. Nitrogen flow rate: 6.3 cm/s; sample weight: 1–100 μg. (b) (1) methane, (2) n-pentane, (3) cyclohexane, (4) 2,4-dimethylpentane, (5) methylcyclohexane, (6) 2,2,4-trimethylpentane, (7) 2,3,4-trimethylpentane, (8) n-octane, (9) 2,2,5-trimethylhexane, (10) diethyl ether at 121°C. $\Delta p = 5.5$ atm. Nitrogen flow rate: 5.5 cm/s.

bonded stationary phases as the chloromethyl terminal groups can be modified in a number of ways. Because selective chemically bonded stationary phases can combine efficiently, and have high thermal stability, the use of these materials is likely to grow in the near future.

Figure 15 shows the separation of some hydrocarbons. The separation was made on a chemically bonded stationary phase (Porasil C with specific surface 250 m^2/g and pore diameter 100–200 Å, treated with 3-hydroxy-propionitrile at 180°C).[520]

Davison and Moore[526] proposed as a stationary phase a poly(ester-acetal). Their poly(ester-acetal) was a copolymer of dimethylterephthalate, ethylene glycol, and 3,9-bis(7-carbomethoxyheptyl)-2,4,8,10-tetraoxa-spiro-5,5-undecane. The last compound is the pentaerythritol-acetal of methyl azela-aldehydate (MA$_z$-PEA) and has the structure I:

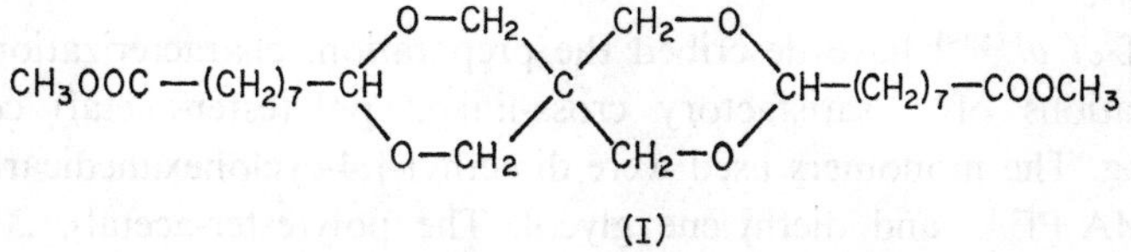

(I)

The linear polymer was found to crosslink in the presence of p-toluene sulphonic acid catalyst at 260°C. Probably the cross-linking reaction occurs between the acetal linkages in two neighbouring linear polymers as shown in II. Reaction of these labile acetal linkages may produce a silicon–oxygen bond on the surface of the solid support.[527]

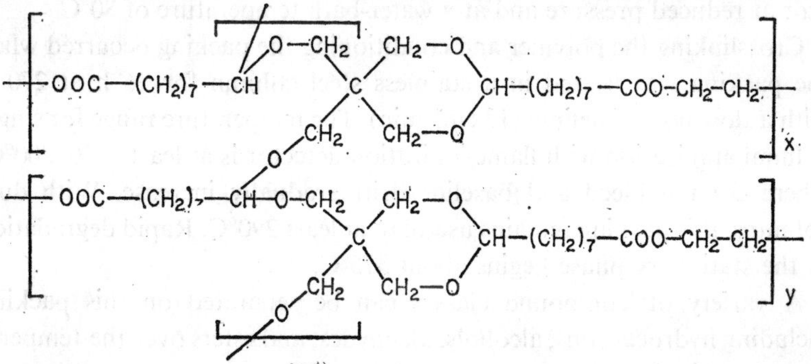

(II)

11*

For chromatographic column packings, the product of this cross-linking reaction has three main advantages:

   (i) it is a non-volatile, semi-polar packing which has essentially no bleeding at all temperatures below polymeric degradation (400°C);
  (ii) it has a deactivated support which is apparently a result of polymer-support interbonding;
 (iii) it provides excellent resolutions which may be carried on with a single column for many compound classes.

In a repeat of their work, however, they were unsuccessful in reproducing the separating ability of the packing; but they were successful in reproducing the thermal stability as measured by thermogravimetric analysis.

Neff *et al.*[528] have described the preparation, characterization, and applications of a satisfactory cross-linked poly(ester-acetal) column packing. The monomers used were dimethyl-1,4-cyclohexanedicarboxylate, MA_zPEA, and diethylene glycol. The poly(ester-acetal), 20% of support weight, dissolved in chloroform in the presence of *p*-toluene-sulphonic acid monohydrate as the crosslinking catalyst (0.01 g in 2 ml of absolute ethanol) was added to the support (Chromosorb PAW 45–60 mesh) composed of a diatomaceous earth calcined without a flux, acid-washed, and producing an acidic solution (pH = 6.50) when slurried in water. The solvent was removed from the mixture in a rotatory evaporator at reduced pressure and at a water-bath temperature of 80°C.

Crosslinking the polymer and conditioning the packing occurred when the packing was heated in a stainless steel column for 16 h at 270°C with a slow flow of helium (13 cm³/min). The temperature range for single column application with flame ionization detector is at least −60–200°C, where column bleed and baseline drift gradually increase. With dual columns, this packing remains useful to at least 290°C. Rapid degradation of the stationary phase begins about 375°C.

A variety of compound classes can be separated on this packing including hydrocarbons, alcohols, aldehydes, and esters over the temperature range indicated.

4.1.3 Concentrated stationary phases

This class includes those stationary phases in which a large proportion of the molar volume is occupied by polar groups. The polar part may be the strongly electronegative atoms F, O, N, or electron attracting groups, such as $-NO_2$, $-CN$, and $-CF_3$, or electron-repelling groups such as $-NMe_2$, $-CMe_3$, and $-OMe$. It is the type and number of polar groups in a molecule relative to the size of the neutral part which determines the polarity or distribution of electrons on the different parts of the molecule. If the molecule contains double bonds with their mobile π-electrons, this allows electron attracting or repelling groups to transmit their effect further through the hydrocarbon part of the molecule than is possible with saturated hydrocarbon chains.

Most stationary phases can act as electron donors or acceptors or both, but usually one effect is predominant. It is the concentration of electrons or the electron-cloud density which governs the potential donor or acceptor property of the stationary phase.

Many stationary phases from this class interact with the solutes to be separated, forming hydrogen bonds. Alkanes and other hydrocarbons have very small retentions. Thus they are generally unsuitable for hydrocarbon analysis unless it is to separate alkanes as a group from other solutes which are polar.

Polar solutes have considerable retentions in these stationary phases, and it is for these that they are principally used.

Water

The use of water as a stationary liquid phase in gas chromatography has been reported by Karger et al.[339, 529] It was used for the rapid separation of hydrocarbons at temperatures 200–300°C below their boiling points. n-Tetracosane elutes at 85°C on a 30 cm column of 45% w/w water on Chromosorb P, 60–80 mesh, in 100 min at a flow rate of 50 ml/min. As its boiling point is 391°C, elution is achieved 306°C below the boiling point.[339] Also, n-octadecane could be eluted in 80 min at a flow rate of 50 ml/min from a 100 cm 46% w/w water column at 35°C. Because water is a volatile solvent it is necessary to pre-saturate the carrier gas with water at the column temperature before introduction into the column.

It is clear that a major retention mechanism for hydrocarbons chromatographed on water is gas–liquid interfacial adsorption.

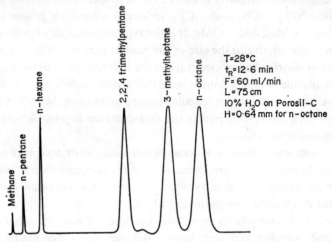

FIG. 16. Separation of octane isomers on a 10% w/w Water-Porasil C column.[339]

Figure 16 shows the separation of octane isomers on a 10% w/w H_2O-Porasil C column.

Glycerol

$$CH_2OH—CHOH—CH_2OH$$

$d_4^{20} = 1.260$ g/cm³; minimum operating temperature 20°C; maximum operating temperature 50°C. Commercial name: Perkin Elmer KA.

Diglycerol

$$(CH_2OH—CHOH—CH_2)_2O$$

$d_4^{20} = 1.26$ g/cm³; minimum operating temperature 20°C; maximum operating temperature 100°C.

Erythritol

$$CH_2OH—CHOH—CHOH—CH_2OH$$

Maximum operating temperature 150°C.

d-Sorbitol

$$CH_2OH—(CHOH)_4—CH_2OH$$

Maximum operting temperature 190°C.

Inositol

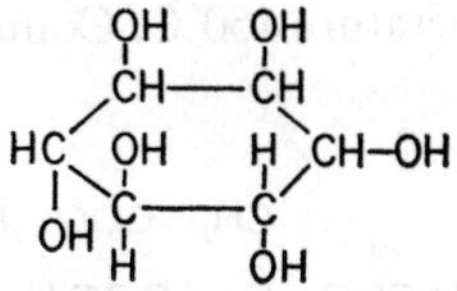

Maximum operating temperature 230°C.

The solutes that can form hydrogen bonds are strongly retained on these stationary phases.[530-532]

Acetonylacetone (2,5-hexanedione)

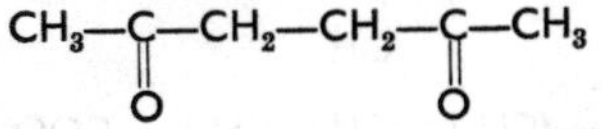

$$CH_3—C—CH_2—CH_2—C—CH_3$$
$$\quad\ \ \overset{\|}{O}\qquad\qquad\ \overset{\|}{O}$$

Minimum operating temperature −10°C; maximum operating temperature 20°C.

Bis(2-methoxyethyl)adipate

$$H_3C—O—CH_2—CH_2—OOC—(CH_2)_4—COO—CH_2—CH_2—O—CH_3$$

Minimum operating temperature: room temperature; maximum operating temperature 150°C.

These stationary phases are especially used for the separation of olefinic and saturated hydrocarbons up to C_4.[533, 534]

Diethylene glycol monoethyl ether (*carbitol*)

$$HO—CH_2—CH_2—O—CH_2—CH_2—OC_2H_5$$

Maximum operating temperature 60°C.

Tributyl citrate

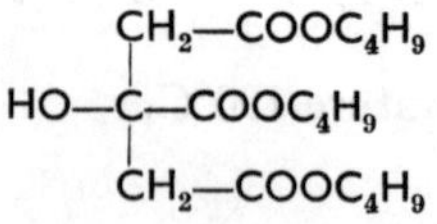

$$CH_2—COOC_4H_9$$
$$HO—C—COOC_4H_9$$
$$CH_2—COOC_4H_9$$

Maximum operating temperature 150°C. Commercial name: Citroflex 4.

Acetyl tributyl citrate

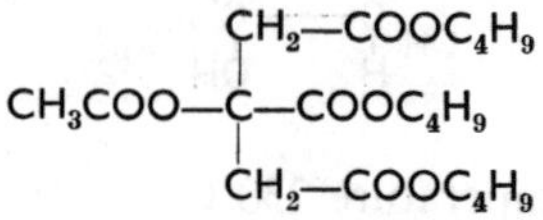

$$CH_2—COOC_4H_9$$
$$CH_3COO—C—COOC_4H_9$$
$$CH_2—COOC_4H_9$$

Maximum operating temperature 180°C. Commercial name: Citroflex A-4.

Hydrogenated castor oil

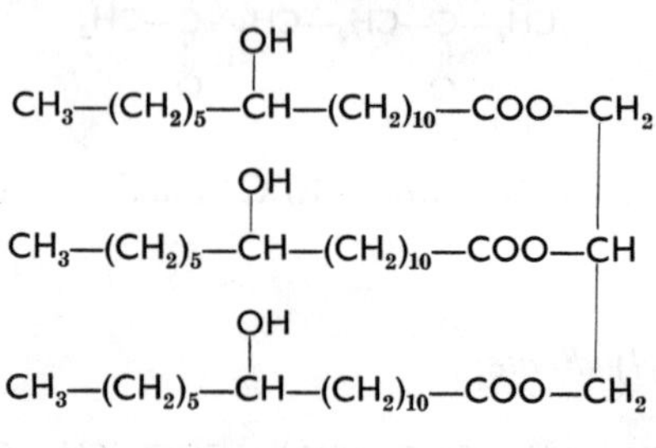

$$CH_3—(CH_2)_5—\overset{\overset{\textstyle OH}{|}}{CH}—(CH_2)_{10}—COO—CH_2$$

$$CH_3—(CH_2)_5—\overset{\overset{\textstyle OH}{|}}{CH}—(CH_2)_{10}—COO—CH$$

$$CH_3—(CH_2)_5—\overset{\overset{\textstyle OH}{|}}{CH}—(CH_2)_{10}—COO—CH_2$$

Maximum operating temperature 200°C. Commercial name: Castorwax.

The cyano derivatives also fall into this class. Nitriles, and especially nitrile-ethers, have only weak interactions with saturated or non-polar solutes, but interact strongly with polar and non-saturated solutes and also with those compounds contain a hydrogen atom capable of hydrogen bonding. The interaction with the polar solutes arises from the strongly polar nitrile groups (the dipole moment is 3.60 D for the alkyl-CN bond and 4.05 D for the phenyl-CN bond).

Because of their polarity, nitriles can induce an electric field in the molecules of unsaturated and polarizable compounds, which results in a certain retention of these compounds. But the donor–acceptor and hydrogen bonding forces are manifested more strongly. Donor–acceptor forces arise because the nitriles act as electron acceptors. Because of the electronegativity of the nitrile group, the compounds with easily ionizable π-electrons (aromatic hydrocarbons) will be retained selectively in the column.

Hydrogen bonds are formed between nitrile ethers and alcohols, phenols, carboxylic acids, primary and secondary amines, etc. High selectivity coefficients are obtained when separating olefins, acetylenes, and cycloparaffins as well as aromatics from n-paraffins, primary alcohols from secondary and tertiary alcohols, ketones and aldehydes from ethers and esters, polar halogenated hydrocarbons from their non-polar or less polar counterparts, and *cis* and *trans* isomers.

The nitrilic combinations most frequently used as stationary phases include:

β, β′-Oxydipropionitrile[315]

$$O\begin{cases} CH_2-CH_2-CN \\ CH_2-CH_2-CN \end{cases}$$

$d_4^{20} = 1.05$ g/cm³. Maximum operating temperature 60°C. Commercial name: Perkin Elmer T.

β, β′-Iminodipropionitrile[535]

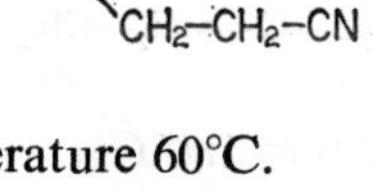

$$HN\begin{cases} CH_2-CH_2-CN \\ CH_2-CH_2-CN \end{cases}$$

Maximum operating temperature 60°C.

β, β'-Thiodipropionitrile[536]

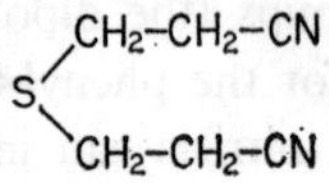

$$S\begin{cases} CH_2-CH_2-CN \\ CH_2-CH_2-CN \end{cases}$$

Maximum operating temperature 85°C.

1,2,3-Tris(5-cyanoethoxy)propane[537]

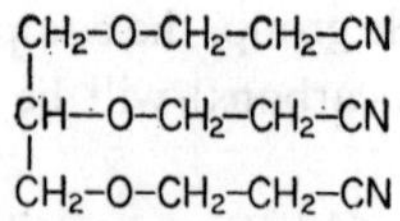

$$\begin{array}{l} CH_2-O-CH_2-CH_2-CN \\ CH-O-CH_2-CH_2-CN \\ CH_2-O-CH_2-CH_2-CN \end{array}$$

Maximum operating temperature 180°C.

1,2,6-Tris(2-cyanoethoxy)hexane[538]

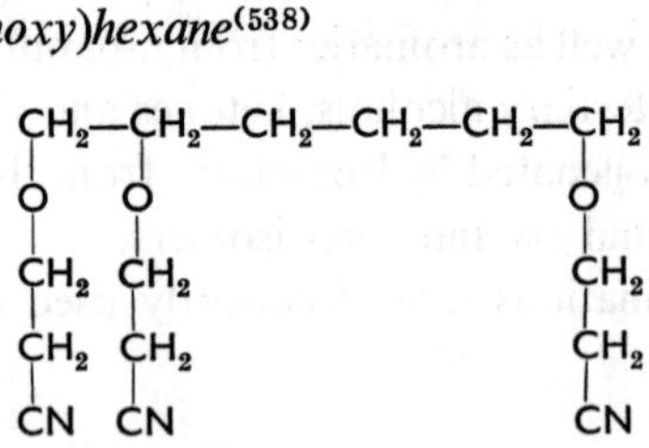

$$\begin{array}{cccccc} CH_2 & CH_2 & CH_2 & CH_2 & CH_2 & CH_2 \\ | & | & & & & | \\ O & O & & & & O \\ | & | & & & & | \\ CH_2 & CH_2 & & & & CH_2 \\ | & | & & & & | \\ CH_2 & CH_2 & & & & CH_2 \\ | & | & & & & | \\ CN & CN & & & & CN \end{array}$$

Maximum operating temperature 190°C.

Tetrakis(β-cyanoethoxymethyl)methane or tetracyanoethyl penta-erythritol[539]

$$C(CH_2-O-CH_2-CH_2-CN)_4$$

Maximum operating temperature 125°C.

2-Methyl-2-propyl-1,3-bis(cyanoethoxy)propane[540]

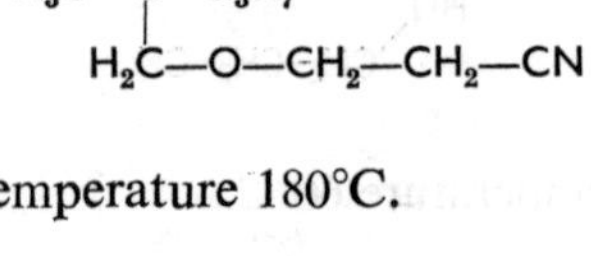

$$\begin{array}{l} H_2C-O-CH_2-CH_2-CN \\ H_3C-C-C_3H_7 \\ H_2C-O-CH_2-CH_2-CN \end{array}$$

Maximum operating temperature 180°C.

TABLE 15. EXPERIMENTAL CONDITIONS FOR PREPARING THE VARIOUS SUBSTANCES TESTED AS STATIONARY PHASES

Compound obtained	Starting material	Solvent for reaction	Excess of acrylonitrile (%)	Catalyst	Reaction temperature (°C)	Yield (%)	Temperature limits (°C)	
							Lower	Upper
Hexakis(β-cyanoethoxy)-1,2,3,4,5,6 hexane (HCEH)	Mannitol	H_2O	0	NaOH	40–50	75	Ambient	90
Hexakis(β-cyanoethoxy)-1,2,3,4,5,6 cyclohexane (HCECH)	Inositol	H_2O	0	NaOH	40–50	74	Ambient	80
Tetrakis(β-cyanoethoxy-methyl)methane (TCEMM)	Pentaerythritol	H_2O	0	NaOH	40–50	47	45	125
(β-Cyanoethoxy)benzene (CEB)	Phenol	C_2H_3CN	43	MeONa	135	40	60	85
Bis(β-cyanoethoxy)-1,2 benzene (BCEB-1,2)	Catechol	C_2H_3CN	200	MeONa	85	24	125	160
Bis(β-cyanoethoxy)-1,3 benzene (BCEB-1,3)	Resorcinol	C_2H_3CN	200	MeONa	85	63	115	155
Bis(β-cyanoethoxy)-1,4 benzene (BCEB-1,4)	Hydroquinone	C_2H_3CN	200	MeONa	85	52	145	165
(β-Cyanoethylthio)benzene (CETB)	Thiophenol	C_2H_3CN	100	Light	20	64	Ambient	60
Bis(β-cyanoethylthio)-3,4-toluene (g) (BCETT)	3,4-Dithiotoluene	C_2H_3CN	100	Light	20	89	70	110

Such compounds may be obtained by action of acrylonitrile on compounds having one or more labile hydrogen atoms in the presence or absence of a catalyst according to the type of reaction (cyanoethylation) and have been intensively studied by Bruson and Riener.[541]

$$R-X-H + H_2C{=}CH-CN \rightarrow R-X-CH_2-CH_2-CN$$

X can be oxygen, sulphur, nitrogen, or carbon.

Chovin and Sannier,[536] using this reaction, prepared a series of compounds that they used as stationary phases. Table 15 gives the experimental conditions for their preparation as well as the temperature limits between which these compounds can be used as stationary phases in gas chromatography. The authors carried out a study on the systems formed by these stationary phases and aromatic hydrocarbons and reached the conclusion that in a compound of the type $R-X-CH_2-CH_2-CN$ replacement of X = O by X = S decreases the retentivity for aromatics. For linear compounds of the type $H-(CH-O-CH_2-CH_2-CN)_n-H$, an increase in the value of n causes practically no change in the retentivity for aromatics.

Cyclization of the carbon chain carrying cyanoethoxy groups, such as in the inositol derivative, does not change the retentivity for benzene by comparison with the open-chain derivative obtained from mannitol. An increase in the number of $O-CH_2-CH_2-CN$ groups on the benzene ring increases the retentivity for benzene. An increase in the number of $S-CH_2-CH_2-CN$ groups on the benzene ring likewise increases the retentivity for benzene.

Among the nitrilic stationary phases the most utilized is 1,2,3-tris (2-cyanoethoxy)propane. Figure 17 shows the chromatogram obtained during the separation of some aromatic hydrocarbons using a column packed with the above stationary phase.[542]

Stationary phases that contain nitro groups behave similarly to those containing nitrile groups (from the point of view of gas chromatography). The same type of donor–acceptor interactions occur as with aromatic compounds such as 1,3,5-trinitrobenzene, picric acid and 2,4,7-trinitro-9-fluorenone.[216, 543–546]

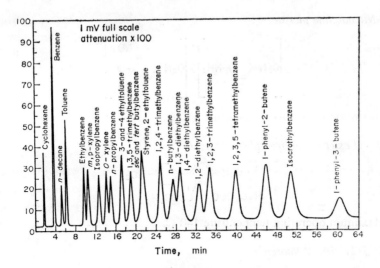

FIG. 17. Chromatogram of a 27-component mixture of aromatic and aliphatic hydrocarbons.[542] Column: 5% 1,2,3-tris(2-cyanoethoxy)propane on 50–60 mesh C-22 firebrick; 11 ft×$\frac{1}{8}$ in. o.d. (3.4 m×3 mm). Helium flow rate: 75 ml/min; temperature, 60°C. Detector: hydrogen flame; 30 ml/ of H_2/min; 350 ml of 30% oxygen in argon/min.

Such a stationary phase may interact through orientation and inductive forces (nitrobenzene has a dipole moment of 4.01 D), while its oxygen atoms are donor atoms and thus they are able to form hydrogen bonds with compounds with protonated functional groups. The nitro groups of these stationary phases are generally attached to an aromatic nucleus. The presence of the phenyl groups means greater solubility of hydrocarbons in these stationary phases.

The specific retention volumes for hydrocarbons, in conditions of similar selectivity, are greater in this instance than with the nitrile- ethers. This property is used for the separation of low boiling-point hydrocarbons. When using these stationary phases, very good selectivity is obtained for the separation of aromatic hydrocarbons from alkanes, substituted aromatic hydrocarbons, and low boiling-point halogenated hydrocarbons.[216, 547]

p-Nitroaniline picrate

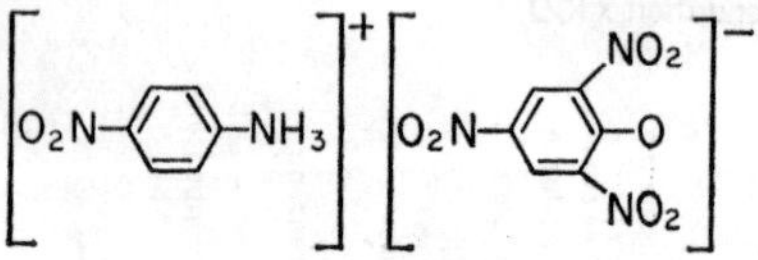

Maximum operating temperature 110°C.

Hexyl ester of dinitrodiphenylic acid

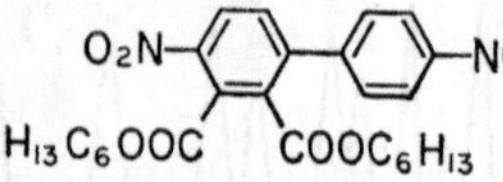

Maximum operating temperature 140°C.

2,4,7-Trinitro-9-fluorenone

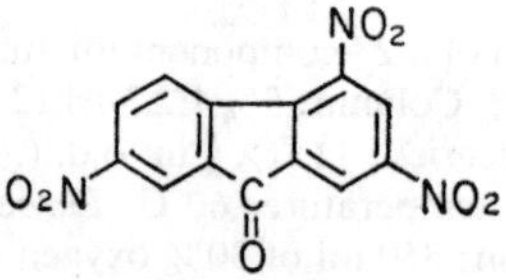

Maximum operating temperature 180°C.

A disadvantage of these stationary phases is that when hydrogen is used as carrier gas at higher temperatures, the nitro groups may be reduced.

Triethanolamine

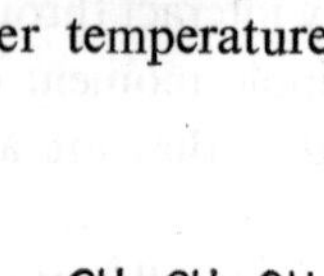

Maximum operating temperature 75°C.

Tetraethylene pentamine

$$H_2N—C_2H_4—NH—C_2H_4—NH—C_2H_4—NH—C_2H_4—NH_2$$

Maximum operating temperature 125°C.

N, N, N′, N′-Tetrakis(2-hydroxyethyl)ethylenediamine

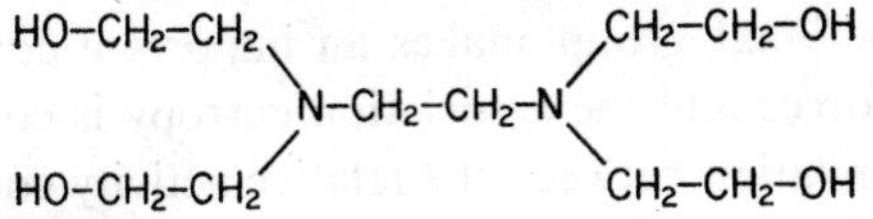

$$HO-CH_2-CH_2 \diagdown \atop HO-CH_2-CH_2 \diagup N-CH_2-CH_2-N \diagup CH_2-CH_2-OH \atop \diagdown CH_2-CH_2-OH$$

Minimum operating temperature 50°C; maximum operating temperature 150°C. Commercial name: THEED.

N, N, N′, N′-Tetrakis(2-hydroxypropyl)ethylenediamine

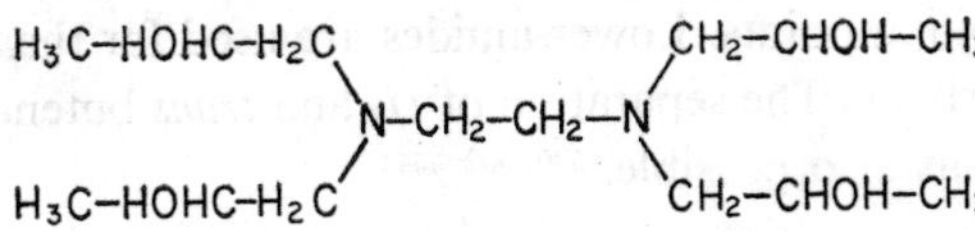

$$H_3C-HOHC-H_2C \diagdown \atop H_3C-HOHC-H_2C \diagup N-CH_2-CH_2-N \diagup CH_2-CHOH-CH_3 \atop \diagdown CH_2-CHOH-CH_3$$

Maximum operating temperature 150°C. Commercial name: Quadrol (Wyandotte Chemical Corp.).

Aliphatic and hydroxy amines are used for the analysis of mixtures of alcohols, glycols, pyridines, piperazines, mercaptans, and thioethers. The presence of hydroxyl groups (as in triethanolamine, tetrahydroxyethylenediamine, and quadrol) increases the tendency to form hydrogen bonds.[544, 548–550]

Carboxylic acids amides are very polar stationary phases. The "polar" properties of these stationary phases results from the large dipole moment (for *N*-methylacetamide it is 3.86 D) as well as from the associative properties of the carbonyl and NH groups. Amides form very strong hydrogen bonds with electron donors and acceptors. Thus chloroform interacts with amides, forming hydrogen bonds with the carbonyl group. (The NH group shows a weaker dipole-dipole interaction, while nitriles interact strongly with the NH group; the forces produced by the interaction with the carbonyl group have only a simple dipole-dipole character.)

Ecknig and Lenz[551] show that the polar contributions of the intermolecular forces that determine the gas chromatographic selectivity of the amides, originate from the association of the polar groups of the stationary phase and the solutes to be separated. These association forces can be described in terms of measurable infrared spectroscopic values.

Thus the different frequency shifts of the carbonyl and NH groups are a measure of the interaction of the stationary phase with the solute.

When the functional group makes an important contribution to the intermolecular forces and the association entropy is constant, the existence of a linear relation between the relative activity coefficients and the frequency shifts is to be expected. Such a relation has been observed for N-methyl acetamide and a series of components for $\Delta v_{C=O}$ from the amide molecule.[551]

The saturated hydrocarbons are less soluble in these stationary phases and thus they are eluted much more rapidly than polar compounds with similar boiling points. Lower amides are used for the separation of C_1–C_5 hydrocarbons. The separation of *cis* and *trans* butene-2, isobutene and butene-1 was also possible.[530, 552–554]

N, N-Dimethylformamide[530]

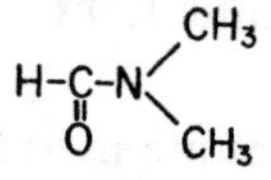

$$H-\underset{\underset{O}{\|}}{C}-N\underset{CH_3}{\overset{CH_3}{<}}$$

$d_4^{22.4} = 0.9484$ g/cm^3. Maximum operating temperature 20°C.

Formamide[552, 340]

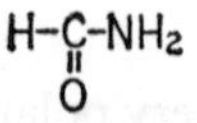

$$H-\underset{\underset{O}{\|}}{C}-NH_2$$

$d_4^{22.4} = 1.1284$ g/cm^3. Maximum operating temperature 20°C.

N, N-Diethylformamide[555]

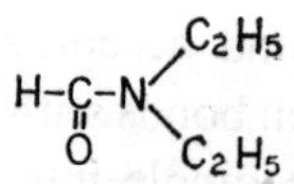

$$H-\underset{\underset{O}{\|}}{C}-N\underset{C_2H_5}{\overset{C_2H_5}{<}}$$

$d_4^{19} = 0.908$ g/cm^3. Maximum operating temperature 25°C.

N, N-Diphenylformamide[315]

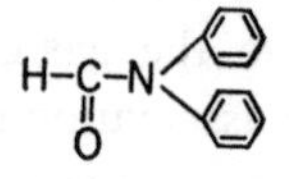

$$H-\underset{\underset{O}{\|}}{C}-N<$$

Maximum operating temperature 100°C.

N-Methylacetamide[551]

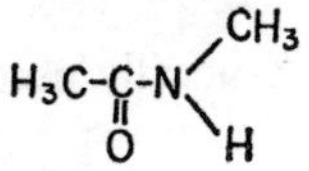

$$H_3C-\underset{\underset{O}{\|}}{C}-N\underset{H}{\overset{CH_3}{<}}$$

Maximum operating temperature 60°C.

N, N-Bis(2-cyanoethyl) formamide[556]

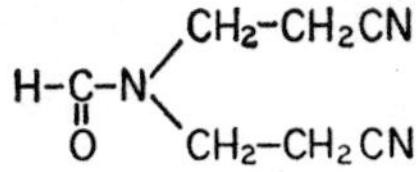

$$H-\underset{\underset{O}{\|}}{C}-N\underset{CH_2-CH_2CN}{\overset{CH_2-CH_2CN}{<}}$$

Maximum operating temperature 125°C.

Hexamethyl phosphoric triamide

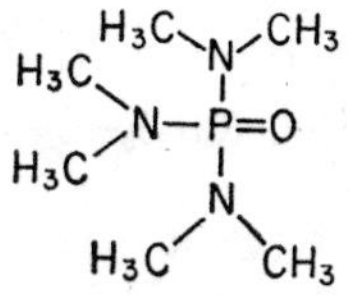

$$\underset{H_3C}{\overset{H_3C}{>}}N-\underset{\underset{N}{\underset{H_3C \diagup \diagdown CH_3}{|}}}{\overset{H_3C\diagdown_N\diagup CH_3}{\overset{|}{P}}}=O$$

Maximum operating temperature 50°C.

N, N-Dimethylstearamide

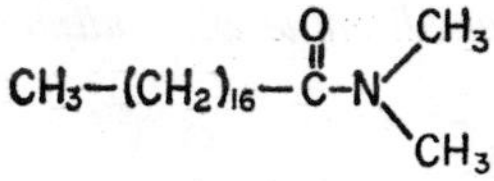

$$CH_3-(CH_2)_{16}-\underset{\overset{\|}{O}}{C}-N\underset{CH_3}{\overset{CH_3}{<}}$$

Maximum operating temperature 150°C. Commercial name: Hallcomid M-18.

N, N-Dimethyloleylamide

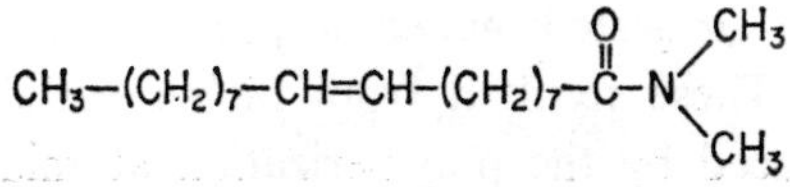

$$CH_3-(CH_2)_7-CH=CH-(CH_2)_7-\underset{\overset{\|}{O}}{C}-N\underset{CH_3}{\overset{CH_3}{<}}$$

Maximum operating temperature 150°C. Commercial name: Hallcomid M-180 L.

These stationary phases are generally used for the analysis of alcohols, ketones, and aldehydes.[557]

Palmitic (I) and stearic (II) acid diethylamides as well as adipic (III), azelainic (IV), and sebacic (V) acid tetraethylamides have also been proposed.[558]

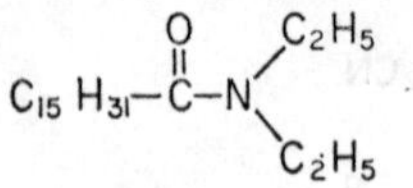

$$C_{15}H_{31}-\overset{\overset{O}{\|}}{C}-N\overset{C_2H_5}{\underset{C_2H_5}{}}$$

(I)

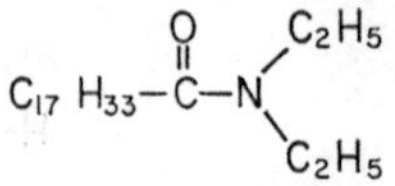

$$C_{17}H_{33}-\overset{\overset{O}{\|}}{C}-N\overset{C_2H_5}{\underset{C_2H_5}{}}$$

(II)

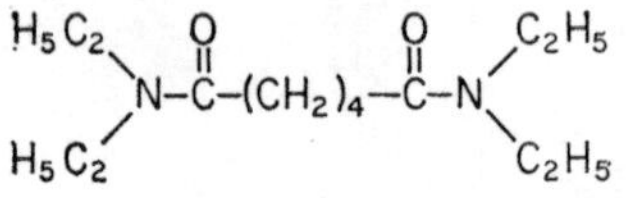

$$\overset{H_5C_2}{\underset{H_5C_2}{}}N-\overset{\overset{O}{\|}}{C}-(CH_2)_4-\overset{\overset{O}{\|}}{C}-N\overset{C_2H_5}{\underset{C_2H_5}{}}$$

(III)

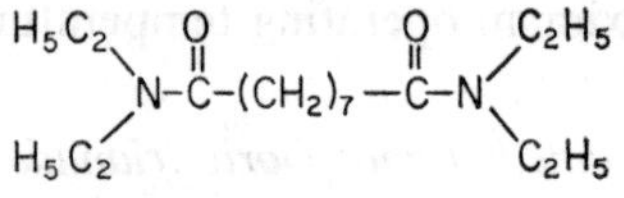

$$\overset{H_5C_2}{\underset{H_5C_2}{}}N-\overset{\overset{O}{\|}}{C}-(CH_2)_7-\overset{\overset{O}{\|}}{C}-N\overset{C_2H_5}{\underset{C_2H_5}{}}$$

(IV)

$$\overset{H_5C_2}{\underset{H_5C_2}{}}N-\overset{\overset{O}{\|}}{C}-(CH_2)_8-\overset{\overset{O}{\|}}{C}-N\overset{C_2H_5}{\underset{C_2H_5}{}}$$

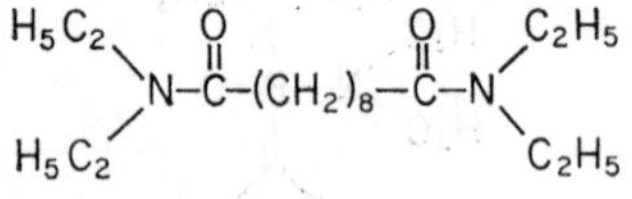

(V)

These stationary phases have been used for the selective retention of aromatic hydrocarbons.

Copolymer of hexymethylenediamine with adipic acid

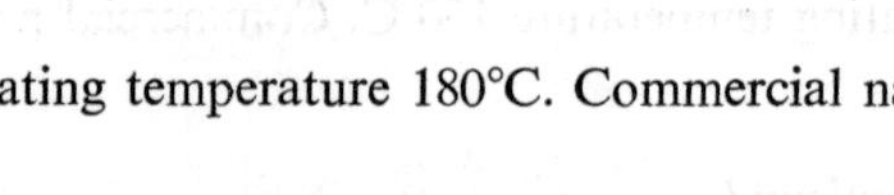

$$-(-NH-(CH_2)_6-NH-\overset{\overset{O}{\|}}{C}-(CH_2)_4-\overset{\overset{O}{\|}}{C}-)_n-$$

Maximum operating temperature 180°C. Commercial names: Nylon 6; Nylon 66.

It has been used for phenol analysis.[559]

It has also been suggested that some polyamides could be used as stationary phases. These resulted from higher carboxylic acids (a 36 dimeric acid produced by the polymerization at mid-molecule of two

unsaturated C_{18} monobasic acids) and secondary diamides of the pipera-
zine and piperidine type.[560] From all these, three polyamides, namely
PZ-103-A, PZ-103, and PZ-101-A, proved useful as stationary phases.
Table 16 gives some characteristics of the above polyamides.

TABLE 16

Polyamide	Monomers	Mole ratio
PZ-103-A	Dimeric acid	1.00
	1,3-Di(4-piperidyl)propane	0.90
	Piperidine	0.20
PZ-103	Dimeric acid	1.00
	1,3-Di(4-piperidyl) propane	0.90
	4-(3-phenylpropyl)piperidine	0.20
PZ-101-A	Dimeric acid	1.00
	1,3-Di(4-piperidyl)propane	0.75
	4-(3-Phenylpropyl)piperidine	0.50

Based on the mole ratios of the reactants, the structure of the PZ-103-A
polymer may be idealized as follows:

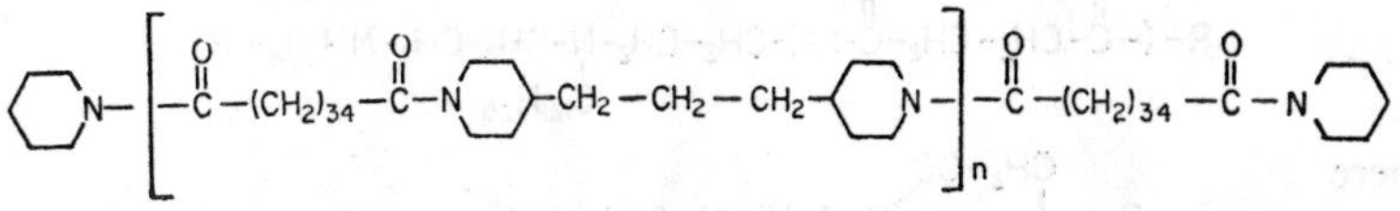

Packed or capillary columns prepared with these stationary phases
provide very good separation of heavy petroleum fractions and are also
useful for the separation of derivatives of amino acids, fatty acids, steroids,
and sugars.

Maximum operating temperature of these stationary phases is 300°C.

Figure 18 shows the temperature-programmed analysis of a wax
standard on PZ-103.[560]

Another polyamide, obtained from 4-dodecyldiethylene triamine and
dimethyl succinate, has been proposed as a stationary phase.[561] The
product thus obtained is believed to be a linear polymer of 4-dodecyl-

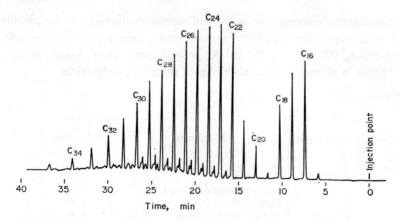

FIG. 18. Analysis of hydrocarbon wax. Column: 100 ft×0.0625 in. o.d.× 0.020 in. i.d. (30.5 m×1.6 mm×0.5 mm) stainless steel coated with 5% PZ-103. Temperature programmed from 125° to 190°C at 10°/min, then at 5°/min to 275°C, then isothermal. Pressure: 10 lb/in² (0.7 kg/cm²) helium.[560]

diethylene triamine succinamide, terminated by succinimide groups:

$$R-(-\overset{O}{\overset{\|}{C}}-CH_2-CH_2-\overset{O}{\overset{\|}{C}}-NH-CH_2-CH_2-\overset{\underset{C_{12}H_{25}}{|}}{N}-CH_2-CH_2-NH-)_n-R,$$

where

$$R = \begin{matrix} CH_2-CO \\ | \\ CH_2-CO \end{matrix} \Big\rangle N-C_2H_4-\overset{\underset{C_{12}H_{25}}{|}}{N}-C_2H_4-NH-$$

This stationary phase has been used for the analysis of amines and amino esters.

Stationary phases with sulphur in their molecules are very widely used for the separation of C_2–C_6 hydrocarbons.

Dimethyl sulphoxide[562]

$$CH_3—SO—CH_3$$

Maximum operating temperature 30°C.

Di-n-propyl sulphone

$$(CH_3—CH_2—CH_2)_2SO_2$$

Maximum operating temperature 60°C.

Tetramethylene sulphone (Sulpholane)

$$(CH_2)_4SO_2$$

Maximum operating temperature 30°C.

Dimethyl sulpholane (3,4-dimethyltetrahydrothiophen-S-dioxide)

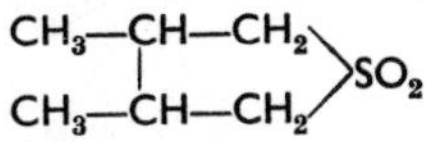

Maximum operating temperature 35°C. Commercial name: Perkin Elmer E.

In admixture with di-n-propyl sulphone, dimethyl sulpholane separates C_2–C_5 hydrocarbons.[563, 493, 356]

Complex mixtures of hydrocarbon gases up to and including *cis*-2-butene can be analysed in 25 min on dimethyl sulphoxide and tetramethylene sulphone as stationary phases.[562] These stationary phases do not react with unsaturated compounds. When acetylenic compounds with active hydrogen atoms are analysed, however, glass or stainless steel tubing is required as they react with copper in the presence of dimethylsulphoxide.

For di-n-propyl sulphone, sulpholane, and dimethyl sulpholane, copper tubing can be used.

Among the heterocyclic compounds used as stationary phases are:

1-Methyl-5-(2-methoxyethyl)tetrazole

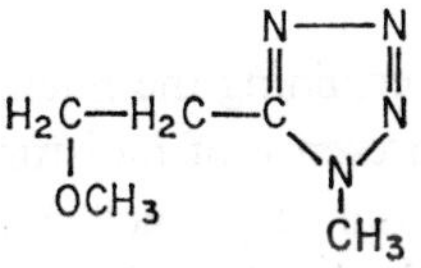

Maximum operating temperature 80°C.

This compound has been used for the separation of C_1–C_7 hydrocarbons.

1-Hydroxyethyl-2-heptadecenylglyoxalidine

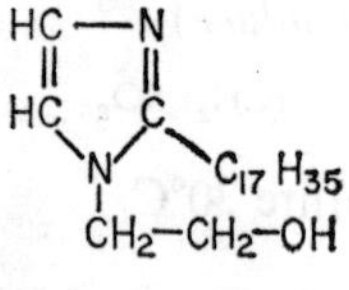

Maximum operating temperature 180°C. Commercial name: Amine 220.

Poly(vinylpyrrolidone)

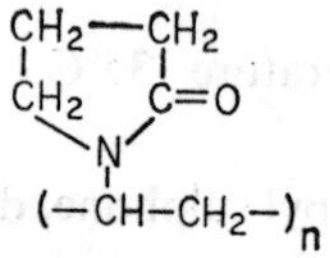

Poly(vinylpyrrolidone) (PVP) introduced first to deactivate support materials, was later used as a stationary phase for the examination of plant sterols.[564] It melts at 200°C. Maximum operating temperature 220°C.

The stationary phases most used in gas–liquid chromatography are the poly(ethylene glycols). The materials used in gas chromatography are almost invariably prepared by polymerization of ethylene oxide.[565] The product of these polymerizations, which are polydisperse in terms of molecular weight, are separated into series of fractions having nominal average molecular weights of 200, 300, 400, etc. (calculated from the chemically determined hydroxyl concentration). For a particular poly(oxyethylene glycol) fraction, these average molecular weights are not all the same; for instance, with polyoxyethylene glycol 400, a tolerance of ±5% is generally accepted.

The principal factors determining the retention characteristics of these stationary phases are the presence of moisture, the concentration of the hydroxyl end groups, and to a very much smaller degree, the molecular weight distribution of the liquid phase.[566] The presence of moisture has been found to cause large changes of absolute retention which may be positive or negative depending on whether the solutes are hydrophilic or hydrophobic, respectively. For the latter, the decreases of retention are almost uniform, so that values for relative retention are unaffected. The explanation afforded for these observations is that the introduction of moisture causes the formation of a hydrogen bonded cross-linked network, involving the water molecules as cross-links between the poly(oxyethylene glycol) molecules, which exclude all but the most strongly hydrophilic solutes.

The most important single factor upon which the retention characteristics of poly(oxyethylene glycol) liquid phases depend has been found to be the concentration of hydroxyl groups available for partitioning purposes. With celite as support, values for the retention index have been found to increase linearly with the hydroxyl concentration. On the other hand, the molecular weight distribution has been found to have an insignificant effect upon retention data. Therefore, it should be possible to produce liquid phases of any desired retention behaviour by the appropriate blending of different poly(oxyethylene glycols), provided the overall molecular weight range is sufficiently small to preclude errors arising from the volatilization of the lower molecular weight species during the preparation and use of the column.

The order of elution of alkanes on these stationary phases is determined mainly by variations in the inductive forces between the solvent and the solute, the methylene groups of the stationary phase exerting only a small influence.

In all aromatic compounds, the inductive forces can interact with the delocalized electrons of the nucleus. It seems probable that the $+I$ effect of the methyl group (as in toluene and the xylenes) is negligible, because if the $+I$ effect were decisive, there would be a more pronounced difference between the three xylenes.[567] The inductive effects with dipole formation operate in o- and m-xylene but cancel out in p-xylene. Halogens directly attached to the ring give rise to a dipole that is stronger than that with the methyl group and has an opposite charge ($-I$ effect). Furthermore, the mesomeric $+M$ effect imparts a partial double bond character to carbon–halogen bonds, particularly when the halogen is fluorine. These circumstances may lead to a stronger interaction between the sample and the stationary phase. The halogens attached to the aromatic ring thus operate in the opposite direction to the methyl group. The fact that ΔH_e is small or negative for halobenzenes indicates a stationary phase–solute interaction stronger than with benzene and its methyl derivatives. Decreasing ΔH_e (stronger solvent–solute interaction) is accompanied by a decrease in ΔS_e (ordering processes in the solution). As the forces involved are basically directional, this behaviour is to be expected.

The concomitant presence of acceptors (hydroxylic and etheric oxygen atoms) and donors (hydroxylic hydrogen atoms) makes possible the interaction of the stationary phases with the compounds that have hydroxyl or primary amino groups in their molecule and at the same time with those having carboxyl or secondary or tertiary amine groups in their molecule.

As hydrogen bonding is the main attractive force in these instances, there are no significant differences between the behaviour of a given stationary phase with any of the above compounds. Thus aldehydes, ketones, and ethers are eluted in the order of their boiling points. Polyglycols with large molecules contain few hydroxylic hydrogen atoms so that the proton acceptor properties conditioned by the increased number of etheric oxygen atoms are stronger. That is, hydrogen bonds are formed especially when the hydrogen atoms are "supplied" by the substance to be analysed. Thus poly(ethylene oxide) selectively separates primary, secondary, and tertiary amines with similar boiling points, the elution being in the order tertiary, secondary, primary.

The maximum allowable operating temperature is not very high, and these stationary phases are oxidized rather easily by air.

The UCON series also belongs to the polyethers. The water insoluble UCONs designated with a suffix H or HB contain 50% poly(ethylene glycol) units. The water insoluble UCONs designated with a suffix LB contain more than 50% poly(propylene glycol) units.

Poly(ethylene glycol)

$$HO-CH_2-CH_2-[-O-CH_2-CH_2-]_n-O-CH_2-CH_2-OH$$

Minimum operating temperature 50–70°C, depending on the molecular weight; maximum operating temperature 100°C ($n \approx 9$; molecular weight ≈ 400); 130°C (molecular weight ≈ 1500); 160°C (molecular weight 2000–6000). Commercial names: polyethylene glycol 400, 600, 1600, 2000, 6000, Carbowax, Ucon 50 HB up to 2000 and 50 HB up to 5100 (Union Carbide Chemicals Co.). With approximately 30 units: Triton X305; Perkin Elmer K (PEG 1500).

Poly(propylene glycol)

$$HO-\underset{\underset{CH_3}{|}}{CH}-CH_2-\left[-O-\underset{\underset{CH_3}{|}}{CH}-CH_2-\right]_n-OCH_2-\underset{\underset{CH_3}{|}}{CH}-OH$$

Maximum operating temperature 100–160°C depending on the molecular weight. Commercial names: Polypropylene glycol 1025; Ucon Polyglycol LB-550-X; Perkin Elmer R.

Poly(styrene oxide)

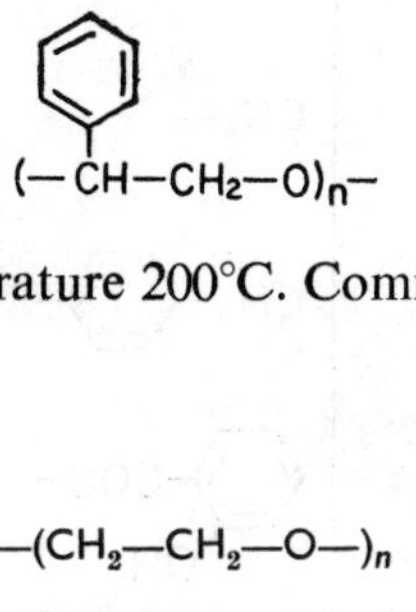

$$(-\underset{\underset{C_6H_5}{|}}{CH}-CH_2-O)_n-$$

Maximum operating temperature 200°C. Commercial name: Dow Polyglycol 174 up to 500.

Poly(ethylene oxide)

$$-(CH_2-CH_2-O-)_n$$

Minimum operating temperature: 30–70°C, depending on the molecular weight; maximum operating temperature: 160°C. Commercial names: Polyethylene glycol 6000, 20,000; Carbowax 6000; Carbowax 20 M; Lubrol MO (I.C.I.); Oxidwachs (Buna).

Poly(tetramethylene oxide)

$$H[-O-(CH_2)_4-]_n-OH$$

Maximum operating temperature 200°C.

The treatment of the free hydroxyl groups of the poly(ethylene glycol) with different reagents led to some compounds that function as polar stationary phases which are relatively thermally stable.[568-570] By etherifi-

TABLE 17. SUBSTITUTED POLY(ETHYLENE GLYCOLS). GENERAL FORMULA
$RO-CH_2-CH_2-(-O-CH_2-CH_2-)_n-O-CH_2-CH_2-OR'_{17}$

No.	R	R'	Name of the substituted compound
1	$C_{16}H_{33}O-$	$HO-$	Poly(ethylene glycol) hexadecylether
2	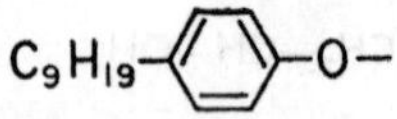 C_9H_{19}—⟨benzene⟩—$O-$	$HO-$	Poly(ethylene glycol) nonylphenylether (Tergitol NPX)
3	$C_{17}H_{35}COO-$	$HO-$	Poly(ethylene glycol) monostearate (Ethofat 60/25)
4	CH_3-	CH_3-	Methoxypoly(ethylene glycol)
5	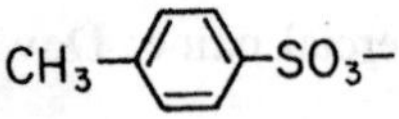 CH_3—⟨benzene⟩—SO_3-	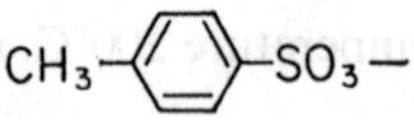 CH_3—⟨benzene⟩—SO_3-	Poly(ethylene glycol) ditosylate
6	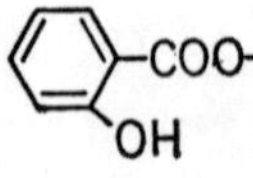 ⟨benzene⟩—$COO-$ with OH	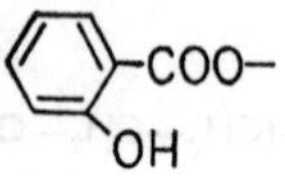 ⟨benzene⟩—$COO-$ with OH	Poly(ethylene gylcol) disalicylate
7	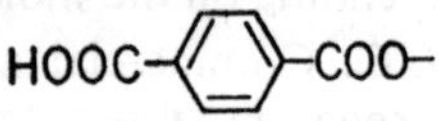 $HOOC$—⟨benzene⟩—$COO-$	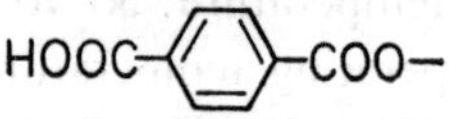 $HOOC$—⟨benzene⟩—$COO-$	Poly(ethylene glycol) diterephthalate
8	$C_{17}H_{33}COO-$	$C_{17}H_{33}COO-$	Poly(ethylene glycol) dioleate
9	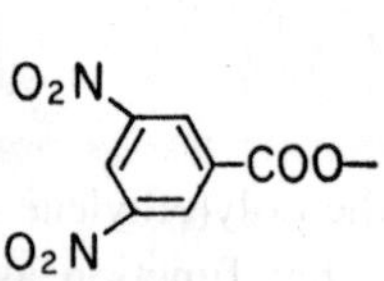 O_2N—⟨benzene⟩—$COO-$ with O_2N	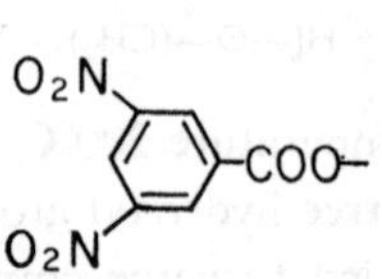 O_2N—⟨benzene⟩—$COO-$ with O_2N	Poly(ethylene glycol) bis 3,5-dinitrobenzoate

TABLE 17 (*cont.*)

No.	R	R'	Name of the substituted compoud
10	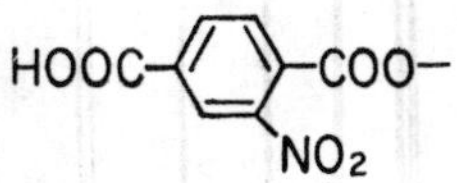 HOOC—⟨⟩—COO— NO₂	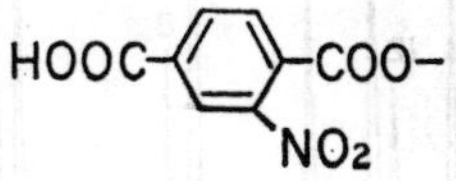 HOOC—⟨⟩—COO— NO₂	Poly(ethylene glycol) di-2-nitroterephthalate (FFAD)
11	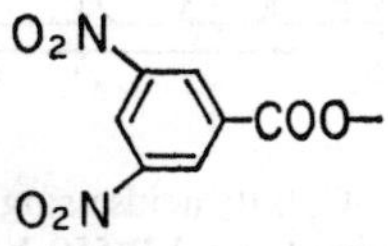 O₂N—⟨⟩—COO— O₂N	$C_{17}H_{35}COO$—	Poly(ethylene glycol) 3,5-dinitrobenzoate monostearate
12	$HOOC—(CH_2)_3—COO$—	$HOOC—(CH_2)_3—COO$—	Poly(ethylene glycol) diglutarate

cation of the terminal hydroxyl groups, variation of the selectivity and the applicability of the poly(ethylene glycols) as a function of the substance to be analysed, may be obtained.

Table 17 is a general list of partly or totally substituted poly(ethylene glycols). Certainly the introduction of long alkyl chains into the molecule of the poly(ethylene glycol) favours the retention of hydrocarbons, whereas alcohols are eluted more rapidly than on unmodified poly(ethylene glycol).

The replacement of both hydroxyl groups with tosyl, salicylic, terephthalic, glutaric, and especially 3,5-dinitrobenzoic groups, favours the retention of alcohols to such an extent that it compensates for the lack of interactions with the free hydroxyl groups from the poly(ethylene glycol).

The most important applications of the polyglycols as stationary phases are for the separation of saturated and aromatic hydrocarbons,[571] and in general for the separation of oxygenated compounds.

Figure 19 presents the separation of saturated C_2–C_6 fatty acids on a column of UCON LB-550-X.[572]

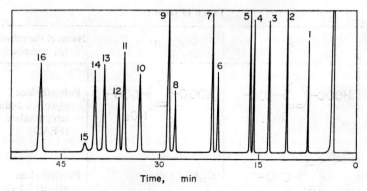

FiG. 19. Capillary gas chromatogram of saturated C_2-C_6 fatty acids using a steel capillary column 45 m×0.25 mm i.d. containing Ucon LB-550-X and phosphoric acid at 125°C. (1) Acetic acid, (2) propionic acid, (3) *iso*-butyric acid, (4) trimethylacetic acid, (5) butyric acid, (6) 3-methyl-butyric acid, (7) (±)-2-methylbutyric acid, (8) n-pentanoic acid, (9) 2,2-dimethylbutyric acid, (10) (±)-2,3-dimethylbutyric acid, (11) 2-ethylbuty-ric acid, (12) (±)-2-methylpentanoic acid, (13) (±)-3-methylpentanoic acid, (14) 4-methylpentanoic acid, (15) 3,3-dimethylbutyric acid, (16) n-hexanoic acid.[572]

For the gas chromatographic analysis of long-chain hydroxylic combinations, polyglycerol has been proposed as a stationary phase.[573] Polyglycerol may be obtained by the reaction of glycerol with 1-hydroxy-2,3 epoxypropane in the presence of basic or acidic catalysts.

$$HO-CH_2-CH-CH_2-OH \atop OH \quad + \quad {}_n \quad CH_2-CH-CH_2-OH \atop O \quad \xrightarrow[(BF_3)]{KOH}$$

$$HO-CH_2-CH-CH_2- \left[-O-CH_2-CH-CH_2- \right]_n -O-CH_2-CH-CH_2-OH \atop \quad\quad OH \quad\quad\quad\quad\quad OH \quad\quad\quad\quad\quad\quad OH \qquad (a)$$

$$HO-CH_2-CH- \left[-O-CH_2-CH- \right]_n -O-CH-CH_2-OH \atop \quad\quad CH_2OH \quad\quad\quad CH_2OH \quad\quad CH_2OH \qquad (b)$$

From viscosimetric determinations it is presumed that polyglycerol is a mixture of structures a and b. Polyglycerol contains proton-acceptor etheric oxygen atoms and the corresponding number of hydroxyl groups (terminal hydroxyl groups are not included in this count) so that they have donor and acceptor properties.

From the point of view of selectivity, polyglycerol behaves very much in the same manner as the carbowaxes. Some differences appear only in the separation of aldehydes and ketones, the aldehydes being retained more strongly. Polyglycerol may be used as a stationary phase in the range 0–225°C.

For the separation of saturated alcohols from unsaturated alcohols with the same number of carbon atoms, cyanoethylated polyglycerol has been proposed as a stationary phase. It is obtained by the interaction of the hydroxyl groups with acrylonitrile in the presence of basic catalysts, the result being the formation of nitrilic ethers.[574]

$$R_1R_2CHOH + CH_2{=}CH{-}CN \xrightarrow{(KOH)} R_1R_2CH{-}O{-}CH_2{-}CH_2{-}CN$$

From infrared spectrophotometric determinations it was concluded that this stationary phase contains etheric groups as well as hydroxyl and nitrile groups. The good results obtained for the separation of unsaturated alcohols from saturated alcohols arise from the presence of the nitrilo groups in the molecules of the stationary phase. These groups have π-electron acceptor properties that lead to the selective retention of unsaturated alcohols.

Figure 20 shows the separation of n-alcohols on a cyanoethylated polyglycerol column.[574] The stationary phase may be used from 0° up to 200°C.

Poly(vinylformal propionitrile)

This compound was proposed as a stationary phase for the separation of polyols.[575] The formation of poly(vinylformal propionitrile) takes place as follows:

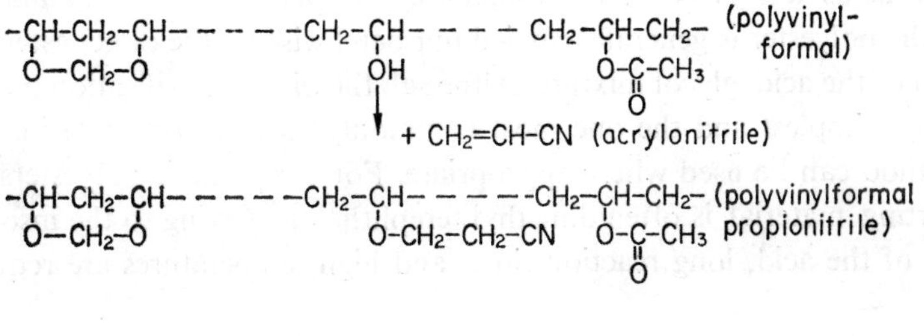

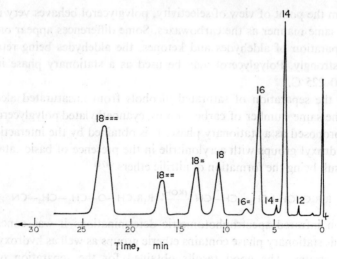

FIG. 20. Separation of n-alcohols (C_{14}, C_{16}, C_{18}, $C_{18}^=$, $C_{18}^{==}$, $C_{18}^{===}$) at 200°C. Hydrogen flow rate: 6.6 l/h; packing: 20% POAN on Sterchamol. Column: 2.00 m×6 mm i.d.; 2.6 μl (400°C).[574]

The temperature range in which this stationary phase may be used is 80–300°C.

The polyesters may be classed as concentrated stationary phases. A polyester, as the name suggests, is a macromolecular compound formed by a series of carboxylic ester linkages.[576]

The term polyester covers a broad range of resinous compositions derived mainly from the reaction of a polybasic acid and a polyhydric alcohol. It is the linear aliphatic polyesters which are normally used as liquid phases for gas chromatography.[577, 578] They are highly polar materials whose structures are somewhat varied depending on the particular acids and glycols from which they are prepared. The manufacture of the polyester is generally carried out batchwise by the direct esterification of the acid–glycol mixture. Although the direct esterification method is the simplest and the one most commonly used, a *trans*-esterification method can be used where appropriate. For terephthalic polyesters, the starting material is often dimethyl terephthalate. Owing to the insolubility of the acid, long reaction times and high temperatures are required for the direct esterification of terephthalic acid. The *trans*-esterification procedure is also advantageous with dicarboxylic acids which are prone to decarboxylation, such as malonic acid. All the *trans*-esterification procedures require a catalyst which is not necessarily needed with a direct esterification method. Impurities present in the polyester are actually very low if pure starting materials are used.

The polyester liquid phases are generally quite stable below 200°C and acceptable at higher temperatures. However, certain compounds and situations should be avoided if long life and consistent analyses are desired.

Polyester resins are particularly susceptible to hydrolysis by water and compounds which are strongly acidic or basic. Hydrolysis basically results in breakage of the ester linkages along the polymer backbone with the formation of additional hydroxyl and carboxylic acid functions. This degradation changes the separation characteristics of the polyester liquid phase and lowers its usable temperature range. Hydrolysis severely limits the use of polyester liquid phases when analysing dilute aqueous solutions containing organic compounds. The relative stability of liquid polyesters towards hydrolysis increases on passing to higher or more hydrophobic glycols and dicarboxylic acids. The temperature at which polyester resins are utilized also drastically affects the rate at which decomposition by hydrolysis takes place. The most hydrolytically stable polyesters decompose instanteneously with water at 280°C.

Sample components containing amine groups cause aminolysis of ester linkages. In this instance, the carboxylic acid amide and a hydroxyl group from the polyol portion are produced.

These reactions also severely change the useful operating temperature of the polyester liquid phase. As polyesters contain residual carboxyl and hydroxyl groups, analysis of compounds which rapidly react with these groups must be avoided. Compounds of this type include isocyanates, epoxides, anhydrides, halo-acids, and acyl halides.

At high temperatures, polyesters are oxidized by atmospheric oxygen. The oxidation is accompanied by degradation of the polymer. Oxidation proceeds initially through the formation of hydroperoxides, which later decompose with the formation of free radicals. The free radicals cause decomposition of the polyester chain with the formation of vinyl-type unsaturation. Along with the fragmentation of the polymer chains, free radicals can cause polymerization to form different macromolecules. All these changes in the composition and molecular weight of the polymer backbone can change the stability and resolution characteristics of the liquid phase. On heating, polyesters are degraded, the extent of which depends on the temperature, the heating time and the presence of catalysts or catalyst residues. Thermal degradation of polyesters in the absence of oxygen takes place rapidly at elevated temperatures, but below certain temperatures, polyesters can be heated for relatively long periods without significant degradation. The temperature at which decomposition begins and the rate at which it takes place depends on the composition and molecular weight of the polymer.

Linear polyesters are unique because they can undergo complete degradation at high temperatures with the formation of cyclic esters. The ease of ring formation depends on the structure of the initial components and the characteristics of the catalyst. Esters of oxalic and carbonic acid are the most readily decomposed; however, no general rule relating degradability to chain length of the acid and polyol has been elucidated.

The presence of compounds such as $SnCl_2 \cdot 2\,H_2O$, $MnCl_2 \cdot 4\,H_2O$, $FeCl_2 \cdot 4\,H_2O$, and $MgCl_2 \cdot 6\,H_2O$ severely increases the formation of volatile products during the thermal degradation of polyester resins. When polyester liquid phases are used to separate components for preparative gas chromatography, it is always possible to contaminate the fractions with the liquid phase.

The general formula of a polyester formed by the esterification of a linear dicarboxylic acid with a diol may be written in the form

$$H\left[-O-(CH_2)_n-O-\overset{\overset{O}{\|}}{C}-(CH_2)_n-\overset{\overset{O}{\|}}{C}-\right]_n-OH$$

The great affinity of the oxygen atoms for electrons makes the interaction with the solutes to be separated more favourable because of the dipoles in the polyester molecules:[579, 580]

$$\overset{..}{O}=\overset{|}{\underset{|}{C}} \longleftrightarrow -\overset{|}{\underset{|}{C}}{}^+ - \overset{..}{\underset{..}{O}}|^{(-)}$$

It was observed that the number of methylene groups between the two carboxyl groups of the acidic fragment of the polyester has a profound influence on the degree of separation. The retention times of hydrocarbons are nearly proportional to the number of methylene groups present. This linearity was found to be valid up to brassylic acid [HOOC—$(CH_2)_{11}$ —COOH] and deviated to some extent only with higher dicarboxylic acid polyesters.[581, 582] The ratio of the specific retention volumes on two different polyesters for a typical hydrocarbon is nearly equal to the number of methylene groups spaced between the two carboxyl groups of the acidic fragments in the polyester. It has also been found that the molecular weight plays a less important part than the number of methylene groups present in the acidic fragment of the polyester.[581]

Among the polyesters used as stationary phases in gas chromatography are:

Poly(ethylene glycol succinate)[577]

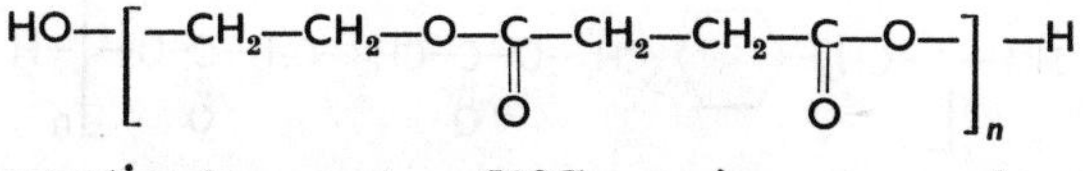

$$HO-\left[-CH_2-CH_2-O-\underset{O}{\overset{}{C}}-CH_2-CH_2-\underset{O}{\overset{}{C}}-O-\right]_n-H$$

Minimum operating temperature 50°C; maximum operating temperature 180°C. Commercial names: HI-IFF-2 B; EGS; LAC 4R 886.

Poly(diethylene glycol succinate)[583]

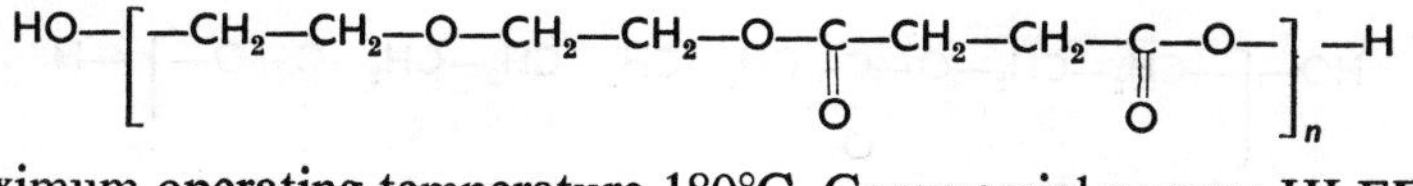

$$HO-\left[-CH_2-CH_2-O-CH_2-CH_2-O-\underset{O}{\overset{}{C}}-CH_2-CH_2-\underset{O}{\overset{}{C}}-O-\right]_n-H$$

Maximum operating temperature 180°C. Commercial names: HI-EFF-1 B; DEGS; LAC 3R 728; LAC 2R 446; Perkin Elmer P; Polyester A.

Poly(1,4-butanediol succinate)[584]

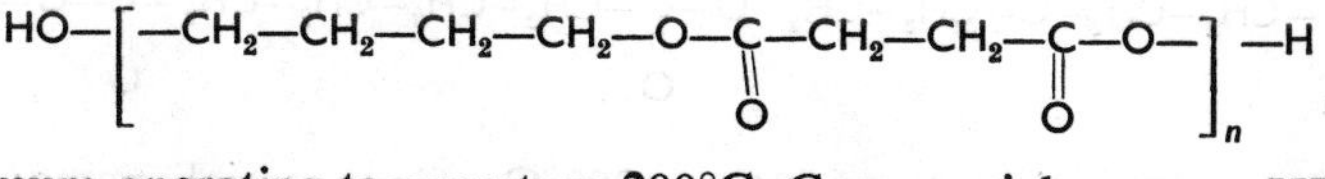

$$HO-\left[-CH_2-CH_2-CH_2-CH_2-O-\underset{O}{\overset{}{C}}-CH_2-CH_2-\underset{O}{\overset{}{C}}-O-\right]_n-H$$

Maximum operating temperature 200°C. Commercial names: HI-EFF-4B; Craig phase.

Poly(phenyldiethanolamine succinate)

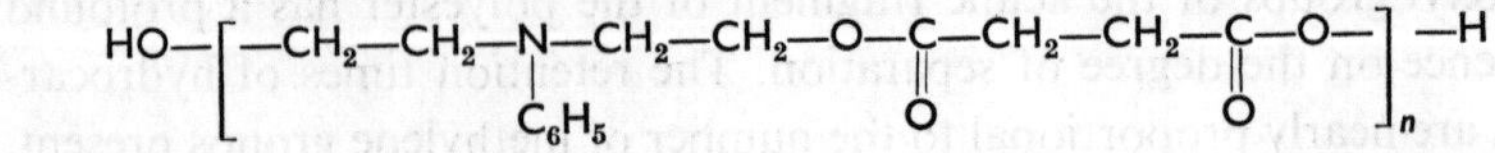

$$HO-\left[-CH_2-CH_2-\underset{\underset{C_6H_5}{|}}{N}-CH_2-CH_2-O-\underset{\underset{O}{\|}}{C}-CH_2-CH_2-\underset{\underset{O}{\|}}{C}-O-\right]_n-H$$

Maximum operating temperature 250°C. Commercial names: PDEAS.

Poly(neopentyl glycol succinate)[585]

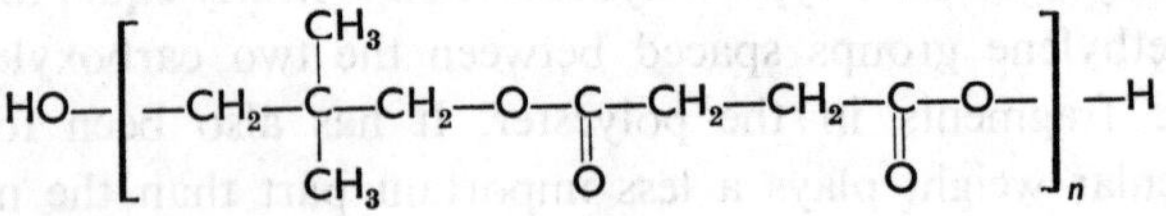

$$HO-\left[-CH_2-\underset{\underset{CH_3}{|}}{\overset{\overset{CH_3}{|}}{C}}-CH_2-O-\underset{\underset{O}{\|}}{C}-CH_2-CH_2-\underset{\underset{O}{\|}}{C}-O-\right]_n-H$$

Maximum operating temperature 200°C. Commercial names: LAC 18R 767; HI-EFF-3 BP.

Poly(cyclohexanedimethanol succinate)[586]

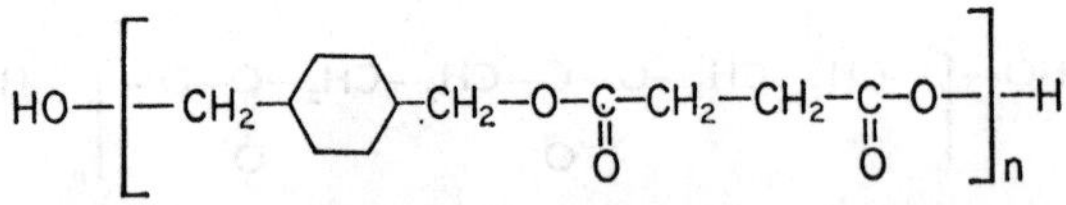

$$HO-\left[-CH_2-\langle\ \rangle-CH_2-O-\underset{\underset{O}{\|}}{C}-CH_2-CH_2-\underset{\underset{O}{\|}}{C}-O-\right]_n-H$$

Maximum operating temperature 250°C. Commercial names: LAC 796; HI-EFF-8 BP.

Poly(ethylene glycol adipate)[584]

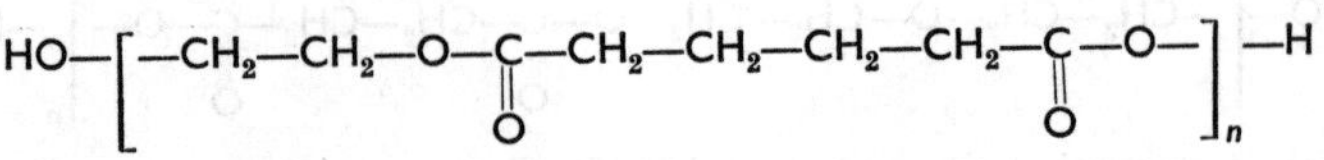

$$HO-\left[-CH_2-CH_2-O-\underset{\underset{O}{\|}}{C}-CH_2-CH_2-CH_2-CH_2-\underset{\underset{O}{\|}}{C}-O-\right]_n-H$$

Maximum operating temperature 200°C. Commercial name: LAC 741.

Poly(diethylene glycol adipate)[584]

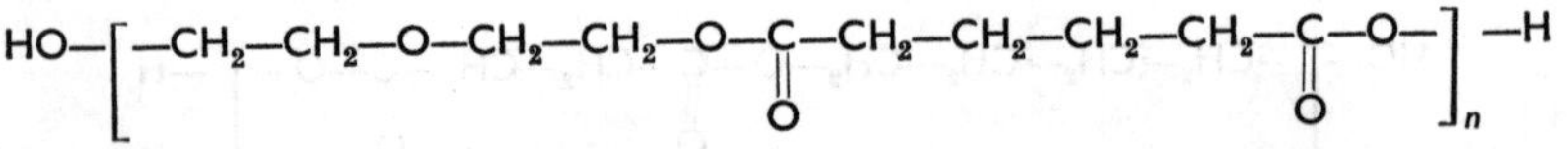

$$HO-\left[-CH_2-CH_2-O-CH_2-CH_2-O-\underset{\underset{O}{\|}}{C}-CH_2-CH_2-CH_2-CH_2-\underset{\underset{O}{\|}}{C}-O-\right]_n-H$$

Maximum operating temperature 200°C. Commercial name: LAC 1R-296.

Diethylene glycol adipate cross-linked with penta-erythritol is found in the literature under the name LAC 2R-446.

Poly(propylene glycol adipate)[587]

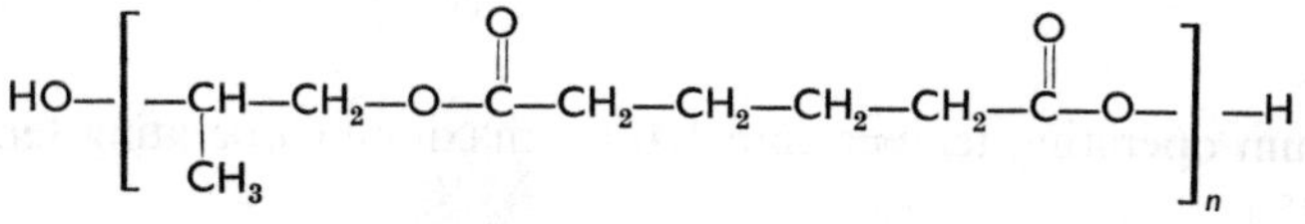

$$HO-\left[-\underset{\underset{CH_3}{|}}{CH}-CH_2-O-\overset{\overset{O}{\|}}{C}-CH_2-CH_2-CH_2-CH_2-\overset{\overset{O}{\|}}{C}-O-\right]_n-H$$

Maximum operating temperature 220°C. Commercial name: Reoplex 400.

Poly(1,4-butanediol adipate)[588]

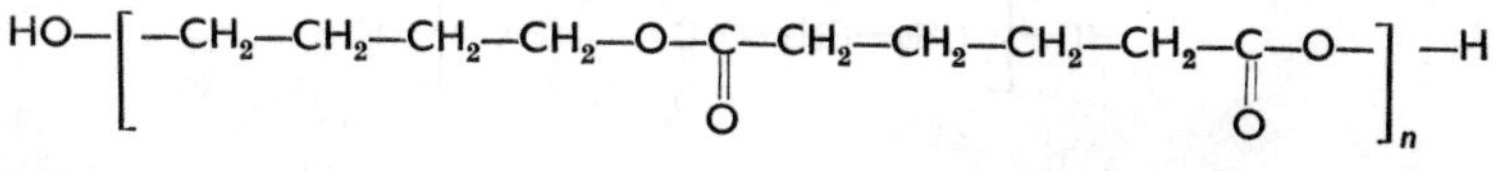

$$HO-\left[-CH_2-CH_2-CH_2-CH_2-O-\underset{\underset{O}{\|}}{C}-CH_2-CH_2-CH_2-CH_2-\underset{\underset{O}{\|}}{C}-O-\right]_n-H$$

Maximum operating temperature 200°C. Commercial name: HI-EFF-4A.

Poly(neopentyl glycol adipate)[589]

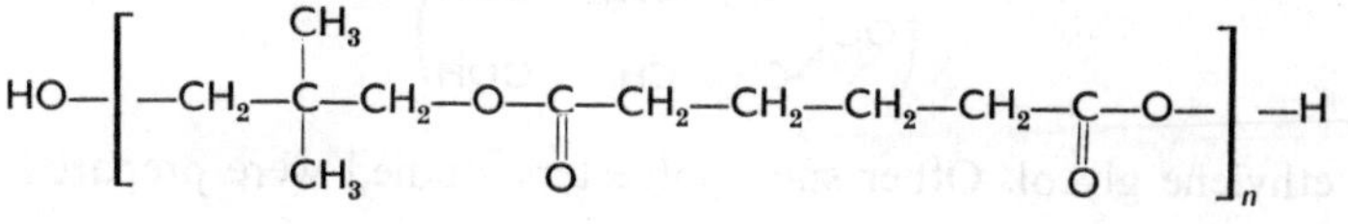

$$HO-\left[-CH_2-\underset{\underset{CH_3}{|}}{\overset{\overset{CH_3}{|}}{C}}-CH_2-O-\underset{\underset{O}{\|}}{C}-CH_2-CH_2-CH_2-CH_2-\underset{\underset{O}{\|}}{C}-O-\right]_n-H$$

Minimum operating temperature 150°C; maximum operating temperature 200°C. Commercial name: LAC 769.

Poly(neopentyl glycol sebacate)[589]

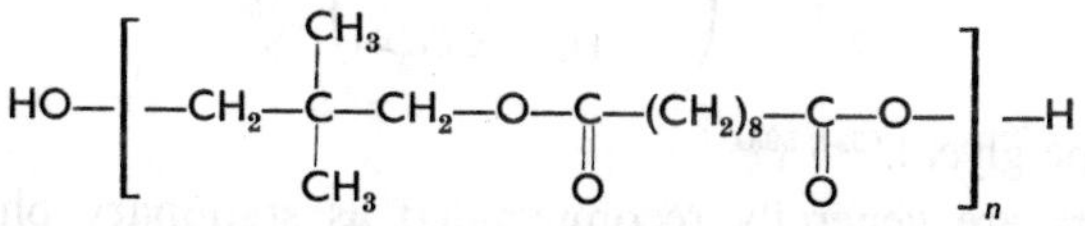

$$HO-\left[-CH_2-\underset{\underset{CH_3}{|}}{\overset{\overset{CH_3}{|}}{C}}-CH_2-O-\underset{\underset{O}{\|}}{C}-(CH_2)_8-\underset{\underset{O}{\|}}{C}-O-\right]_n-H$$

Maximum operating temperature 225°C. Commercial name: LAC 17R-770.

Poly(ethylene glycol phthalate)[590, 591]

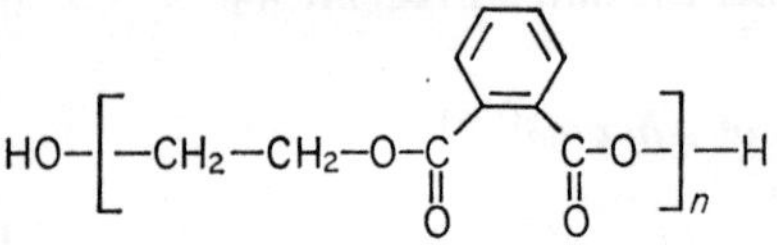

$$HO-\left[-CH_2-CH_2-O-\underset{O}{\overset{}{C}}-\underset{O}{\overset{}{C}}-O-\right]_n-H$$

Minimum operating temperature 100°C; maximum operating temperature 225°C.

Poly(ethylene glycol tetrachlorophthalate)[592]

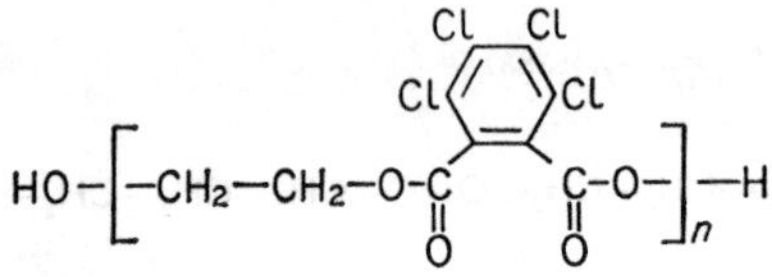

$$HO-\left[-CH_2-CH_2-O-\underset{O}{\overset{}{C}}-\underset{O}{\overset{}{C}}-O-\right]_n-H$$

Minimum operating temperature 120°C; maximum operating temperature 250°C.

Falk[593] recommends as a stationary phase with increased polarity a polyester of γ-ketopimelic acid

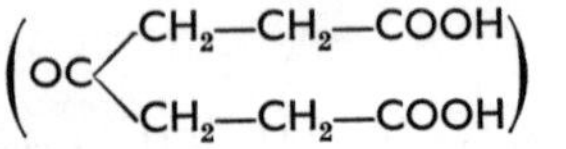

$$\left(OC\underset{CH_2-CH_2-COOH}{\overset{CH_2-CH_2-COOH}{\big\langle}}\right)$$

and ethylene glycol. Other such polyesters studied were prepared from maleic acid

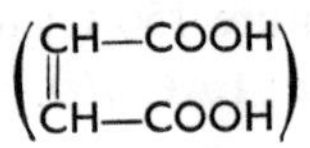

$$\left(\underset{CH-COOH}{\overset{CH-COOH}{\big|}}\right)$$

or malic acid

$$\left(\underset{H_2C-COOH}{\overset{HO-HC-COOH}{}}\right)$$

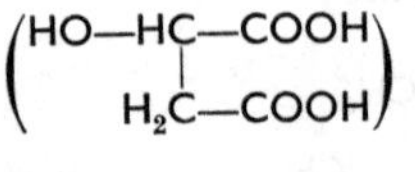

and ethylene glycol.[594, 595]

Polyesters are generally recommended as stationary phases for the separation of fatty acid methyl esters, essential oils, steroids, amino acid derivatives, etc.

Polycarbonate resins

Polycarbonates are linear polyesters of carbonic acid in which the carbonate groups recur in the polymer chain according to the general formula

$$H-\left[-O-R-O-\overset{\overset{\displaystyle O}{\|}}{C}-O-R-\right]_n-OH$$

Depending on the nature of R in the above formula, the polycarbonates can be subdivided into aliphatic, aliphatic–aromatic, or aromatic polycarbonates.[596] Maximum operating temperature 250°C. Commercial name: Lexan.

Another type of stationary phase with many ester groups, esters of cyclodextrin, has been applied to fatty acid analysis.[597–599] Sand and Schlenk[597] studied the possibility of using β-cyclodextrin acetate (cyclohepta-amylose heneicosa-acetate; m.wt. 2018; melting point 199–201°C), β-cyclodextrin propionate (m.wt. 2312; melting point 169°C), α-cyclodextrin acetate (cyclohexa-amylose octadeca-acetate; m.wt. 1730; melting point 243–245°C), and their mixtures as stationary phases in gas chromatography. Their use is justified by the authors by the fact that the polyesters commonly used as polar stationary phases for the separation of fatty acid esters have carbon to oxygen ratios similar to those of carbohydrate esters. For the separation of fatty acid esters, cyclodextrin esters behave very similarly to the polyesters. The polyesters are prepared more easily, but the cyclodextrin esters yield more reproducible data.[598]

The cyclodextrins include:

β-Cyclodextrin propionate

$$(C_{105}H_{154}O_{56})$$

Minimum operating temperature 170°C; maximum operating temperature 210°C.

Sucrose diacetate hexaisobutyrate (SAIB)[600]

Maximum operating temperature 200°C.

Octakis(2-hydroxypropyl)sucrose[601]

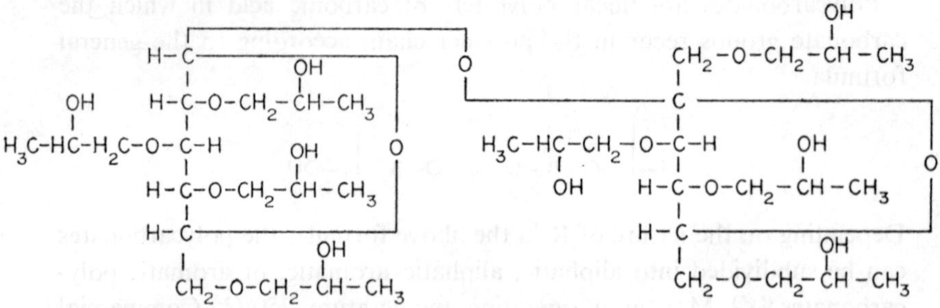

Maximum operating temperature 225°C. Commercial name: HYPROSE-SP 80.

Sorbitan monostearate

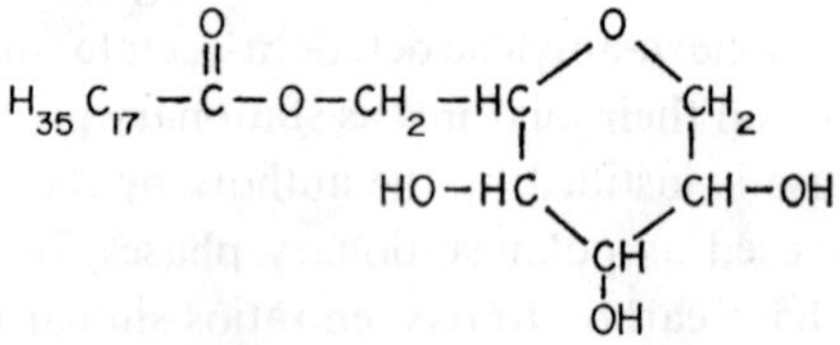

Maximum operating temperature 150°C. Commercial names: SPAN-60; ATPET 80 (monoester of sorbitan and fatty acids).

Sorbitan monooleate

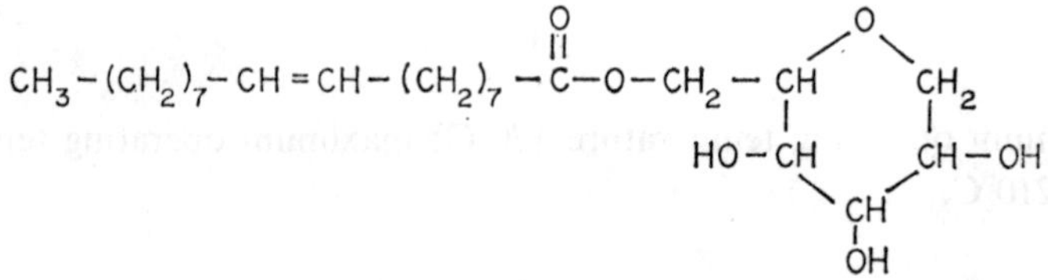

Maximum operating temperature 150°C. Commercial name: SPAN-80.

Sorbitan polyglycol ether

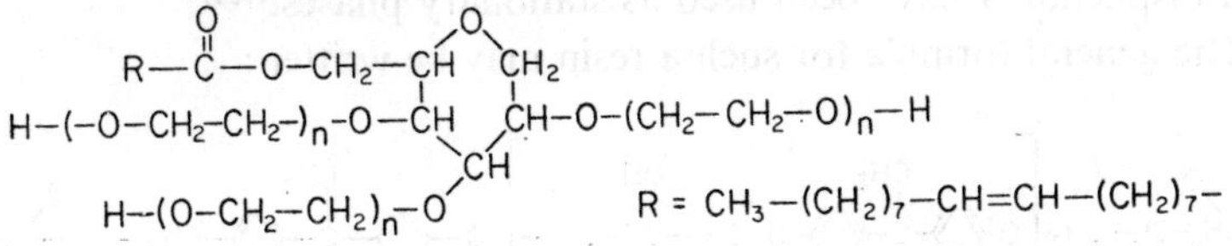

$$R-\overset{\overset{O}{\|}}{C}-O-CH_2-CH \overset{O}{\diagdown} CH_2$$
$$H-(-O-CH_2-CH_2-)_n-O-CH \quad CH-O-(CH_2-CH_2-O)_n-H$$
$$H-(O-CH_2-CH_2)_n-O \overset{CH}{\diagup}$$

$$R = CH_3-(CH_2)_7-CH=CH-(CH_2)_7-$$

Maximum operating temperature 150°C. Commercial name: Emasole 4130.

Octylphenoxypoly(ethyleneoxy)ethanol

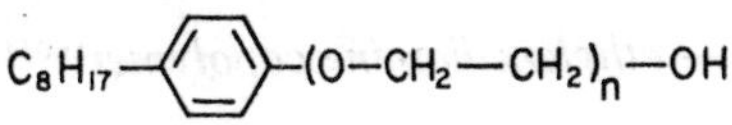

$$C_8H_{17}-\langle\rangle-(O-CH_2-CH_2)_n-OH$$

Maximum operating temperature 190°C. Commercial names: Triton X-100; Triton Y-305.

Octylphenoxyhexaethyleneoxyethanol

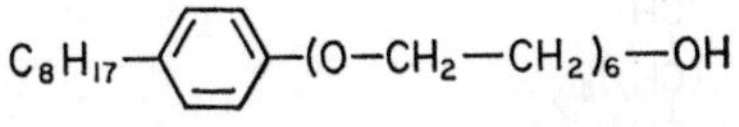

$$C_8H_{17}-\langle\rangle-(O-CH_2-CH_2)_6-OH$$

Maximum operating temperature 180°C. Commercial name: Marlophen 87.

The compounds of the Triton series are recommended for the separation of inorganic gases, mercaptans and alkyl sulphides.[602]

Nonylphenoxypoly(ethyleneoxy)ethanol

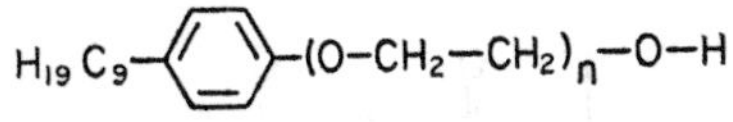

$$H_{19}C_9-\langle\rangle-(O-CH_2-CH_2)_n-O-H$$

Maximum operating temperature 200°C. Commercial names: Igepal CO 880 (n = 30); Igepal CO 890 (n = 40); Dowfax 9N9 (n = 9) (maximum operating temperature 150°C).

Igepal is recommended as a stationary phase for the separation of aromatic amines.[603]

Some commercial epoxy resins obtained by reacting epichlorohydrin and bisphenol A have been used as stationary phases.[604]

The general formula for such a resin may be written:

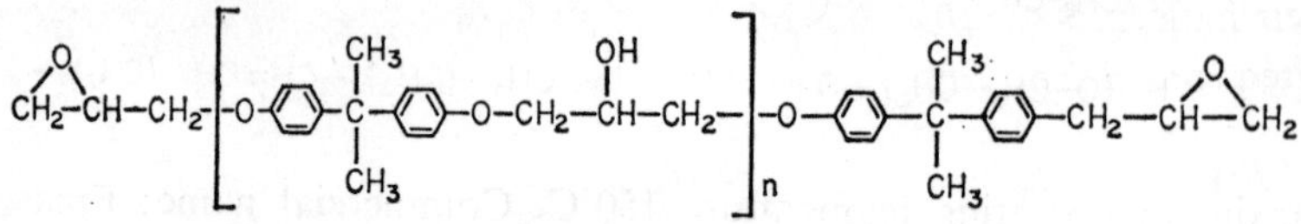

Maximum operating temperature 250°C. Commercial names: Epon Resin 1001; Epikote 1001 (Shell Chemical Co.).

This stationary phase was proposed for the analysis of steroids.[605]

Linoleic acid dimer + ethylenediamine copolymer[604]

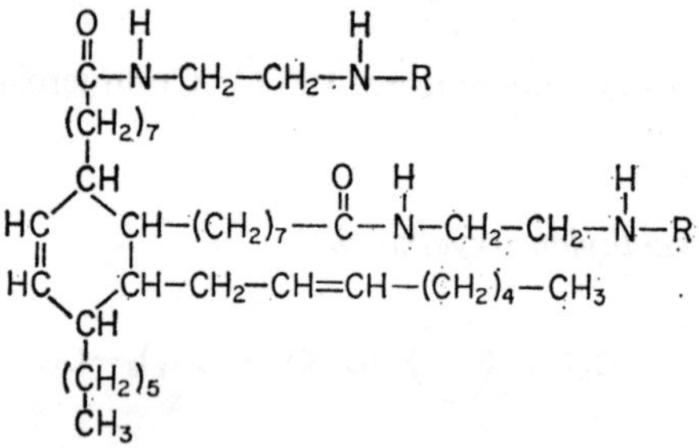

where R may be hydrogen or another linoleic acid dimer group. Maximum operating temperature 250°C. Commercial name: Versamid 900 (General Mills).

Versamid is preferentially used for amines and amides.

Mathews *et al.*[606] prepared and used a polyimide as a stationary phase. This compound has the formula:

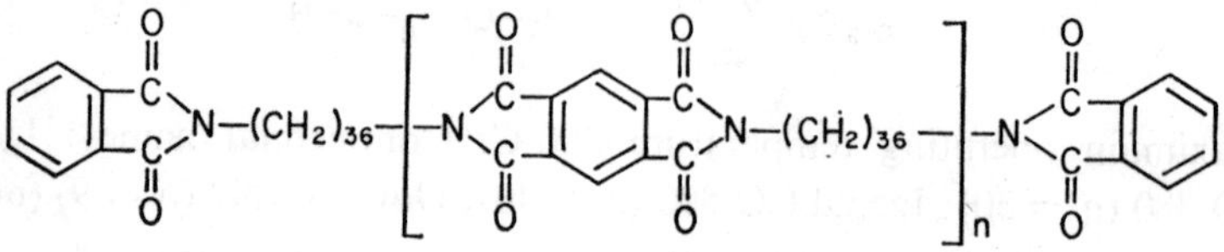

Maximum operating temperature 300°C. Commercial name: PZ-109 polyimide.

This stationary phase shows a marked selectivity for the polar groups of steroid molecules.

A series of stationary phases used in gas chromatography contain in their molecules an aromatic ring, ester or hydroxyl groups, and a fluorinated side chain. They include.

1,3-Bis(2-hydroxyhexafluoro-2-propyl)benzene

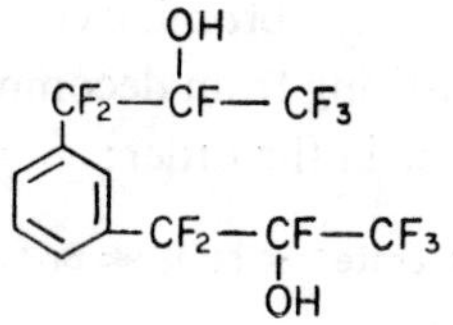

Maximum operating temperature 150°C.

4,4-Bis(2-hydroxyhexafluoro-2-propyl)diphenyl ether

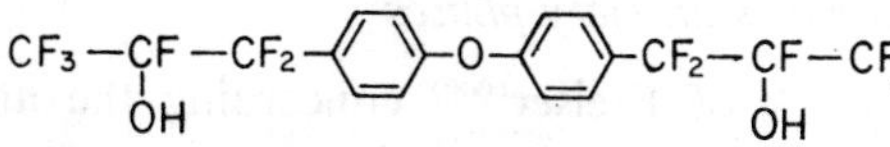

Maximum operating temperature 170°C.

Zonyl E 7

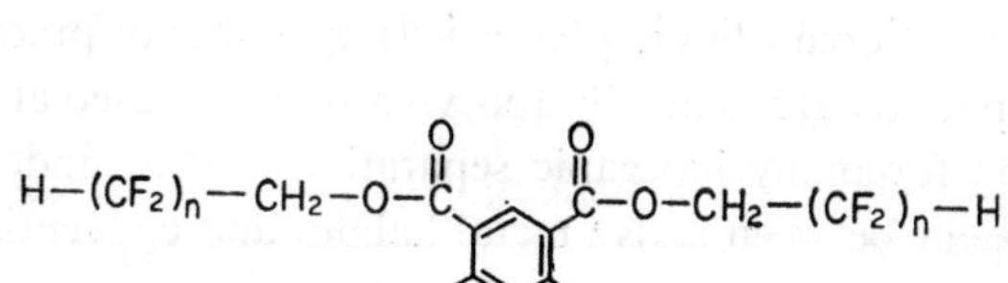

Maximum operating temperature 200°C.

Gordon *et al.*[607] studied the behaviour as stationary phases of molten tetra-n-pentylammonium picrate (I) and bromide (II) and tetra-n-hexylammonium nitrate (III).

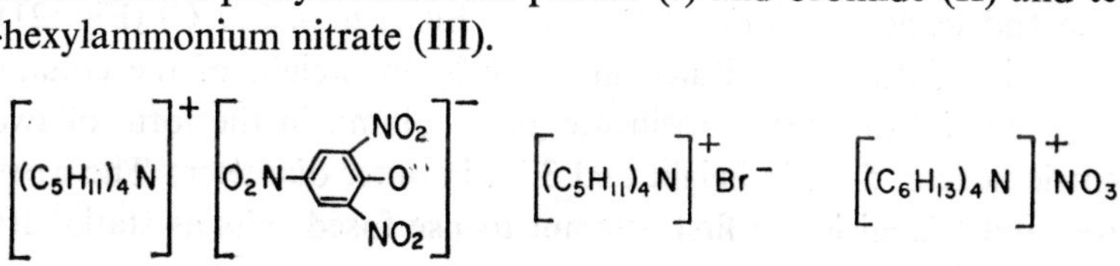

(I)   (II)   (III)

Retention indices (I) are reported for twenty-five solutes including n-alkanes, RX, with X = $-CH=CH_2$, $-C\equiv CH$, $-OC_2H_5$, $-Br$, $-I$, $-NH_2$, $-OCOCH_3$, $-CHO$, $-COCH_3$, $-CN$, and $-OH$.

The molten nitrate and bromide show a pronounced selective retention of ROH and to a lesser extent of $RC\equiv CH$. The molten picrate shows a small but apparently genuine selectivity toward π-donor molecules. Chemical transformations of RBr and RI on the molten nitrate column and of RCl and RI on the molten bromide column were observed.

The background signals arising from decomposition products from the molten salt columns decrease in the order:

$$\text{picrate}^- < NO_3^- \ll Br^-$$

It is suggested that the fused picrate on a deactivated diatomaceous support would constitute a practically useful polar column up to at least 140°C.[608]

Fused inorganic salts as stationary phases

Discussing the work of Freiser[609] concerning the analysis of low boiling tin tetrahalides on titanium using a n-hexadecane column at 102°C, Juvet and Wachi[610] show that the separation of metal halides using organic stationary liquid phases is frequently not practical because organic compounds are generally too volatile to be used at the temperatures required for many inorganic separations. Also, undesirable reactions often occur between active metal halides and conventional organic liquid phases.

Metal halides may be separated by partition gas chromatography using fused salts as the stationary liquid phases. A mixture of titanium(IV) chloride saturated at room temperature with antimony(III) chloride was separated on a column of a eutectic mixture of anhydrous bismuth(III) chloride and lead(II) chloride (89 mole% $BiCl_3$ m.p. 217°C) (Fig. 21). The eutectic mixture constituted about 70% by weight of the column packing material (a 12-ft borosilicate glass column in the form of two concentric helices, 5 in. in height and 3 in. in outer diameter). The work of Juvet and Wachi is the first attempt to use fused salts as stationary liquid phases in gas chromatography.

Tadmor[611] separated mixtures of tin(IV) chloride, bromide, and iodide at 150°C using aluminium bromide as the liquid phase, and eluted iron(III) chloride and mercury(II) chloride at 290°C from a column containing bismuth(III) chloride as the liquid phase (a silicone grease column was found to react with several metal halides).

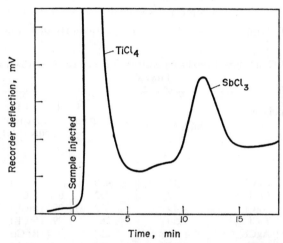

FIG. 21. Gas chromatographic separation of titanium tetrachloride solution saturated at 27°C with antimony trichloride. Column: BiCl₃–PbCl₂ eutectic on C-22 firebrick; column temperature: 240°C±1°C; carrier gas: dry nitrogen at 30 ml/min.[610]

The elution characteristics of eleven transition metal chlorides on twelve inorganic fused salt mixtures are reported by Zado and Juvet.[612] The compounds studied were Al_2Cl_6, $AsCl_3$, Fe_2Cl_6, $HfCl_4$, $MoCl_5$, $NbCl_5$, $SbCl_3$, $SnCl_4$, $TaCl_5$, $TiCl_4$ and $ZrCl_4$.

The composition and physical properties of the fused salt liquid phases are given in Table 18.

Most of the solutes studied are capable of forming chloro-complexes with available chloride ions in the liquid-phase melt according to the equilibrium

$$MCl_n + Cl^- \rightleftarrows MCl_{n+1}^-$$

The vapour pressure of the solute over its chloro-complex is, of course, less than that of the uncomplexed solute. An increase in temperature shifts the equilibrium to the left and increases the partial pressure of the solute above the melt, allowing its elution from the chromatographic column. The ratio of the stabilities (dissociation pressures) of two chloro-complexes is, in general, different from the relative volatility ratio of the uncomplexed solutes. Therefore, separation of solutes based on this difference in complex stability would be particularly desirable for situa-

TABLE 18. COMPOSITION AND PHYSICAL PROPERTIES OF THE FUSED SALT LIQUID PHASES[612]

Component		Mole% of A	Melting point (°C)	Estimated temperature at which vapour pressure = 0.1 mmHg (°C)
A	B			
NdCl$_3$	NaCl	41.2	430	748 (NaCl)
KCl	LiCl	49	352	680 (LiCl)
InCl$_3$	NaCl	49	272	292 (InCl$_3$)
InCl$_3$	TlCl	50	262	292 (InCl$_3$)
RbCl	AgCl	40	253	680 (RbCl)
BiCl	PbCl$_2$	89	215	160 (BiCl$_3$)
ZnCl$_2$	TlCl	52	213	365 (ZnCl$_2$)
PbCl$_2$	FeCl$_3$	37	177	175 (FeCl$_3$)
BiCl$_3$	FeCl$_3$	48	171	155 (FeCl$_3$ and BiCl$_3$)
NaFeCl$_4$		100	158	300 (NaFeCl$_4$)
NaAlCl$_4$		100	152	450 (NaAlCl$_4$)
TlCl	BiCl$_3$	32.5	150	160 (BiCl$_3$)

tions in which solutes have similar boiling points. The situation is more complicated when one of the components of the melt itself forms chloro-complexes. The solute then competes with the solvent for available chloride ions.

For the NdCl$_3$–NaCl system, the eutectic mixture at 41.2 mole% of NdCl$_3$ consists simply of a mixture of the uncomplexed species. This liquid phase is particularly promising for use at high temperatures in the separation of ZrCl$_4$–HfCl$_4$ mixtures via their hexachloro complexes.

KCl–LiCl

Most of the solutes studied in this liquid phase are eluted with no retention, presumably because of lack of solubility in this very polar melt.

InCl$_3$–NaCl; *InCl$_3$–TlCl*

At the eutectic composition, the sodium chloride and thallium chloride are completely associated with the indium chloride as Na_3InCl_6 and Tl_2InCl_5. A column only 5 cm long completely resolves a $TaCl_5$–$NbCl_5$ mixture on the $InCl_3$–$TlCl$ melt with a relative retention of 4.34 at a column temperature of 329°C. On the $InCl_3$–$NaCl$ melt, the components are not resolved on a longer column (23 cm). This retention behaviour indicates no special interaction of the $InCl_3$–$NaCl$ melt with the $NbCl_5$–$TaCl_5$ mixtures. Although the complexes $NaNbCl_6$ and $NaTaCl_6$ are known to exist, they do not form under these conditions because Na_3InCl_6 is apparently more stable. For the $InCl_3$–$TlCl$ eutectic mixture, $TaCl_5$ was selectively retained by the liquid phase; apparently a thallium tantalum chloro-complex is formed that is more stable than the corresponding indium complex.

From the other molten salts presented in Table 18, $NaFeCl_4$ and $NaAlCl_4$ show interesting properties as stationary phases. These materials are chemically inert, stable compounds of low melting point and high boiling point. Although $ZrCl_4$ and $HfCl_4$ were completely retained and $SnCl_4$, $TiCl_4$ and $AsCl_3$ were eluted with air, the three solutes $SbCl_3$, $NbCl_5$ and $TaCl_5$ had good retention characteristics on these liquid phases.

The stability of antimony, niobium, and tantalum complexes with the alkali metal tetrachloro-aluminate and -ferrate melts increases as the alkali metal cation radius increases.[116]

The use of molten salts as stationary phases raises some instrumental problems resulting from the high temperature and corrosive influence of the molten salts. These difficulties have mainly been overcome and their applications to the ultrapurification of semiconductor materials and of elements difficult to separate by other means will lead to greater use of inorganic gas chromatography.[613]

The choice of fused salt as stationary phase is dependent upon several factors:

(a) The fused salt should be a good solvent for volatile inorganic compounds. In general, the effectiveness of the molten salt medium as a solvent is increased as the temperature of the column is decreased.

(b) The fused salt should have a low vapour pressure at the temperature of the column.

(c) The fused salt should, in general, possess an ion in common with that of the solute molecules. This minimizes the possibility of undesirable reactions in the column.

The first successful separation of organic compounds using non-volatile inorganic fused salts as liquid phases was reported by Hanneman et al.[614] The column packing was an eutectic mixture of sodium, potassium, and lithium nitrates (18.2 : 54.5 : 27.3 wt.%) deposited 28.6 wt.% on GC-22 firebrick, 42–60 mesh. This column could be operated between 150°C (the melting point of the eutectic) and 400°C and was used to separate polyphenyl compounds and of a mixture of phenyl-m-terphenyls.

Gas chromatographic columns using molten inorganic salts as a stationary liquid phase apparently behave in the same manner as columns with conventional organic liquid phases. This seems to fit the observed variation of HETP with flow rate, the linear plots of log retention time vs. carbon number for a homologous series and the symmetrical peaks.

Also, the use of molten salt stationary phases makes possible separations of high boiling organic compounds, limited only by the thermal stability of the organic compounds. If the separation of metal chlorides on eutectics as liquid stationary phases is based on their dissolution in the stationary phase,[615] the separation of organic compounds is based on adsorption phenomena.

4.1.4 Specific stationary phases

In certain instances, the stationary phase can consist of, or contain, a chemical reagent which can react specifically with a particular class of solute in a specifical chemical manner. It is obvious that the interactions between stationary phases and solutes must, on the one hand, be strong enough to provide the specific properties mentioned above, and, on the other hand, be sufficiently labile easily to remove the components to be analysed from the stationary phase. Thus these interactions cannot be localized as in C–C, C–N, C–O, etc., and must be very sensitive to the thermodynamic and chemical conditions of the medium.

Such specifical interactions may occur if transition metal complexes are used as stationary phases. Consider a square-planar complex ML_4 where the central metal ion participates in co-ordination with nd, $(n+1)s$, and $(n+1)p$ orbitals, where n is the principal quantum number. The spatial orientation of the metal orbitals are:

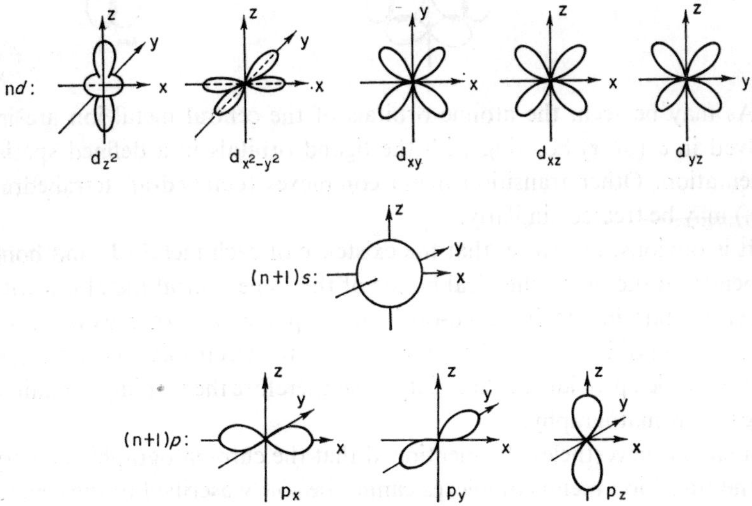

Depending on the spatial orientation, the metal orbitals may form σ- and π-bonds with the ligand orbitals, as follows:

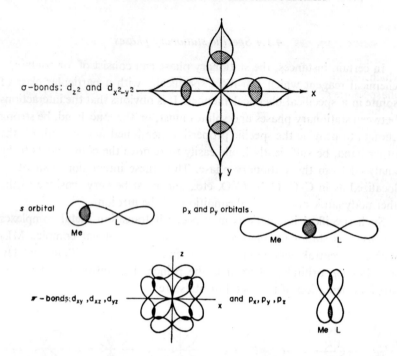

As may be seen, the atomic orbitals of the central metal ion are involved in σ (or π) bonding with the ligand orbitals in a defined spatial orientation. Other transition metal complexes (octahedral, tetrahedral, etc.) may be treated similarly.

It is obvious, therefore, that the existence of each metal—ligand bond depends on the other chemical bonds of the same central metal ion with other ligands in the inner co-ordination sphere. In other words each bond is conditioned by the other bonds by electronic and/or steric factors, which provide the "mobility" and therefore the "lability" required in gas chromatography.

It should nevertheless be mentioned that the chromatographic activity of the transition metal complexes cannot be solely ascribed to the central metal atom. The ligands which apparently do not participate in the selection activity of the central atom may control the interaction between the stationary phase and solutes through steric factors or inductive factors (like the trans effect for square-planar complexes, etc.) and may play a role in the stabilization of a certain oxidation state of the metal.

Besides the "unsaturated" co-ordination of the central metal ion in which originates its chromatographic activity, selectivity in a stationary phase-solute system is also a function of the intimate molecular structure of the solutes.

The coordination ability of the solutes in the sample to be analysed is also governed by many factors (electronic, energetic, steric, polarizability, etc.).

As an example, consider the rhodium(I)-dicarbonyl-β-diketonate complexes used as stationary phases in squalene for the separation of olefins as recently reported by Gil-Av and Schurig.[616] The complexes are square-planar:

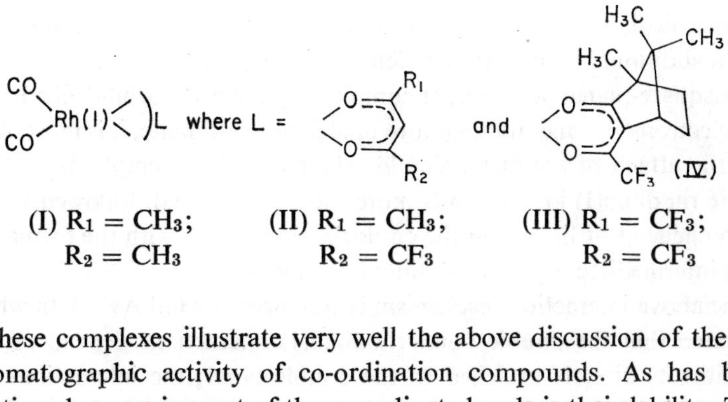

(I) $R_1 = CH_3$;   (II) $R_1 = CH_3$;   (III) $R_1 = CF_3$;
    $R_2 = CH_3$ $R_2 = CF_3$ $R_2 = CF_3$

These complexes illustrate very well the above discussion of the gas chromatographic activity of co-ordination compounds. As has been mentioned, a requirement of the co-ordinate bonds is their lability. This property is assured by the presence of eight electrons in the d-shell of rhodium(I), which undoubtedly make the d-orbitals less able to form chemical bonds as strongly as rhodium(III) with a $4d^6$ electronic configuration.

The low oxidation state of rhodium is stabilized by the strong back-donation to form metal–ligand π-bonds with the two carbon monoxide molecules in the coordination sphere.[617] This illustrates how ligands which do not participate directly in the gas chromatographic activity of the complex can play a role in the stabilization of a given oxidation state of the central metal ion. All $4d^8$ rhodium(I) complexes are square-planar, like the four-co-ordinated platinum(II) or palladium(II) complexes.

The gas chromatographic activity of these complexes was considered by Gil-Av and Schurig[616] as a SN_2 reaction of the four-coordinate rhodium(I) complex with the olefinic double bond, as follows:

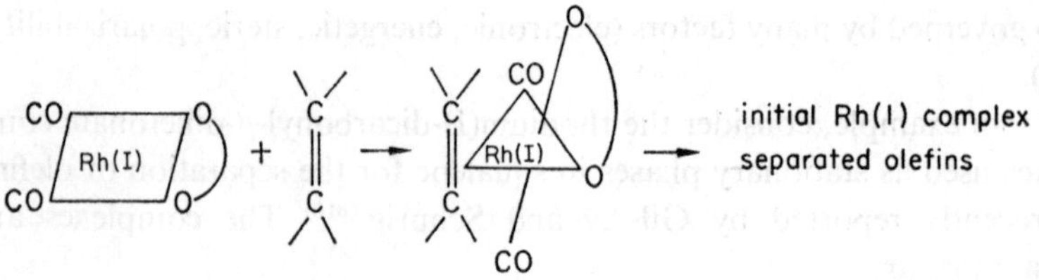

the intermediate complex being a five-coordinate trigonal bipyramid having an additional bond with the olefinic π-orbital.

As square-planar rhodium(I) complexes have a d_{z^2} orbital filled with two electrons[618] and are therefore unable to co-ordinate, it is most likely that the attack of the olefin should take place via an empty $5p_z$ orbital on the rhodium(I) ion, which is more radially extended, followed by the rearrangement of ligands in the co-ordination sphere with the formation of an intermediate trigonal bipyramid complex (D_{3h}).

The above interaction mechanism is supported by Gil-Av's data which indicated that the double bond–rhodium(I) interactions depend on the substituents directly attached to the double bond; the interactions decrease, going from $H_2C{=}CH_2$ to $RCH{=}CHR'$ or $RR'C{=}CH_2$ or from cis to trans isomers, which emphasizes the influence of the substituents on these, primarily through steric factors.

The best chromatographic activity studied by the above authors for the separation of olefins has been obtained with the complex IV (dicarbonyl-rhodium(I) trifluoroacetylcamphorate). It may be assumed that the inequivalence and the weakness of the coordinate bonds with the two oxygen atoms owing to the presence in the ligand molecule of the —CF$_3$ group, may favour the rearrangement of the ligands in order to stabilize the intermediate D_{3h} complex by the mechanism discussed above.

As has been discussed above, the gas chromatographic activity of complexes may therefore take place via a SN_2 mechanism as in the olefin–rhodium(I) complex system.

Another mechanism should be direct ligand substitution, that is a SN_1 reaction between the stationary phase and the solutes, as has been suggested in the separation of olefins using palladium(II) complexes of the form:[619]

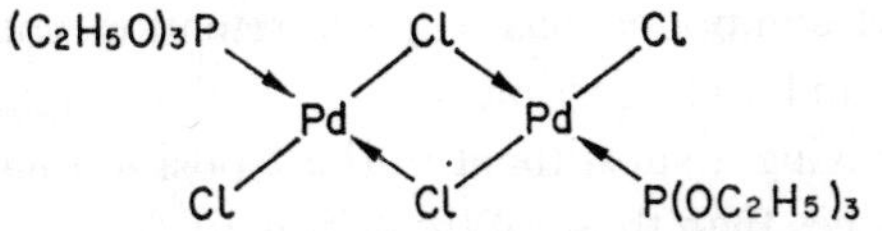

$$(C_2H_5O)_3P\diagdown\underset{Cl}{\overset{}{Pd}}\diagdown\underset{Cl}{\overset{Cl}{Pd}}\diagdown P(OC_2H_5)_3$$

Owing to the strong *trans* effect of the chloride ion, the $P(C_2H_5O)_3$ group is easily substituted by olefins *via* a SN_1 mechanism as follows:[620]

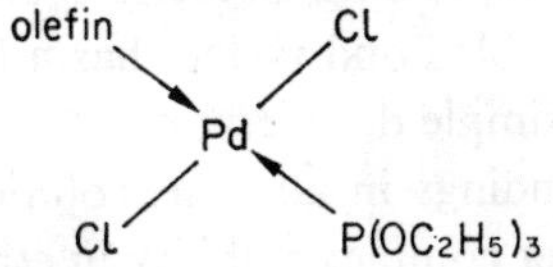

$$\underset{Cl}{\overset{olefin}{\diagup}}Pd\underset{P(OC_2H_5)_3}{\overset{Cl}{\diagdown}}$$

The olefin retention depends on the olefin–palladium(II) bond strength.

The most data in the literature concerning the use of charge-transfer interactions with metals in the gas chromatographic separations of olefins conerns the use of silver ions. Bradford *et al.* were the first to report the high selectivity of silver nitrate–ethylene glycol as a stationary phase for the separation of butenes.[621] The striking efficiency of this complex-forming phase for the analysis of olefins with similar boiling points was subsequently confirmed by many investigators.[622–629]

On the basis of retention data for 83 hydrocarbons (saturated and unsaturated) on a silver nitrate–ethylene glycol column, the analytical potential of such columns was evaluated by Smith and Ohlson[629] on the basis of the degree, type, geometry, and position of unsaturation. The selectivity of silver-nitrate-containing phases in general results from the marked effect of relatively small structural changes in olefins on the stability constants of the complexes. According to Dewar,[227] the bond between the olefinic ligand and the metal ion is formed by donation of electrons from the ethylenic bond to the vacant s-orbital of the silver ion and back donation of d-electrons from the metal to the antibonding orbitals of the unsaturated compound. The following general rules, correlating structure with retention volume and the stability constants of the Ag^+-olefin complexes, have been established:[624, 626]

   (i) Substitution at the double bond decreases the retention volume.

  (ii) A 1-alkyl compound has a lower retention volume than those of the 3- and 4-alkyl isomers.

 (iii) Olefins having a substituent in the 3-position have higher retention volumes than those of the 4-isomers.

 (iv) Cyclobutenes have less tendency to form complexes than the corresponding five- and six-membered cyclo-olefins.

  (v) Cyclopentene derivatives have higher retention volumes than those of the corresponding isomeric cyclohexenes.

 (vi) A conjugated double bond system has a lower complex-forming capacity than a simple double bond.

(vii) In contrast to findings in aqueous solutions, conjugated dienes do not show lesser complex stability in ethylene glycol.

(viii) Compared with their behaviour in aqueous media, unconjugated dienes co-ordinate strongly in ethylene glycol, probably as a result of chelation.

One of the most interesting features in chromatography on silver nitrate containing stationary phases is the secondary isotope effect. The pronounced secondary deuterium isotope effects on the stability of silver–olefin complexes have been studied in detail by gas chromatography.[104] Experiments with specially deuterated propylenes showed that the isotope effect is much larger when deuterium is directly bound to the unsaturated carbon atoms ("α-effect") than when it is in the β-position to the double bond ("β-effect"). The progressive increase in retention time on silver nitrate–ethylene glycol columns by deuterium substitution has been used as a basis for the separation and analysis of the various isotopic olefins differing only in the number and positions of deuterium atoms.[630–633]

A serious drawback of the silver nitrate–ethylene glycol column, which limits its application to low-boiling compounds, was pointed out by Keulemans.[634] According to him, the temperature for such chromatography should be kept below 40°C, because above this temperature the adducts do not form and the stationary phase is not stable.

To improve the stability and selectivity of silver nitrate columns, several solvents were investigated. These include poly(ethylene glycol) and glycerol,[635] di and triethylene glycol,[315, 636] benzyl cyanide,[637, 638] ethylene glycol bis(cyanoethyl) ether,[639] and tetra-ethylene glycol.[640-642] None of the above solvents was able significantly to improve the stability and selectivity of silver nitrate–ethylene glycol columns.

In this connection, it must be mentioned that dicarbonylrhodium-trifluoroacetylcamphorate has a greater stability than the silver nitrate phases and, contrary to the solvents used for silver nitrate, the apolar stationary phase (squalane) which dissolves the rhodium complex permits non-olefinic hydrocarbons to be separated according to their boiling points. On the other hand, because the interaction of olefins with dicarbonylrhodium(I) complexes greatly exceeds that with silver ions, the complex concentration in the liquid phase can be reduced by a factor of 10 and more. Some analytical applications of this stationary phase, such as the analysis of small amounts of ethyl acetate in vinyl acetate and the separation of C_2H_4 and C_2D_4 have been reported.[643]

Recently, Wasik and Tsang[644] used aqueous solutions of silver nitrate as stationary phases. The strongly ionizing properties of water (thus increasing the silver ion activity) make such columns more efficient than the conventional silver nitrate–ethylene glycol columns. These columns have interesting analytical applications, particularly in the field of isotope separation.

Wasik and Tsang[645] also demonstrated the possibility of separating hydrocarbons such as olefins, alkanes, and aromatic compounds by using aqueous solutions of silver and mercury(II) salts as stationary phases. Mercury(II) ions complex very strongly with olefins and act essentially as a barrier to the elution of any olefins, but do not complex with alkanes or aromatic compounds; silver ions complex weakly with aromatic compounds. Thus aromatic compounds are eluted after longer times than alkanes while the olefins are retained completely. In spite of extensive investigations, the problem of operating the silver nitrate column at elevated temperatures still remains, which restricts its application to low boiling compounds.

To improve on the stability of silver nitrate columns at higher temperatures, Banthorpe et al.[646] replaced the silver by thallium(I) nitrate in diethylene glycol or poly(ethylene glycol) as the stationary phase, without loss of selectivity. However, the separation of cis- and trans-2-butenes was poorer on the thallium nitrate column than on the silver nitrate–ethylene glycol column.

Cartoni et al.[619] proposed as liquid stationary phases the N-dodecyl-salicylaldimines of nickel, palladium, platinum, and copper (MeSal$_2$) (I) and the methyl-n-octyl glyoximes of nickel, palladium, and platinum (MeGly$_2$) (II).

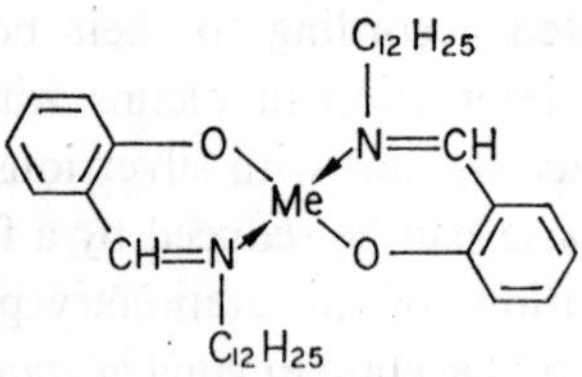

(I)

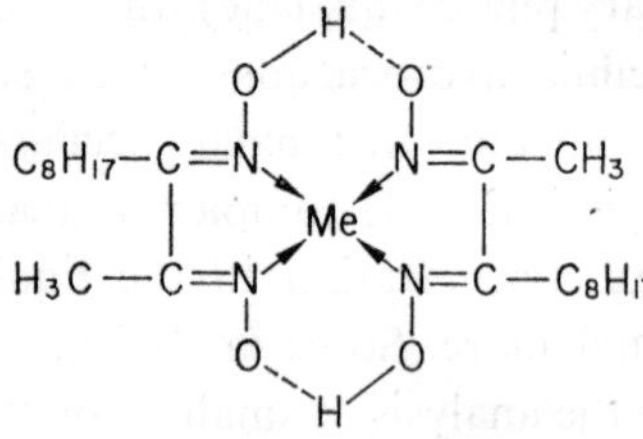

(II)

Owing to the weak interaction that occurs at the unoccupied co-ordinating positions, these complexes, when used in liquid phases, specifically retard those molecules which act as ligands, especially amines, ketones, alcohols, and molecules containing double bonds. The strong interaction of nickel with amines was expected, as a large number of square-planar diamagnetic complexes of nickel are known to react with two molecules of amine to give octahedral, paramagnetic complexes.[647] No such favourable rearrangement of electrons is possible in copper(II) complexes.

The temperature limits between which these stationary phases may be utilized are: NiSal$_2$, 54–180°C; PdSal$_2$, 100–160°C; PtSal$_2$, 116–135°C; CuSal$_2$, 60–180°C; NiGly$_2$, 105–180°C; PdGly$_2$, 110–180°C; PtGly$_2$, 120–190°C.

Barber *et al.*[648] suggested the use of the stearates of manganese(II), cobalt(II), nickel(II), copper(II), and zinc(II) $(CH_3—(CH_2)_{16}—COO)_2Me$, as liquid phases. A very strong interaction was observed when amines were passed through the columns of metal salts. Primary amines decomposed the stearate column. For secondary and tertiary amines on manganese, cobalt, and zinc stearates, the retardation factors were proportional to the basicity of the amines. The special interaction with metal stearates suggested their use for effecting difficult separations. For example, use of a column with 20% manganese stearate on Celite at 156°C enabled β-picoline, α-picoline, and 2,6-lutidine, compounds with the same boiling point (143°C), to be separated.

The selective retention of specific organic donors on metal(II) caproates was reported by Bayer.[649] Nickel caproate in silicone high-vacuum grease was used for the separation of amino-acid methyl esters. Among the metal ions studied (Ni, Fe, Cu, Co), nickel caproate showed the greatest retention volumes and the most stable bonding between the NH_2 group of the amino-acid methyl ester and the metal.

Cartoni *et al.*[650] suggest the use of copper, beryllium (m.p. 53–53.5°C), aluminium (m.p. 40–40.5°C), nickel (m.p. 48–49°C), and zinc (liquid at room temperature) n-nonyl-β-diketonates as liquid stationary phases. The operating temperature is 120°C. The retention of organic compounds on columns packed with these liquid phases showed alcohols to be effective ligands for complex formation with metal diketonates. The complexing action increases from the tertiary to the primary alcohols and is quite effective for the zinc and beryllium chelates and more pronounced for the nickel chelate.

A similar complexing action occurs when zirconium $[ZrO_{0.5}(NO_3)(DEHP)_2]_n$, cobalt $[Co\ en_3(DEHP)_3]$, and thorium $[Th(NO_3)(DEHP)_3]$ di-2-ethyl-hexyl-phosphates[320] are used. The complexing power increases from tertiary alcohols to primary alcohols and is more pronounced for the zirconium compound.

Castells and Catoggio[119] found a very anomalous sequence in the interactions between aliphatic amines and metal ions by gas chromatography using different metal stearate solutions in quadrol [N, N, N', N'-tetrakis(2-hydroxypropyl)ethylene diamine] as stationary phases.

Greater interaction with cadmium and zinc, which have completely filled atomic orbitals, than with nickel, which has incompletely filled orbitals, cannot be explained exclusively on the basis of amine–cation interaction. Possibly, the presence of quadrol in the stationary phase has complicated the system.

Another example of specific interaction is met with in the separation of amino-acid enantiomers on optically active stationary phases. Amino acids are used in the form of their volatile lower aliphatic (C_1-C_4) trifluoroacetyl (TFA)-alcohol esters. At present, three types of optically active stationary phase have been successfully tested:

(I) N-TFA-amino-acid esters of higher aliphatic alcohols (C_6-C_{12}).[323, 324]

(II) Ureide of L-valine isopropyl ester.[651, 652]

(III) N-TFA-dipeptide esters of lower aliphatic alcohols (C_1-C_6).[653–663]

The best separation was achieved on N-TFA-dipeptide cyclohexyl ester phases. Mechanisms for the separation of D, L-amino acids have been proposed, which assume the formation of diastereoisomeric complexes held together by hydrogen bonding.[324, 654, 658]

For phases of types (I) and (II) (esters and ureides), complex A can be illustrated as follows:[662]

Complex A

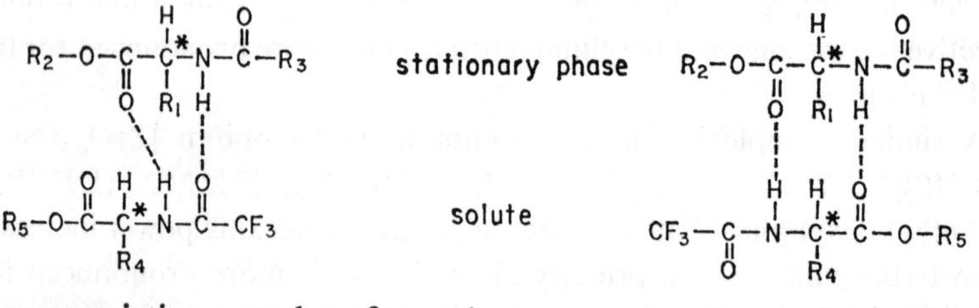

In an association complex of type A, no more than two hydrogen bonds can be formed as only one hydrogen donor is available in each of the molecules considered. For the dipeptide phases, two complexes (B and C) can be proposed. In complex B, the ester part of the peptide is involved in the formation of the hydrogen-bonded diastereoisomeric association complex, whereas in complex C, both amide functions are responsible for association.

Complex B

```
        H       H                                              H      H
        |*      |*                                             |*     |*
C₆H₁₁O-C-C-N-C-C-N-C-CF₃    stationary phase    C₆H₁₁O-C-C-N-C-C-N-C-CF₃
        ‖ | | ‖ | | ‖                                          ‖ | | ‖ | | ‖
        O R₁ H O R₂ H O                                        O R₁ H O R₂ H O

        O H H O                                                O H H O
        ‖ | |*‖                                                ‖ |*| ‖
CF₃-C-N-C-C-OR₅              solute               R₅O-C-C-N-C-CF₃
        R₄                                                     R₄
```

Complex C

```
        H      H                                               H      H
        |*     |*                                              |*     |*
C₆H₁₁O-C-C-N-C-C-N-C-CF₃    stationary phase    C₆H₁₁O-C-C-N-C-C-N-C-CF₃
        ‖ | | ‖ | | ‖                                          ‖ | | ‖ | | ‖
        O R₁ H O R₂ H O                                        O R₁ H O R₂ H O

              O H H O                                          O H H O
              ‖ |*| ‖                                          ‖ | |*‖
      R₅O-C-C-N-C-CF₃          solute              CF₃-C-N-C-C-OR₅
              R₁
```

R_1, R_2, R_3, R_4, and R_5 are aliphatic radicals.

It was shown that complex C is statistically predominant in the dipeptide phases.[661, 662]

In the conformation in which the solutes are hydrogen-bonded to the stationary phase, they cease to be mirror images of each other and become "conformational" diastereomers. Interactions of such diastereomers with the stationary phase will, in general, be different. This aspect of the association of the dipeptide phase with the enantiomeric N-TFA-amino-acid esters helps in the understanding of the very high resolution factors found. Because the resolution factors have such large values, separation of enantiomers can be achieved by dipeptide phases not only on capillary columns but also on packed columns.[653]

The relative sizes of R_1, R_2, and R_4 are clearly a steric factor in the resolution of D, L pairs. When R_1 and R_2 are isopropyl groups, the dipeptide phase is N-TFA-L-valyl-L-valine cyclohexyl ester (val-val). If R_1 is [—CH₂—CH(CH₃)₂] and R_2 is a benzyl group, then the dipeptide phase is N-TFA-L-phenylalanyl-L-leucine cyclohexyl ester (phe-leu). A molecule having a [—CH₂—CH(CH₃)₂] group in the R_1 position and an isopropyl group [—CH(CH$_3$)$_2$] in the R$_2$ position is N-TFA-L-valyl-L-leucine cyclo-hexyl ester (val-leu).

It is interesting to note that the resolution factor for D, L-proline on a val-leu phase (1.026)[660] lies between the values observed on val-val (1.041)[655] and phe-leu (1.000) phases.[657] These values indicate that D, L-proline is resolvable on val-val, partially so on val-leu and unresolva-ble on phe-leu. A satisfying explanation for this pattern can be formulated by considering the hydrogen-bonded diastereomeric association com-plexes formed as a model.[660] In proline, only two of the three theoretic-ally possible hydrogen bonds can be formed. The sizes of R$_1$ and R$_2$ now become a much greater steric factor in the formation of complexes C shown earlier. When R$_1$ and R$_2$ are small, as in val-val, there is still sufficient space for the formation of hydrogen bonds. When the size of R$_1$ is increased, spatial orientation decreases hydrogen bond formation. Hence the better resolution on val-val, in which both possible hydrogen bonds can develop, than on val-leu, where one of the two available hydrogen bonds is crowded by R$_1$ and only partial resolution is achieved. When R$_1$ and R$_2$ are both lengthened, as in phe-leu, neither hydrogen bond can be readily formed and no resolution occurs.

It has been reported that when the trifluoroacetyl group in N-TFA-L-phenylalanine cyclohexyl ester is replaced by an isobutyryl group, the phase becomes almost incapable of resolving TFA-α-amino acid es-ters.[324] This result has been explained by the increased ability for hydro-gen bonding by the —NH group attached to a trifluoroacetyl group. It has been shown that this indirect effect may be only a minor factor in other systems.[654] However, an important effect of the trifluoroacetyl group is the decreased melting point of the dipeptide phase, which permits the phase to operate at lower temperatures with correspondingly higher resolution factors. The best separations of enantiomers of amino-acid derivatives, using peptide stationary phases, would be achieved using a dipeptide stationary phase possessing a TFA group, bulky side groups, and a similarly bulky ester group, with a column operated at the lowest feasible temperature.[661]

Somewhat greater chromatographic temperature-stability can be achieved, with a small sacrifice in the resolution factor, by going to derivatised tri- or higher peptides. Such a peptide, N-TFA-(L-val)$_3$-O-iso-propyl, has been synthesized by Feibush and Gil-Av.[654] The compound has a melting point of 202°C. The tripeptide derivative tends to depress slightly the resolution of N-TFA-alanine t-butyl ester; similar observations were made with the methyl, ethyl, n-propyl, n-butyl, and cyclopentyl esters of N-TFA-alanine.

Nevertheless, it should be pointed out that for some binary compositions the melting point is below 80°C, and very high resolution factors can be obtained at this lower temperature. For instance, N-TFA-alanine t-butyl ester has a resolution factor as high as 1.29 at 80°C on a binary mixture of the di- and tripeptide, containing a 0.42 mole fraction of the latter.

The derivatives of the volatile solutes should also possess a bulky ester group (such as t-butyl or *iso*propyl) and a N-TFA-group.[657, 661]

Parr *et al.*[659] investigated the influence of perfluoroacyl derivatives of amino-acid esters other than the trifluoroacetyl group with respect to retention times and resolution factors. They synthesized N-pentafluoro-propionyl (PFP), N-heptafluorobutyryl (HFB), and N-pentadecafluoro-octanoyl (PDFO) -D, L-leucine esters, and studied their gas chromatographic behaviour on the optically active stationary phase N-TFA-L-phenylalanyl-L-leucine cyclohexyl ester. The new derivatives, and especially PFP amino-acid isopropyl esters, have two advantages. Owing to their higher volatility, the PFP derivatives allow the resolution of lower volatile amino-acid enantiomers. On the other hand, a reduction in column temperature can be achieved in order to give a better resolution of enantiomers. However, in a comparison of the chromatographic properties of an N-PFP stationary phase with its corresponding N-TFA derivative, no advantages were apparent.[661]

The resolution of enantiomers is not restricted to α-amino-acid esters because other classes of racemic compounds may be separated on optically active stationary phases.[651, 652] A particularly interesting aspect of resolution on optically active stationary phases is that the determination of optical purity is absolute. If the asymmetric reagent used for resolution is not pure, the only effect will be to reduce the resolution factor, but the relative peak areas will not change. On the other hand, it seems possible that the method will permit the determination of the absolute configuration of certain compounds. As an example, the separation of some N-TFA-isopropyl esters of the relatively volatile amino acids on a capillary column with N-TFA-L-valyl-L-valine cyclohexyl ester as stationary phase is shown in Fig. 22.[655]

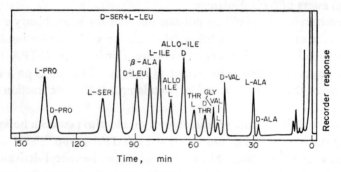

Fig. 22. Chromatogram of the N-TFA-*iso*propyl esters of the relatively volatile amino-acids on a 500 ft×0.02 in. stainless steel column, coated with N-TFA-L-valyl-L-valine cyclohexyl ester. Helium pressure: 20 lb/in² (1.4 kg/cm²); column temperature: 110°C; injector temperature: 180°C; detector temperature: 290°C; air pressure: 25 lb/in²; hydrogen pressure: 14 lb/in².[655]

Liquid crystals must also be included in the class of specific stationary phases. The term liquid crystal was first suggested by Lehmann[664] for those substances which, within definite temperature intervals, are liquid in mobility and crystalline in optical properties. These materials do not pass directly from the solid to the normal or isotropic liquid phase but go through one or more discreet phase transformations involving liquid crystal intermediate phases (mesophases). More than 500 organic compounds are known to exhibit liquid crystal phases under some conditions of temperature or solution. Their general physical properties and classifications have been compiled in detail.[665, 666] Liquid crystals are divided into categories depending on the nature of the liquid crystal formed (smectic, nematic, or cholesteric) as was discussed previously (p. 92).

Gas chromatographic investigations of liquid crystals as stationary phases have been carried out for isomers of a variety of compounds, including aliphatic and aromatic hydrocarbons, halogenated hydro-carbons, phenols, etc.,[667-672] and have been reviewed by Kelker and von Schivizhoffen.[354]

The first investigation in which a liquid crystal was used as a stationary phase was described in 1963 by Kelker.[667] The compound studied as the stationary phase was p,p'-azoxyphenetole, which has a nematic phase between 138–168°C.

$$C_2H_5O-\!\!\langle\bigcirc\rangle\!\!-N\!\!=\!\!NO-\!\!\langle\bigcirc\rangle\!\!-OC_2H_5$$

It was found that this compound in its nematic form separates the m- and p-xylenes.

For each of the compounds studied, including cyclohexane, toluene, and m-, p-, and o-xylenes, there is a uniform decrease in retention volume up to the nematic–liquid transition temperature. At the transition tem-perature of 168°C there is an abrupt increase in retention volume and solubility. Some of this undoubtedly corresponds to an increase in entropy of solution consistent with the formation of a complex, less-ordered phase capable of a variety of interactions. The decrease in liquid phase orientation would result in a loss of local molecular selectivity and there-fore affords a greater volume of solvent stationary phase wherein solu-bility can take place.

Dewar and Schroeder[350, 670] studied a number of liquid crystals from the point of view of their behaviour as stationary phases in gas chromato-graphy. The compounds studied together with their melting points and transition temperatures are given in Table 19.

These stationary phases were studied in connection with the possibility of the separation of m- and p-substituted benzenes. On p-azoxyanisole(I), m- and p-xylene cannot be separated, but the relative retention times showed an interesting variation with temperature. In the normal liquid region, the retention time for the p-isomer was less than for the m-isomer, as has been observed for other liquids as stationary phases. However, this order was reversed in the nematic region owing to selective retention of the linear isomer by the meso-phase.

Excellent separations of the xylenes would be effected by operating at 75°C in the smectic region with 4,4′-di-n-hexyloxyazoxybenzene(II) and

4,4'-di-n-heptyloxyazoxybenzene (III) as stationary phases. Relative retention times for the three xylenes on a column of 4,4'-di-n-hexyloxy-azoxybenzene decrease at the smectic to nematic transition temperature (80°C). This is unexpected, for the decrease in order in passing from a smectic to a nematic phase should lead to an increase in solubility and thus in retention time. A possible explanation seems to be that smectic stationary phases may not, like normal liquids, operate under equilibrium conditions. The viscosity of a smectic liquid crystal is extremely aniso-tropic, being very great for shear across the planes of the two-dimensional liquid. This is a consequence of the layer structure, the mechanical pro-perties of a smectic phase being similar to those of graphite. It therefore seems possible that diffusion through a smectic stationary phase may be slow enough to affect the residence time. If this is so, two factors would operate in the transition from the smectic to the nematic phase; the decrease in order would lead to an increase in retention time, while the viscosity effect would lead to a decrease. The observed result, a small decrease in retention time, could be explained as a sum of these opposing effects, the latter slightly predominating.

Good separation properties of *m*- and *p*-isomers are also observed on the other stationary phases in Table 19 in comparison with the separation of the same compounds on conventional stationary phases (silicone oil and Apiezon M). Thus for nematic 4,4'-bis(*p*-methoxybenzylideneamino)-3,3'-dichlorobiphenyl(IV), it was found that the relative retention of 4-methylbiphenyl (3-methylbiphenyl, 1.00) was 1.42 to 1.52, compared with 1.06 on Apiezon M grease. The ester (V) (benzoic acid *p*-(n-heptyl-oxy)-*p*-phenylene ester) is highly selective in the lower part of its smectic range. At 68°C, the relative retention of *p*-xylene (*m*-xylene, 1.00) is 1.10; at 88°C, *p*-chlorotoluene on *p*-methylanisole had very high relative reten-tions as compared to their *m*-isomers, i.e. 1.17 and 1.33 respectively. The separation of these isomers on columns of ester (VI) (benzoic acid *p*-butoxy-4,4'-biphenylene ester) and (VII) (*p*,*p*'-diphenylene bis(*p*-n-heptyl-oxybenzoate)) is not better than on derivative (V). The same observation may be made in connection with the behaviour as stationary phases of ethyl-4-(*p*-ethoxybenzylideneamino)cinnamate (VIII) and diethyl *p*-azoxy-cinnamate (IX). The best results with these stationary phases were ob-

TABLE 19. MESOMORPHIC STATIONARY LIQUID PHASES

Com-pound	Structure	Smectic range (°C)	Nematic range (°C)	Cholesteric range (°C)
I	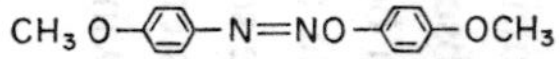 CH_3O—⬡—$N=NO$—⬡—OCH_3	—	121–135	—
II	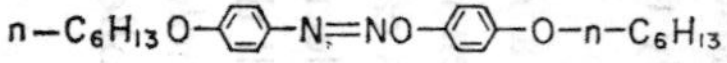 $n-C_6H_{13}O$—⬡—$N=NO$—⬡—$O-n-C_6H_{13}$	71–80	80–130	—
III	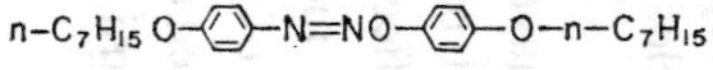 $n-C_7H_{15}O$—⬡—$N=NO$—⬡—$O-n-C_7H_{15}$	75–95	95–127	—
IV	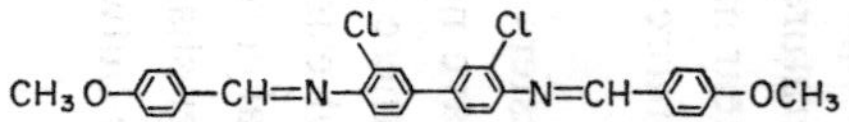 CH_3O—⬡—$CH=N$—⬡(Cl,Cl)—$N=CH$—⬡—OCH_3	—	154–334 dec.[a]	—
V	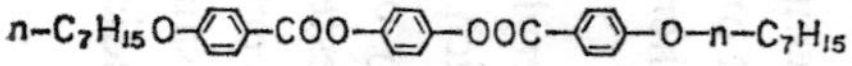 $n-C_7H_{15}O$—⬡—$CH=N$—⬡—OOC—⬡—$O-n-C_7H_{15}$	83–125	125–206	—
VI	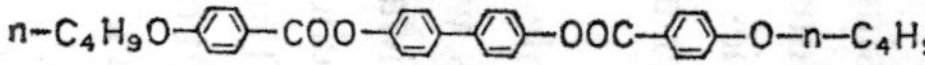 $n-C_4H_9O$—⬡—COO—⬡—OOC—⬡—$O-n-C_4H_9$	171–184	184–358 dec.	—
VII	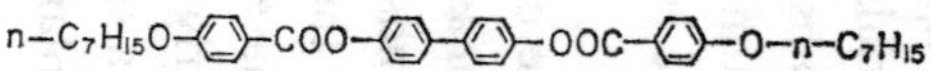 $n-C_7H_{15}O$—⬡—COO—⬡—OOC—⬡—$O-n-C_7H_{15}$	150–211	211–316	—
VIII	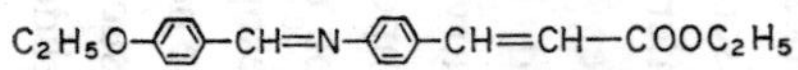 C_2H_5O—⬡—$CH=N$—⬡—$CH=CH$—$COOC_2H_5$	78–118[b] 118–157	157–160	—
IX	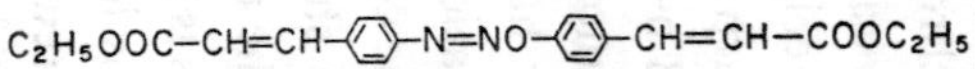 C_2H_5OOC—$CH=CH$—⬡—$CH=N$—⬡—$NO=N$—⬡—$CH=CH$—$COOC_2H_5$	140–251 dec.	—	—
X	Cholesteryl benzoate	—	—	149–180
XI	Cholesteryl nonanoate	(75)[c]	—	79–90

[a] Nematic phase decomposes before transition to isotropic liquid.
[b] This compound has two distinct phases.
[c] Monotropic (cholesteric→smectic) transition which is observed at 75°C only on cooling.

tained for the separation of dihalobenzene isomers. The relative retentions of *o*-, *m*-, and *p*-xylene on cholesteryl benzoate (X) and cholesteryl nonanoate (XI) proved to be similar to those obtained with conventional liquid phases.

From the studies of Dewar and Schroeder, it may be concluded that:
(a) Smectic phases are more effective than nematic phases.
(b) The solubility of solutes in liquid crystals depends on factors other than shape (e.g. dipole–dipole interactions may counterbalance the effects of geometry).
(c) The performances obtained with compound (V) as stationary phase seem superior to any previously tested compounds for separating benzene positional isomers.

Barall *et al.*[671] studied the behaviour as stationary phases of the acetate, n-valerate, and n-nonanoate esters of cholesterol. Cholesteryl acetate undergoes a solid–cholesteric transition at 94.5°C and a cholesteric–isotropic liquid transition at 116.5°C. Cholesteryl valerate is in the cholesteric phase from 83° to 89°C; above 89°C it is an isotropic liquid.

The authors showed that the cholesteric esters, when used as stationary phases in gas chromatography, have several features in common with other liquid crystal materials similarly employed. For example, the log retention time–reciprocal absolute temperature relationships exhibit slope changes and sharp discontinuities at or near mesophase transition points. The acetate and n-valerate ester adhere more closely to this observation than the n-nonanoate.

On the other hand, the cholesteryl esters have a number of unique properties as stationary phases which have not been observed for other liquid crystalline compounds. Many of these effects can be explained by considering the shape of the cholesteryl ester molecule and its special orientation in the mesophase. The central core of all the solid and mesophase structures is the nearly flat or saucer-shaped cholesteryl ring system. The ester groups in such a stack radiate outward following both a horizontal and a screw glide plane.

When the ester groups are long enough (n-nonanoate), organization probably takes place in this side-chain portion of the molecule as well. For a longer ester group, a greater degree of interaction between it and other long-chain molecules would be expected. Thus the retention time–boiling point curves for both aromatic compounds and alkanes in both the mesophase and isotropic liquid of cholesteryl acetate are linear. On the other hand, the boiling point–retention time isotherms of the n-nonanoate ester are sharply curving for alkanes at temperatures below the isotropic liquid transition point, whereas little curving is noted for aromatic compounds under corresponding conditions. Specific interactions between the paraffin chains and the oriented ester tails can account for this. When the orientation of the cholesteryl stack is destroyed at the isotropic transition, a simple boiling point relationship is observed.

Further evidence for this specific interaction between alkanes and oriented ester chains is shown by the generally higher retention times of alkanes on the nonanoate column compared to the acetate and valerate ester columns below the isotropic liquid transition temperature and the nearly equal retention times above the isotropic liquids. For example, at 120°C all of the esters are in the isotropic liquid phase, and the elution times for all alkanes and aromatic compounds are the same, irrespective of the ester on the column. Long-chain cholesteryl esters in the mesophase range interact specifically with paraffinic hydrocarbons.

An interesting study in connection with the effects of alternating- and direct-current electric fields on the gas chromatographic behaviour of a liquid crystal stationary phase was made by Taylor et al.[673] The liquid crystals studied ("product A", a mixture of cinnamate esters and cholesteryl myristate) have a well-defined nematic mesophase over a broad temperature range, including room temperature (20–81.2°C). The use of an electrical potential resulted in improved resolution of the chromatographic peaks because retention times increased and peak widths decreased, as shown in Fig. 23.

This effect was explained in the following way. In the absence of an electric field the normal condition in the nematic meso-phase is a swarming of molecules into regions of average longitudinal order, i.e. molecules oriented with their long axes parallel, with no particular order between ends. These swarms orientate on a surface so that the molecules are predominantly parallel to the surface. However, in an electric field, the swarms reorientate with the long axes parallel to the electric field gradient.

In a gas chromatographic column this would correspond to a change in orientation from wall-parallel to wall-vertical upon application of a field. In the wall-parallel configuration, the aromatic rings are exposed to the solute vapours; in the wall-vertical configuration the aliphatic ends of the rod-like molecules are exposed. The effect of applying the field is to shift the nature of the stationary phase from a polar, aromatic one to an aliphatic, less polar one. As these aromatic molecules display an induced dipole in an electric field, an interaction between field frequency and the molecules is to be expected. The direct-current or low-frequency alternating-current field should give the most stable orientation.

In addition to orientation, another field effect reported for nematic liquid crystals is streaming. Swarms of molecules of the same orientation undergo a turbulent disturbance in the field and are transported in the field gradient. This effect would make more of the viscous stationary phase accessible to the moving solute and, thereby, increase the apparent capacity of the stationary phase. Streaming would also tend to disorder the swarm alignment and thus decrease the selective effect of the field. Such behaviour suggests that there may be an optimal potential gradient at which swarms are aligned but not undergoing significant streaming.

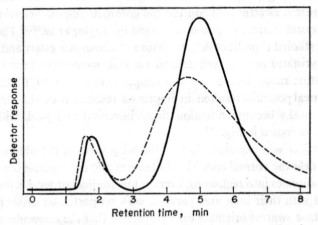

FIG. 23. Chromatogram at 54°C of a diethyl ether–benzene mixture using a coiled capillary column coated with 50% solution of Vari-Light A: (------) potential = 0 V; (———) potential = 500 V.[673]

These observations agree substantially with the effects observed in the gas chromatographic column found by Taylor *et al.*[673] The percentage increases in capacity ratios appeared to depend on a combination of molecular shape and dipole moment of the solute. Smaller or more nearly linear molecules may be retained longer because they fit more easily between the molecules of the nematic phase. In addition, those molecules having significant dipole moments may interact more strongly with the ordered dipoles of the liquid crystals. The dipole moments (in Debye units) of the solutes in the gaseous state are: benezene, 0; heptane, 0; chloroform, 1.02; diethyl ether, 1.15; n-propanol, 1.69; acetone, 2.89.

Diethyl ether and acetone, which are small and have large dipole moments, showed the greatest increases in capacity ratios, while heptane, which is somewhat larger, and chloroform, which has only a small dipole moment, showed smaller changes. Benzene, which is moderately large and has no dipole moment, showed the smallest change in capacity ratios. n-Propanol, which is delayed, to a large extent, by hydrogen bonding, showed little or no effect resulting from changes in polarity of the stationary phase. It is believed that an area that deserves exploration is the use of programmed potential changes. By applying a ramp potential it might be possible to achieve effects analogous to those obtained using programmed temperature changes.

The fact that a saturated solution of tri-*o*-thymotide is such that narrow, straight-chain molecules are able to enter holes or channels in its structure, whereas broad, branched-chain molecules are excluded from such holes or channels and are consequently less retarded, was used by Maczek and Phillips[674] for selective separations based on molecular shape:

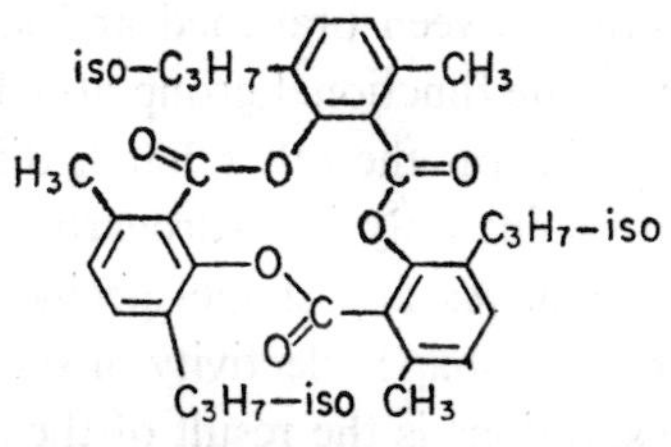

tri-*o*-thymotide 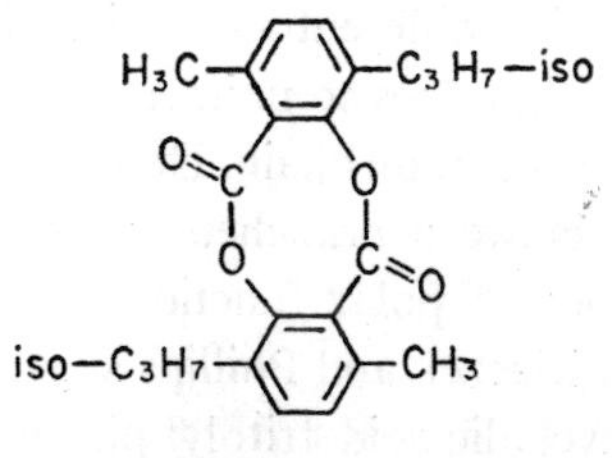   di-*o*-thymotide

The behaviour of tri-o-thymotide dissolved in tritolyl phosphate as a stationary phase was compared with di-o-thymotide, which is chemically similar but which does not present such properties.

It has been found that the branched hydrocarbons, tertiary and secondary alcohols, are similarly retained on both columns, whereas the linear saturated and unsaturated hydrocarbons and primary alcohols are retained 1.3 times more strongly on the column with tri-o-thymotide; the halogenated and aromatic hydrocarbons are 1.4 times more strongly retained.

A similar study was carried out concerning the behaviour as stationary phase of desoxycholic acid dissolved in tritolyl phosphate in the ratio 1 : 1.[675] Desoxycholic acid is known to form crystalline inclusion compounds with a variety of molecules of restricted width but not with broad molecules (branched-chain hydrocarbons).

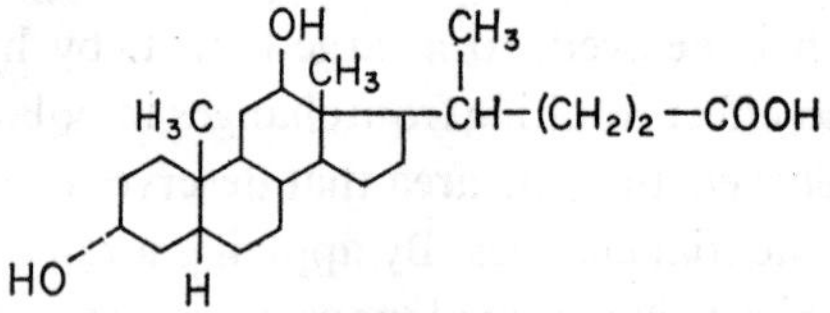

desoxycholic acid

From the gas chromatographic study it was concluded that a saturated solution of desoxycholic acid in tritolyl phosphate strongly retards n-alkanes and to a lesser extent n-alkenes; this retention is accompanied by a loss of column efficiency and asymmetrical peaks.

There is a quite marked difference between desoxycholic acid dissolved in tritolyl phosphate and comparable tri-o-thymotide solutions. Tri-o-thymotide differentiates quite distinctly between branched and straight chains, but does so uniformly, whatever the functional group attached to the hydrocarbon chain. Desoxycholic acid, on the other hand, differentiates between branched and straight chains only in hydrocarbons; the presence of polar functional groups removes any shape-specific selectivity. Maczek and Phillips suggest that the shape selectivity of saturated desoxycholic acid–tritolyl phosphate solutions is the result of the ability of straigth-chain hydrocarbons to form inclusions in the open desoxycholic acid crystal structure, inclusions for which branched chains are too broad.

A very selective stationary phase for the separation of o-, m-, and p-xylene is Bentone 34. It is obtained by treating bentonite (a clay, Al_2O_3 $(SiO_2)_4H_2O$) with dimethyldioctadecylammonium salts. This modified clay is suspended in benzene to give a thixotropic gel and is then mixed with a normal support material. This material was first used for gas chromatography by White and Cowan.[193] The separation factor for m- and p-xylenes on dimethyldioctadecyl ammonium bentonite at 70°C is such that a 99.9% separation could be achieved using only 800 theoretical plates. The organo-clay compound is so selective in its retention of the m-isomer that the elution order for the xylenes is p-, o-, m-. The special role of Bentone 34 can be attributed to its layer structure; the ease of adsorption of a foreign material between the layers of the expanded clay presumably depends on the geometrical shape of the adsorbed molecules, and this enables the clay to discriminate between benzene positional isomers.[350, 670]

Mortimer and Gent[398] found that the peak symmetry is improved by using an organo-clay modified with a liquid solvent of the type used in conventional gas chromatography. Grant[676] showed that a high degree of selectivity is obtained for m- and p-aromatic isomers, anthracene and phenanthrene, quinoline and isoquinoline, β- and γ-picolines, and cis- and trans-p-methyl cyclohexanols on a 150 cm$\times$3 mm i.d. chromatographic column packed with 100-mesh glass beads coated with 0.8% by weight of Bentone 34. The general utility of Bentone 34 columns for the separation of aromatic hydrocarbons, especially of polyphenyls, has been demonstrated by Versino et al.[677] The most used modifying solvent for Bentone 34 is silicone oil.[678, 679] Figure 24 shows a chromatogram, obtained using such a stationary phase.[678]

After this short review of some of the most commonly encountered stationary liquid phases, it must be underlined that the dividing lines between different classes of stationary liquids cannot be absolute. Thus the strong retention of aromatic amines, which are electron donors, by 2,4,7-trinitrofluorenone, an electron acceptor,[215] or the selective retention of aromatic compounds by 2,4,6-trinitrophenetole,[680] could be

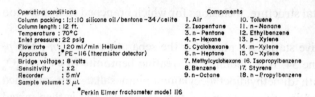

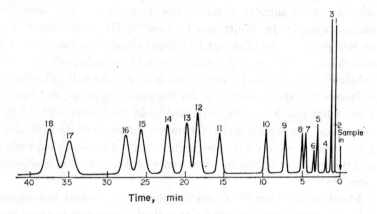

FIG. 24. Separation of hydrocarbons, including xylenes.[678]

considered as resulting from specific interactions. At the same time, these two stationary phases contain a great number of polar groups that favour these interactions. There are many examples of this type, so that in some instances it is difficult to include a compound in one or another class of stationary phase; it must be concluded that, at present, any classification of stationary liquid phases is inevitably imprecise.

The difficulties encountered in the classification of the stationary phase also arise from the extremely great number of stationary phases for proposed use in gas chromatography. Mann and Preston[681] catalogue over 700 liquid phases in the *Gas Chromatographic Data Compilation DS 25 A*, an ASTM publication.

In this respect it should be noted that the Rohrschneider or McReynolds constants for a number of stationary phases indicate equivalent retention behaviour. On this basis, Preston[682] has spoken of "stationary phase pollution" and suggested a drastic reduction in the number of stationary phases required in gas chromatography. The idea of using a limited number of "preferred" or "standard" liquid phases has been promoted over the last 3 years in several editorials and letters in the *Journal of Chromatographic Science*.[683-690]

The various promoters of standard liquid phases seem to be proposing the following:

(i) A selection of only 6–25 standard liquid phases should be used. These should be manufactured specifically for gas chromatographic use.

(ii) The standard liquid phases should be defined by Rohrschneider or McReynolds constants.

(iii) Retention data should be reported as retention indices.

(iv) Only retention data from these phases should be published or, at least, compiled.

Some problems concerning the standardization of the chromatographic process are discussed by Keller[690] and Henly.[689] In order to ensure that the phase alone is being defined, i.e. that support effects are eliminated, the constants mentioned in (ii) need to be determined at high phase concentration, e.g. 20 wt.% on high quality silane-treated supports. But many analyses are run with liquid phase loadings of 3–10 wt.%, and others cannot be done at loadings greater than 3 wt.%. It has been proposed to relate these constants to the specific packing or column in use,[681] but Henly[689] found them most useful only for defining the liquid phases.

Difficulties also arise from accuracy of column temperatures, effect of tailing reducers, and column ageing effects. On the other hand, the accuracy of the identity of two stationary phases with the same Rohrschneider constant may be questioned, e.g. for the separation of a particular mixture. It also must be established what difference in the constants is necessary in order to state that two liquid phases are significantly different. Keller[690] has observed that neither Rohrschneider nor McReynolds constants offer much help in separating positional isomers of nearly the same boiling points or with polyfunctional group solutes.

Gas chromatography is basically and primarily a separation process, and the number of possible combinations of solutes to be separated is infinite. We consider that standard gas chromatographic procedures and materials can be established for specific analyses, but to what extent the whole gas chromatographic process can be standardized remains to be seen.

There is little doubt that most gas chromatographic analyses could be done with significantly fewer liquid phases than are presently in use; their number could and should be decreased and limited. The problem is to decide which and how many standard liquid phases will cover most gas chromatographic analyses.

4.2 SOLID STATIONARY PHASES

Chromatography through gas–solid elution was the first gas chromatographic technique applied to the separation of a mixture of components. Chromatography using gas–liquid partition may practically be considered as being initiated by the James and Martin paper[691] in 1952, but at that time gas–solid chromatography was already well established.[692–695] Gas-solid chromatography was mainly used in the separation of mixtures of the fixed gases and the lower hydrocarbons.

Compared with the rapid development of gas–liquid chromatography, gas–solid chromatography was subsequently largely neglected. There were a number of reasons for its relative unpopularity. Firstly, adsorption isotherms in gas–solid systems are frequently non-linear, even at low column loadings, and are occasionally irreversible. This leads to several detrimental effects: solute retention volumes which vary with sample size, band tailing, and incomplete recovery of sample from the column. Secondly, retention volumes are generally excessively large, particularly for large polar molecules. Thirdly, adsorbents are in general difficult to standardize and reproduce compared with liquid phases. Fourthly, active adsorbents at elevated temperatures frequently induce catalytic alteration of the sample. Finally, the number of different, commercially available adsorbents which appear useful is relatively small compared with the large number of different liquids available for gas–liquid chromatography.

The chemical composition of the surface of the usual commercial adsorbents is heterogeneous and difficult to control. In addition, the geometry of conventional adsorbents varies widely. The adsorbents contain pores of various sizes, including some of molecular dimensions, in which adsorption is usually augmented. Such heterogeneity greatly lowers the effectiveness of the column, making it impossible to realize in full the selectivity of an adsorbent of any given chemical composition.

Giddings[696] has pointed out that, despite the above limitations, gas–solid chromatography has greater future analytical potential than gas–liquid chromatography. The column efficiencies theoretically attainable in gas–solid chromatography appear to be several orders of magnitude higher than are possible in gas–liquid chromatography. Giddings has also called attention to some examples of unique selectivity provided by adsorption separation. The replacement of a liquid stationary phase by a solid adsorbent should allow the mass transfer term to be reduced in magnitude, and so allow a gas–solid column to operate more efficiently than an otherwise similar gas–liquid column.

Furthermore, many of the foregoing limitations of the gas–solid technique are today more apparent than real. The highly sensitive detectors that have been developed for gas–liquid chromatography permit gas–solid separations of extremely small samples, which in turn means linear adsorption isotherms and constant solute retention volumes. In addition, there are a number of ways in which homogeneous surfaces may be prepared. Thus, symmetrical elution peaks have been obtained with graphitized carbon black.[66, 697] Elimination of the geometric heterogeneity of adsorbents increases not only the effectiveness of adsorption columns, but their selectivity as well. The use as adsorbents of inorganic salts, metal complexes, and porous polymer beads opens extensive possibilities for selecting and controlling the nature of the adsorbent surfaces and, therefore, for controlling the selectivity of gas-adsorption columns.

Thus, gas–solid chromatography offers in some applications unchallenged advantages over the gas–liquid technique. The advantages for separating gases and vapours of low-boiling substances are universally known, and further comment is unnecessary. For the analysis of compounds having higher boiling points, nowadays there are some stationary liquid phases that can be used above 300°C when working with ionization detectors. The adsorbents are much more thermally stable and are often insensitive to attack by oxygen. In gas–liquid columns, bleeding from the column liquid occurs continuously during operation, and thus the columns have limited lives. The small amounts of oxygen which almost inevitably occur in the carrier gas slowly oxidize the column liquid, changing its characteristics.

Gas–solid chromatography is preferable to gas–liquid for the temperature programmed operation, and also for determining micro-impurities in quantities near the limit of sensitivity of ionization detectors. Gas–liquid columns cannot be used to determine 0.1–0.01 ppm of impurities; the gas-adsorption method must be employed.[698]

Gas–liquid columns are commonly suited only to elution chromatography: gas–solid columns may also be used for displacement. Thus, although the displacement technique is generally inferior to the elution technique for direct analysis, it does have very distinct advantages in preparative work and for trace analysis. The selectivity of gas–solid columns is in general much superior to that available with gas–liquid columns. Also, owing to rapid mass exchange, gas–solid chromatography offers great possibilities for fast analyses. Despite the great advantages of gas–solid over gas–liquid chromatography in many instances, these methods should supplement one another rather than compete.

The most frequently used solid stationary phases and some of their applications are described below.

4.2.1 Non-polar solid stationary phases

Carbonaceous solid phases

Charcoal has been used extensively as an adsorbent in gas chromatography since its inception. It is simplest to use any of a variety of types of "activated charcoal" which are usually prepared by suitable steam treatment of vegetable charcoals at moderate temperature. The usual way of obtaining this active state is to heat the appropriate size fraction in a vacuum oven at about 150°C for a few hours. Such materials have large surface areas (800–1000 m^2/g), and are extremely heterogeneous.

There are, of course, many varieties of activated charcoal, characterized by differences in activity stemming from differences in porosity and available area, the extent of oxygen and water adsorption, and possibly of surface hydroxylation. Habgood and Hanlan[183] have made an extensive study of the effect of any of these parameters on the level of activation and, in turn, on the gas chromatographic performance. They found that the highest column efficiency was obtained with the most active charcoal, a result which they attributed to reduced resistance to mass transfer to and from the solid surface. This result conflicts with the experience of most workers who, in fact, attempt to reduce the activity either by chemical or physical pre-treatment or by the use of combined gas–solid–liquid columns.

Because of their large surface area, retentions tend to be very great on such materials, so that such columns are normally used at temperatures very much higher than the boiling points of the substances to be separated. Thus they tend to be used for the separation of permanent gases at or near room temperature, and low-boiling hydrocarbons.[699–705] The order of elution of gases and C_1 and C_2 hydrocarbons on a charcoal column at room temperature is: H_2, O_2, N_2, CO, CH_4, CO_2, C_2H_2, C_2H_4, and C_2H_6.[706, 707]

If charcoal or other carbons are treated at about 3000°C, structural changes take place converting the carbon into polyhedra with homogeneous basal graphite faces. Carbon of this form, called "graphitised carbon black", has a relatively small surface area (6–30 m^2/g), is almost homogeneous, and is much more suitable for use in gas chromatography. Graphitized carbon black has been studied extensively by Kiselev et al.[232, 249, 697]

The principal characteristics of gas–solid separation on graphitized thermal carbon black and the energies of the corresponding non-specific molecular interactions are the values of the retention volumes per unit surface area of the adsorbent at several temperatures and the differential heats of adsorption at low coverages. The absolute values of the retention volumes increase with increasing total polarizability of the molecules which determines the energy of dispersion interaction.

As interaction on graphitized carbon black occurs solely through dispersion forces, retentions are not a function of solute dipole moment (the dipole moment of the molecule, which causes some polarization of the carbon atoms in the graphite during adsorption, does not make any significant contribution to the energy of adsorption and therefore does not affect the retention volumes). Also the role of the molecular weight is less important than the geometric differences of the molecules, which affect their possible orientation and energy of adsorption on the basal graphite plane. The absolute retention volume values, or the corresponding relative values with respect to n-alkanes, can be used for solute identification in gas–solid chromatography on graphitized carbon black. The plots of the logarithm of the absolute retention volume values and of the differential heats of adsorption at low surface coverages on graphitized carbon black against the number of carbon atoms in the molecule are straight lines for homologous series. It is possible, therefore, to identify unknown solutes from these dependencies and to predict retention volumes within a homologous series.

The adsorption heats of individual molecules are the sum of terms for their structural components, for n-alkanes and their unbranched derivatives. The increments Q are: $Q_{CH_3} = 2.1$; $Q_{CH_2} = 1.6$; $Q_{OH} = 2.1$; $Q_{O(ethers)} = 1.3$; $Q_{COOH} = 5.2$ kcal/mole.[232] The additivity of the dispersion interaction and the fact that the molecules of organic compounds are made up of a few atoms or groups, allows one to express the potential energy of interaction with the basal face of graphite by means of a few potential functions of atom–atom interaction or group–atom interaction. These potential functions can be calculated satisfactorily by means of the theory of dispersion interaction.[708-710] The energy of the dispersion interaction between graphitized carbon black and different solutes depends greatly on the distance between the adsorbent surface and the force centres of the linkages of the adsorbed molecule.

In this way, graphitized carbon black is a unique non-polar adsorbent, and the energy of interaction of all the force centres of the molecule with this adsorbent can be considered to be additive. The adsorption on such an adsorbent is especially sensitive to the distance of the atoms or molecular bonds from the plane surface of the adsorbent, i.e. to the geometry of the molecule.

This property can be used successfully for the determination of the geometrical structure of the molecules by their emergence sequence and by the values of the retention volumes, which is of interest in structural chemistry problems such as the identification and investigation of the structure of *cis*- and *trans*-isomers. For example, the connection between the structure of molecules and their adsorption properties on graphitized carbon black can be demonstrated by gas chromatography. It is thus known that in a series of unsaturated hydrocarbons the *cis*-configuration, having higher boiling points, are retained less strongly than the *trans*-isomers with the same carbon number. Even slight differences in boiling points yield rather different retention times. This is connected with the diverse orientations of the molecules of these isomers. The *trans*-configurations of the olefins have an energetically more advantageous position on the surface of graphitized carbon black as compared with the *cis*-isomers.

Thus gas chromatography has been used for the investigation of the structure of the *cis*- and *trans*-configurations of 3-methylpentene-2 and 3,4-dimethylpentene-2.[254] In the literature there is contradictory information about the configuration of these compounds, and it is difficult to determine which has a *cis*-configuration and which a *trans*. Investigation of the retention volumes and heats of adsorption of these compounds showed that 3-methylpentene-2, with a boiling point of 70.4°C, is the first to elute from the column at 65°C (in 9 min 40 s) and 3-methylpentene-2, with a boiling point of 67.6°C is the second (in 11 min 34 s). Thus the first compound has a *cis*-configuration and the second a *trans*-configuration. The heat of adsorption calculated from the chromatographic values is 9.6 kcal/mole for the *cis*-isomer and 9.9 kcal/mole for the *trans*-isomer.

For 3,4-dimethylpentene-2, the mixture of these compounds with boiling point 87°C gives two peaks at 65°C. The first (emerging after 15 min 25 s) is the *cis*-configuration, the second (eluting in 16 min 36 s) is the *trans*-configuration. The heat of adsorption of the *cis*-isomer is 10.0 kcal/mole, and that of the *trans*-isomer is 10.6 kcal/mole.

In the same study, a number of hydrocarbons with six-membered rings (cyclohexane, cyclohexene, cyclohexadiene-1,3, and benzene) were investigated. The molecules differ geometrically and in the number of hydrogen atoms. Retention volumes obtained experimentally and the heats of adsorption of these substances show that an increase in these parameters is observed in the series from cyclohexane to benzene. Nevertheless, benzene, although it has both the lowest boiling point and the lowest molecular weight, is retained more strongly on the surface of carbon black than the other substances. This confirms that the chromatographic behaviour of the molecules on the surface of the carbon black depends mainly on their geometrical structure and on their orientation at the surface of the basal graphite face. The cyclohexane molecule, having a preferential chair configuration, only makes contact with the carbon black surface with three of its carbon atoms, the other three carbon atoms being some distance from the surface. As cyclohexene and cyclohexadiene-1,3 have flatter molecules, the distance between the carbon atoms and the surface of carbon black becomes less, so that the energy of interaction with the surface of the adsorbent, which is characterized by the heat of adsorption, increases as the retention volume increases. The differences in retention volumes of cyclohexene and cyclohexadiene-1,3 are not sufficient for their separation (these compounds have similar structural configurations).

Some of the analytical applications of graphitized carbon black have already been mentioned in Chapter 3. These were the separation of isomeric butylbenzenes of the *cis-* and *trans-*isomers of 1,2-, 1,3-, and 1,4-methylcyclohexanols and of terpenes. The separation of saturated aromatic hydrocarbons and their perfluoro analogues is another example.[249] With benzene, the substitution of hydrogen by fluorine takes place in the plane of the aromatic ring and therefore does not lead to an increase in the distance between the carbon atoms of the molecule and the basal plane of the graphite. The polarizability of the fluorinated molecule is somewhat higher, and thus the energy of dispersion interaction with graphitized carbon black increases. Thus perfluorobenzene elutes after benzene.

When methyl or methylene groups are fluorinated, their radius increases and the carbon atoms of fluorinated alkanes are at a greater distance from the basal plane of the graphite. Consequently, the energy of dispersion interaction of fluorinated alkanes and cycloalkanes with the basal plane of the graphite is smaller than that of the unfluorinated analogues. Thus n-perfluoroheptane elutes before n-heptane.

As a chemically inert non-polar adsorbent, graphitized carbon black effectively separates bielement-organic compounds of silicon, germanium, and tin.[711] The efficiency of this separation exceeds that obtained in gas chromatographic columns filled with various non-polar and polar liquid phases.

Gas–solid chromatography using graphitized carbon black was tried by Zane[712] for separating several three- and four-ring polynuclear aromatic hydrocarbons. This adsorbent was found to have a unique ability of being able to separate a mixture of anthracene and phenanthrene on a short column (stainless steel tubing 70 cm×2.3 mm i.d. with 100–120 mesh graphitized carbon black at 415°C). At a column temperature of 470°C, injected samples of chrysene and benz[a]anthracene produced no peaks on the chromatogram after 2 h, even when sample sizes of 20 μg were used. It appears that these polynuclear aromatic hydrocarbons were irreversibly adsorbed on the graphitized carbon black.

The difficult elution of polynuclear aromatic hydrocarbons from graphitized carbon black in conventionally packed columns has been facilitated by depositing this adsorbent on an inert porous support.[713] Thus graphitized carbon black on Chromosorb W (80–100 mesh, 15 : 85) makes possible the separation of benzo[e]pyrene from benzo[a]pyrene and dibenz[a,c]anthracene from dibenz[a,h]anthracene on a column 95 cm×2 mm i.d. at 460°C. The separation factors for benzo[a]pyrene/benzo[e]pyrene and dibenz[a,h]anthracene/dibenz[a,c]anthracene on graphitized carbon black are markedly higher than those on organic liquid phases[475, 714, 715] or on alkali metal and alkaline earth metal chlorides.[716, 717] Also, in order to separate these pairs by gas–liquid chromatography, it usually is necessary to use capillary columns,[718] but for graphitized carbon black deposited on Chromosorb W, a short, packed column is satisfactory (Fig. 25). The separation occurs 125°C lower than

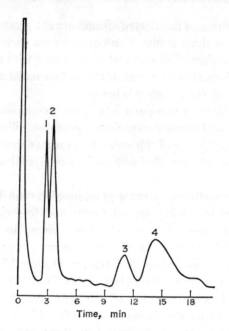

Fig. 25. Chromatogram of pairs of isomeric PAH on a column of graphitized carbon black on Chromosorb W (80–100 mesh) (15 : 85) at 460°C; carrier gas: argon, 20 ml/min. (1) Benzo[e]pyrene, (2) benzo[a]pyrene, (3) dibenz[a,c]anthracene, (4) dibenz[a,h]anthracene.[713]

on a capillary column coated with a thin layer of graphitized carbon black;[719] the peaks remain symmetrical.

Pope[720] has prepared a chromatographic column with polythene moulding powder (100–120 mesh) coated with graphitized carbon black. The performance of these columns compares favourably with that of a n-hexadecane coated capillary for the separation of hexane isomers and allows the resolution of a sample of $C_7F_{15}H$, from which only one peak was obtained with the latter column, into four components.

Goretti et al.[721] prepared a capillary column with a thick layer of graphitized carbon black by drawing out glass tubing loosely packed with adsorbent; the thickness of the layer may vary from 50 to 100 μm. These columns have a number of interesting features, combining the advantages offered by a non-polar adsorbent with the length of an open tubular system, high permeability, small pressure drop, low value for effective plate height, and high selectivity for specific systems. They exhibit outstanding properties for the separation of geometrical and structural isomers such as o-, m-, and p-cresols, the xylenes, and polar compounds (alcohols, amines). A specific application of the thick layer graphitized carbon black open tubular columns has been the separation of isotopic molecules; a number of polar and non-polar compounds have been investigated and separated from their deuterated homologues.

The use of graphitized carbon in gas–solid chromatography for the separation of some hydrocarbons from their deuterated homologues was described by Yashin.[722] In all instances an inverse isotopic effect was found. The deutero species was eluted first, as it is usually observed to do when working with a non-polar stationary phase.

Tailed peaks are usually obtained with these columns; to obtain symmetrical peaks it has been found necessary to make the elution gas (nitrogen) bubble through a water trap; water vapour acts as an effective reducer of tailed peaks. As the support is a non-porous material it seems that the function of the water vapour is to eliminate the active sites on the glass capillary.

An open, tubular, thick layer, graphitized carbon black column was effective for the separation of the following isotopic species:[723] benzene-deuterobenzene (d_6-benzene); toluene-deuterotoluene (d_8-toluene); acetone-deuteroacetone (d_6-acetone); pyridine-deuteropyridine (d_5-pyridine). It seems that the different sizes of hydrogen and deuterium and the greater mobility of the deuterium compounds should be the determining factors which affect the different adsorption of the isotopic molecules. Figure 26 shows the separation of benzene from the deuterobenzene and of toluene from the deuterotoluene.

Di Corcia and Bruner[724] mention, however, that the graphitized carbon blacks have some disadvantages, common to other adsorbent, which can be summarized as follows. They have very high retention volumes and, because of the presence of some active sites on the adsorptive surface, peak tailing occurs even for scarcely polar compounds and, for hydrogen-bonding compounds, almost irreversible adsorption and "ghosting"

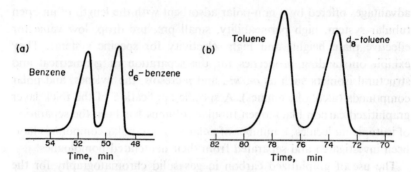

FIG. 26. Chromatographic separation on a 9.6 m×0.15 mm i.d. open tubular thick layer graphitized carbon black glass column of (a) C_6D_6—C_6H_6. Column temperature: 69.5°C; nitrogen pressure: 1.75 atm; (b) C_7D_8—C_7H_8. Column temperature: 98.4°C; nitrogen pressure: 1.9 atm.[723]

phenomena are observed. The authors have shown that by heating the graphitized carbon blacks in a hydrogen stream at 1000°C, very polar compounds could be eluted and peak tailing and "ghosting" phenomena were avoided. This procedure has the purpose of eliminating the active centres constituted by surface oxygen complexes which are always present on an adsorbent surface. In this way, carbon black becomes a very inert adsorbent, allowing the elution of small amounts of very polar compounds such as aliphatic amines and free carboxylic acids. These materials were also used as supports for gas–liquid chromatography of highly reactive gases such as hydrogen halides, chlorine, and boron trifluoride.[725]

Graphitized carbon black is a fine powder. The literature frequently gives commercial names, such as Sterling FT, Sterling MT, Spheron 6 (Cabot Co.), and TeG-10.

The granular particles obtained after shaking and passing the material through a sieve are irregular and have a low mechanical strength. This has a detrimental effect on the hydrodynamic conditions inside the column and decreases the column efficiency. Kiselev[249] investigated the optimum conditions for reinforcing the carbon black particles by treatment with an adhesive polymer (Apiezon L, 0.01% by weight). This treatment gave mechanical strong particles without affecting the properties of the carbon black as a non-polar adsorbent having a homogeneous plane surface. This new material has been called Carbochrom. It has been observed that in the separation of C_5-C_{10} n-alcohols, the efficiency of the Carbochrom column is twice that of a column of the same length filled with untreated graphitized carbon black.

Boron nitride

The lamellar structure of boron nitride crystals is analogous to that of graphite except that at the interaction of the hexagonal basal planes there are alternate boron and nitrogen atoms instead of carbon atoms. When the composition of the crystal corresponds exactly to BN and the lattice is practically free of defects, the basal plane exhibits the properties of a non-polar adsorbent. However, such laminated lattices are rather difficult to obtain in a sufficiently homogeneous form.[726]

Crowell and Chang have calculated the potential energy of adsorption and measured the heat of adsorption of noble gases.[727] Curthoys and Elkington have done the same for hydrocarbons.[252] The calculated values agreed with those found by gas chromatography.

4.2.2 Ionic solid stationary phases

Non-porous ionic crystals of lamellar or simple cubic structure, e.g. crystals of the alkali metal halides are of great interest for gas chromatography. The use of alkali metal halides as adsorbents dates from 1964, when these salts, deposited 25% by weight on Chromosorb P, were proposed as stationary phases suitable for the separation of *o*-, *m*-, and *p*-terphenyls.[728] Solomon[729] also studied *o*-, *m*-, and *p*-terphenyls, quaterphenyl and hexaphenyl isomers, and a variety of other compounds displaying various polarities. The most promising were 20% LiCl, CsCl or $CaCl_2$ coated non-acid-washed Chromosorb P (60–80 mesh). A 10 ft× $\frac{1}{4}$ in. (3 m×6.3 mm) 20% LiCl Chromosorb P (60–80 mesh) column was used for analyses of quaterphenyl and hexaphenyl mixtures. The peaks obtained were very sharp and tailed only slightly. Limited work has shown that compounds other than polyphenyls may be analysed on lithium chloride columns. Those which eluted satisfactoryily are fused-ring aromatic hydrocarbons, arylalkyl hydrocarbons, haloaromatics, aromatic ketones, quinones, aromatic amines, and nitro-aromatics.

Simple aliphatic alcohols, ketones, aldehydes, and acids gave broad peaks and eluted above 200°C. Salt columns, in general, gave better separation of aromatic compounds. Shifts in the elution order of poly-phenyl isomers on various salt columns were associated with certain types of linkages in the molecule. For example, for m-quaterphenyl and 1,3,5-tri-phenylbenzene, the branched compound eluted before the linear one from a caesium chloride column, the two compounds almost coincided on a lithium chloride column, and the branched compound eluted after the linear one on a calcium chloride column. These elution orders were also found to hold for higher molecular weight polyphenyls if these isomeric structures were present.

Solomon concluded that separation by solids is effected by weak bonding between the inorganic salt and the organic molecules. Some generalizations from his studies are:

(i) The melting point of the solid has little effect upon column per-formance.
(ii) If two salts are mixed, retention properties are the average of the two, adjusted for concentration.
(iii) Anion–cation effects are significant. For example, sulphates cause elution temperatures to increase as compared to chlorides.

Figure 27 illustrates the chromatogram obtained with a mixture of aromatic hydrocarbons having from two to seven rings and boiling from 182°C (indene) to about 600°C (coronene). The substrate was 20% LiCl on Chromosorb P $\left(8 \text{ ft} \times \frac{1}{8} \text{ in.}\right.$ (2.4 m×3 mm) stainless steel column); the temperature was programmed from 80° to 320°C at 2°C min, and flame ionization detection was employed.[716] No baseline shift is observed throughout the programming.

Figure 28 shows the separation of some polyphenyls on a column (100 cm×5 mm i.d.), packed with Chromosorb P (100–120 mesh) coated with 10% caesium chloride.[730] The separation was performed using a

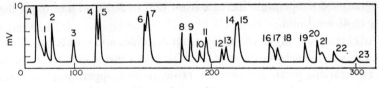

Temperature, °C

FIG. 27. Separation of polynuclear aromatic hydrocarbons on lithium chloride columns.[716] The entire chromatogram, from a 8 ft × 1/8 in. column was programmed from 80°C at 2°C/min. (1) Napththalene, (2) indene, (3) 2-methylnaphthalene, (4) 2,6-dimethylnaphthalene, (5) acenaphthene, (6) 9-methylfluorene, (7) anthracene and phenanthrene, (8) fluoranthene, (9) pyrene, (10) 9,10-dimethylanthracene, (11) 2-methylpyrene, (12) benzo[b]fluorene, (13) 7-H-benzo[c]fluorene, (14) benzo[b] phenanthrene and triphenylene, (15) benz[a]anthracene, benz[b]anthracene, and chrysene, (16) benzo[e]pyrene, (17) benzo[a]pyrene, (18) benzo [b]-fluoranthene, (19) 3-methylcholanthrene, (20) dibenz[a,c]anthracene and dibenz[a,j]anthracene, (21) dibenz[a,h]anthracene, (22) coronene, (23) dibenzo[a,i]pyrene.

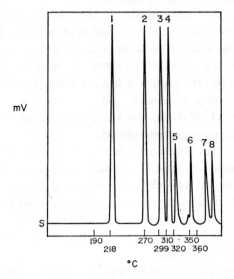

°C

FIG. 28. Chromatogram of the separation of polyphenyls from a linear temperature programmed run (125–360°C) at 12.5°C/min.[730] (1) Biphenyl, (2) o-terphenyl, (3) m-terphenyl, (4) p-terphenyl, (5) triphenylene, (6) o,m-quaterphenyl, (7) m,p-quaterphenyl, (8)p,p'-quaterphenyl.

programmed temperature increase from 125° to 360°C at a nitrogen flow rate of 48.8 ml/min.

An extensive study of the alkali metal chlorides and nitrates as column packing was made by Grob *et al.*[731] The alkali metal salts were ground by a mortar and pestle and sieved to 60–80 mesh. Copper columns, 320 cm ×6 mm i.d. were filled with these materials. The retention data for different non-polar, polar and aromatic compounds for the nitrate columns and heats of adsorption for the chloride columns were extensively discussed. For n-alkanes at 30°C, the retention volumes on different columns decrease in the order $CsNO_3 > RbNO_3 > LiNO_3 > KNO_3 = NaNO_3$.

Some generalities can be drawn from the data regarding the alcohols. The boiling point primarily determines the elution order. However, when chain branching occurs, the more highly branched species elutes before a straight chain alcohol with an identical boiling point. For example, n-propanol and s-butanol have approximately the same boiling point, yet n-propanol is adsorbed more strongly.

The xylenes are retained somewhat by the columns, but this interaction appears to be non-specific, similar to that of the alkanes. The adsorption of *m*- and *p*-chlorotoluene on these columns is also very non-specific.

The retention volume of nitrobenzene is equal to or smaller than that of the cresols, even though it has a boiling point 8°C higher; the nitro-group is electron withdrawing and the cresols have two electron-releasing groups which result in a significantly richer electron ring, so that their retention volumes are greater.

On the other hand, *o*-toluidine boils 15°C higher than aniline. However, its retention volume is at least 50% greater. Perhaps the best explanation is that *o*-toluidine has two electron-releasing groups on the ring. For picolines, the adsorption increases from the 2-position to the 3-position and is most pronounced in the 4-position (the methyl group found in the three picolines is electron releasing and its effect is most pronounced when it is located in the 4-position).

The heat of adsorption for all alkanes is less than the heat of adsorption for the alcohols with the same number of carbon atoms. For alcohols, the straight-chain isomers have strongly negative heats of adsorption on all columns. The authors consider that for the alcohols, purely physical adsorption cannot constitute the only explanation for the adsorption heat data and retention volumes. It appears that some reaction occurs after adsorption and before desorption. One possible explanation is that some type of intermolecular hydrogen bonding occurs after adsorption which decreases the strength of the bond between the metal halide and the oxygen of the hydroxy group. Hence, less energy would be required in the desorptive process. Another explanation could be the formation of alcoholates the heat of formation of which is sensitive to the stereochemistry of the alcohol. For all the compounds studied, the heat of adsorption increases as the polarity or electron density increases.

From the data of Grob *et al.*, some generalizations can be made:

   (i) Alkali metal nitrate columns retard elution greater than alkali metal chloride columns.

  (ii) The retention volume generally follows the boiling point. If two solutes have the same boiling point, the more polar solute has the greater retention volume.

 (iii) Chain branching, causing decreased polarity, reduces retention volume.

 (iv) Electron-releasing groups enhance adsorption; electron withdrawing groups hinder it.

  (v) *Ortho* substituents cause stronger effects than *meta* substituents.

Guran and Rogers have made a gas-chromatographic study of the selectivity of some alkali metal halides toward certain organic isomers and homologues.[188] Each salt (NaCl, NaBr, NaI, KCl, KBr, and KI) was ground in a mortar and sieved to obtain an 80–120 mesh fraction which was packed in coiled, 6 mm i.d. glass columns about 100 cm long. The authors showed that, because values of distribution ratios were very sensitive to column conditioning, they did not represent a reliable basis for comparing the adsorption behaviour of different columns. However, when separation factors were calculated for isomeric pairs, the values for a single salt were almost independent of conditioning, and for this reason the use of these factors could constitute a base for comparing columns of different sodium and potassium halides. Generally, the values obtained under a given set of conditions were reproducible to within ± 0.02 and often within ± 0.01.

The ratios for isomeric ketones (3-pentanone/2-pentanone; 3-heptanone/2-heptanone) at 86.6°C were surprisingly similar on the chloride and bromide salts of each metal, but were significantly different for each cation. Thus the isomeric ketones were better separated on the potassium chloride and bromide columns than on the corresponding sodium halides. In contrast, the ratios taken from the two iodide salts appeared to be independent of the alkali metal ion. The data indicated that, if the anion was not disproportionately larger than the cation, the interaction of the carbonyl dipole in the ketones depended not only on the steric configuration of that dipole but also on the cation. For the iodides, however, the greater polarizability of the anion appeared to have been the govering factor.

Separation of isomeric esters was not influenced by a change in cation, and was poorer compared to the ketone isomers, even though the esters differed much more in their boiling points. This was probably a result of the more open molecular structure around the dipole in the esters and, consequently, less steric influence on the dipole interaction.

The change in free energy for the addition of one CH_2 group to a ketone or an ester was independent of the column material and the adsorbate molecule to which the group is added.

The retention of alcohols on the six columns depends to a significant degree on their abilities to form hydrogen bonds. For example, on the sodium halide columns, n-propanol was retained much longer than t-butanol and was eluted at about the same time as di-isoropyl carbinol. In contrast, on the potassium halide columns, n-propanol eluted with t-butanol, but was very nicely separated from di-isopropyl carbinol. The ratio changes for alcohols between the sodium and potassium halides may be attributed to a greater hydrogen-bonding contribution on the sodium halide columns. As expected, the hydrogen-bonding effect was strongest for the primary alcohol, n-propanol, and consequently it was retained longer on the sodium halides. It is interesting to note that the hydrogen-bonding abilities of di-isopropyl and triethyl carbinols can be differentiated from the ratios of the distribution ratio. The larger ratios on the potassium halides suggest that triethyl carbinol hydrogen bonded more strongly.

In order to obtain reproducible results with these columns, partly deactivated columns must be used. The separation factors, including those for the alcohols, were most nearly constant when the column was "dried" under mild conditions (125°C or less). Another reason for using partly deactivated columns is that tailing becomes more serious as the activity of the column is increased.

The possibility of adduct formation between alkali metal fluorides and oxyfluorides and non-metal oxides, suggested the use of these materials as adsorbents in gas chromatography.[732] For example, on a column, 12 ft × $\frac{1}{4}$ in i.d. (3.6 m × 6.3 mm), containing 10% CsF and 90% CaF$_2$, the following components eluted in the following order: $CF_2{=}CF_2$; $(CF_3)_2C{=}O$; CF_3CFO; $F_2C{=}O$, and were thus separated. With a programmed temperature increase from 30° to 250°C, this separation takes 15 min. The compounds were eluted in the reverse order of their boiling points. Another column containing caesium fluoride was also used in order to separate SF_4 from SOF_2, COF_2 from CO_2, and SOF_4 from SO_2F_2.

The use of columns containing alkali metal fluorides as adsorbents is nevertheless limited by the presence of water and hydrofluoric acid. Owing to the deliquescent nature of the alkali metal fluorides, the adsorption of water vapour from the atmosphere during preparation and filling of the chromatographic column must be avoided. The adsorption of moisture on the fluoride lowers the chromatographic column performance. Hydrofluoric acid interacts rapidly with alkali metal fluorides leading to very stable difluorides (MHF_2). The regeneration of columns poisoned in this way with hydrofluoric acid is not yet possible.

The chlorides, bromides, and iodides of strontium and barium have been investigated as column packings by Grob et al.[733] The alkaline earth halides, with the exception of strontium iodide, were dehydrated in a muffle furnace at 400°C using a nitrogen purge. Strontium iodide was dehydrated at 325°C because higher temperatures cause slight decomposition and discoloration. After cooling in a vacuum desiccator, the halides were transferred to a dry box where they were ground with a mortar and pestle, sieved to 60–80 mesh, and packed into 6 ft×0.23 in. i.d. (1.8 m×6 mm) glass columns using a vacuum and gentle tapping. The retention data were determined at 170°, 200°, and 250°C.

Various groups of organic compounds (alcohols, carboxylic acids, and isomeric aromatic compounds) were studied, their heats of adsorption calculated and models presented to explain their behaviour on these columns. Retention volumes generally followed the boiling points; for two solutes having the same boiling point, a greater retention volume was noted for the more polar solute. Chain branching reduced the retention volume, but the latter increased as the size of the anion in the packing increased.

The sample size required to produce overloading generally increased with increasing anion size in the packing.

The inversion of the heats of adsorption of methanol and ethanol on the barium halide columns seems to be the combined result of the increased electropositive character of the larger barium ion and the difference in polarity between methanol and ethanol. The absence of any anomalous results for formic and acetic acids, which might compare with the heats of adsorption for methanol and ethanol, indicates the possibility of gas phase dimer formation.

Carboxylic acids were eluted from strontium and barium iodide and barium bromide columns only under the most severe conditions. The resulting peaks were characterized by excessive tailing.

Aromatic solutes exhibited a variety of different interactions. For example, as electron-releasing groups were added to the ring, the heat of adsorption increased, whereas the presence of electron-withdrawing groups, which decrease the electron density in the ring, leads to a decreased heat of adsorption. Thus the halobenzenes are not as strongly adsorbed as the alkyl substituted aromatic compounds of comparable boiling point.

Grob and McGonigle[734] studied the behaviour as adsorbents of the anhydrous chlorides of vanadium(II), manganese(II), and cobalt(II). These compounds differ primarily in their number of available $3d$ electrons (V(II)-$3d^3$, Mn(II)-$3d^5$ and Co(II)-$3d^7$). It was hoped that they would be effective as packings for separating various unsaturated compounds. The columns were glass (6 ft$\times$0.23 in. i.d.) (1.8 m$\times$6 mm). Each was packed with 60–80 mesh salt using a vacuum and slight vibration. Useful chromatograms with reasonable retention times and good reproducibility were obtained for acetylenic and olefinic compounds when prepared in 20.0% (v/v) solutions using 1 μl of sample. Initial injections of unsaturated samples exhibited a sequential decrease in retention time per injection until a constant value was obtained. This was interpreted as arising from gradual saturation of irreversible sorption sites. Usually, three or four 1-μl injections of sample solution (20.0%) were required to "saturate" the column. In all instances, the heats of adsorption increased in the order: $3d^5$, $3d^7$, $3d^3$, i.e. $MnCl_2$, $CoCl_2$, VCl_2. This is predictable in that VCl_2 has the greatest number of available d-orbitals, and $3d^5$ ($MnCl_2$) is a particularly stable state. Temperatures as high as twice the boiling point were required to elute unsaturated samples from VCl_2.

It is interesting to note that the difference between the heats of adsorption of benzene and cyclohexane is much less on all columns than that between 1-pentyne and 1-pentene or 1-hexyne and 1-hexene, with the greatest difference noted for VCl_2. This is best explained by the fact that conjugation will tend to weaken the interaction of a π-bond with a metal ion field.

The authors considered that the transition metal salts studied offer a practical basis for quantitative separation of linear isomeric hydrocarbons with an increasing degrees of unsaturation and of cyclic isomeric hydrocarbons with an increasing degrees of unsaturation, compounds having isolated π-bonds being the more strongly adsorbed.

Altenau and Rogers[735, 736] consider that it is possible to create useful adsorbents for gas chromatography by careful elimination of volatile substances from a crystal so as to avoid the collapse of the crystal framework into a very fine powder. When powdering can be avoided, the resulting solid should have fairly uniform porosity in addition to a much higher specific surface area as compared to the crystals of the starting material.

The following adsorbents have been prepared by heating in an oven at the indicated temperature:

$Cu(Py)_2(NO_3)_2$ from $Cu(Py)_4(NO_3)_2$ at 85°C (Py = pyridine)

$Cu(NH_3)_2(NO_3)_2$ from $Cu(NH_3)_4(NO_3)_2$ at 150°C

$Cu(Py)SO_4$ from $Cu(Py)_4SO_4$ at 100°C

$CuSO_4 \cdot H_2O$ from $CuSO_4 \cdot 5H_2O$ at 125°C

$CdSO_4$ from $3CdSO_4 \cdot 8H_2O$ at 165°C

$MgCl_2 \cdot 2H_2O$ from $MgCl_2 \cdot 6H_2O$ at 100°C

Usually the required number of molecules had been volatilized from the starting material after about 4 h. Once the adsorbent had been produced it could be stored indefinitely in a dessicator over calcium sulphate or calcium chloride.

The determinations were made on copper columns (6 ft $\times \frac{1}{8}$ in. o.d., 1.8 m $\times$ 3 mm) filled with 50–60 mesh adsorbent. At 38°C, a column packed with $Cu(Py)_2(NO_3)_2$ successfully separated alcohols, esters, ethers, and ketones up to boiling points of approximately 160°C, alkylated and halogenated benzenes up to boiling points of around 185°C, and aliphatic hydrocarbons starting from n-butane up to those having boiling points of about 210°C. All these classes of organic compounds gave symmetrical peaks with virtually no tailing. The same column also separates isomeric ketones (2- and 3-pentanone and 2-, 3-, and 4-heptanone), n-amyl alcohol and 2-hexanol, and ethyl butyrate and n-butyl acetate.

The retention data for columns of $Cu(Py)_2(NO_3)_2$, $Cu(Py)SO_4$, and $Cu(Py)_4SO_4$ show that the number of groups co-ordinated around the copper(II) ion have a major effect on the retention characteristics of the adsorbent. The more groups co-ordinated around the coper(II) ion, the less the retention of sample molecules. Also, $CuSO_4 \cdot H_2O$ exhibited greater retention for sample molecules (aromatic hydrocarbons and oxygen-containing compounds) than $Cu(Py)SO_4$. This behaviour agrees with the postulate that as the number of groups co-ordinated by the coper(II) is decreased, retention is markedly increased (it is known that the water molecule in $CuSO_4 \cdot H_2O$ is hydrogen-bonded to the sulphate ion and not bound to the copper(II) ion). Comparison of the retention data on columns of $Cu(Py)_2(NO_3)_2$ and $Cu(NH_3)_2(NO_3)_2$ showed that the latter exhibited much stronger adsorption. Ethyl ether was the only oxygen-containing compound that eluted from a column of $Cu(NH_3)_2$ $(NO_3)_2$ at 38°C within 60 min.

Using anhydrous cadmium sulphate, aliphatic hydrocarbons and some aromatic compounds were eluted at 38°C without tailing. Oxygen-containing compounds were not eluted. Retention of aromatic hydrocarbons and oxygen-containing compounds is clearly much stronger on a column of $CuSO_4 \cdot H_2O$ than on a column of $CdSO_4$. Without the molecule of water, $CuSO_4$ would undoubtedly be an even stronger adsorbent. At 38°C, aromatic hydrocarbons had smaller retention times on $MgCl_2 \cdot 2H_2O$ than on $CuSO_4 \cdot H_2O$ or $CdSO_4$.

The adsorption is considered to be governed primarily by the interaction between the metal ion in the adsorbent and the electronegative part of the sample molecules, i.e. the π-electrons of the aromatic ring or an unshared pair of electrons on an oxygen atom. Because oxygencontaining compounds do not elute from $CuSO_4 \cdot H_2O$ and $CdSO_4$ columns until much higher temperatures than comparable aromatic compounds, the interaction between the metal ion and the non-bonded pairs of electrons on the oxygen atom may be much stronger than that between the metal ion and the π-electrons of aromatic rings. Because amines did not elute from any of the columns, the interaction between the metal ion and the non-bonded electrons of the nitrogen must be even stronger than that between the metal ion and the corresponding electrons on oxygen atoms.

The degree of interaction of the metal ion with π-electrons is affected by substituents. For example, on $CuSO_4 \cdot H_2O$, chlorobenzene (b.p. 132°C) had a smaller retention time than toluene (b.p. 111°C); chlorobenzene had a symmetrical peak while toluene showed extensive tailing.

The fact that cadmium sulphate shows weaker retention for aromatic hydrocarbons and oxygen-containing compounds than does $CuSO_4 \cdot H_2O$ may be explained on the basis that, although copper and cadmium each carry two positive charges, cadmium has nineteen more electrons shielding its charge. Furthermore, the $4d$ orbitals of cadmium(II) are completely filled, whereas those of copper(II) in $CuSO_4 \cdot H_2O$ are vacant.

The most probable reason why a column of $Cu(NH_3)_2(NO_3)_2$ showed stronger adsorption for oxygen-containing compounds than either $Cu(Py)_2(NO_3)_2$ or $Cu(Py)SO_4$ is the added effect of hydrogen bonding between the hydrogen of the ammonia molecule and the non-bonded pairs of electrons on oxygen. Hydrogen bonding, as evidenced by tailing, was severe when $Cu(NH_3)_4(NO_3)_2$ was used as an adsorbent. The most severe tailing was encountered using $CuSO_4 \cdot 5H_2O$. For the last two column materials, even some aromatic compounds tailed badly. This may be the result of hydrogen bonding with the aromatic ring.

The stronger retention of aromatic hydrocarbons and oxygen-containing compounds on $Cu(Py)SO_4$ as compared to $Cu(Py)_2(NO_3)_2$ may be rationalized on the basis that the copper(II) ion in $Cu(Py)SO_4$ has one more bonding orbital available. There may also be some effect of a change in steric factors and of the different anion.

The separation of isomeric ketones on a column of $Cu(Py)_2(NO_3)_2$ may be explained on the basis of an interaction between the copper(II) ion and the carbonyl oxygen. The ethyl ketones (3-pentanone and 3-heptanone) have smaller retention times compared to the respective methyl ketones. In the latter, the oxygen atom is more exposed and can interact more strongly with the copper(II) ion. With 4-heptanone, the oxygen atom is even less exposed than with 3-heptanone, in addition to which 4-heptanone has a boiling point 6°C lower.

An ion-induced dipole interaction between the ions of the adsorbent and the aliphatic hydrocarbons is probably a primary contributing factor to the separation of aliphatic hydrocarbons. The columns of $Cu(Py)_2(NO_3)_2$, $Cu(NH_3)_2(NO_3)_2$, $MgCl_2 \cdot 2H_2O$, $CdSO_4$, and $CuSO_4 \cdot H_2O$ exhibit about the same affinity for aliphatic hydrocarbons but $Cu(Py)SO_4$ has a much greater affinity. This might be because amorphous $Cu(Py)SO_4$ (all adsorbents were crystalline expect $Cu(Py)SO_4$) has a larger surface area. Also, it may be functioning, in part, as a molecular sieve, as it was able partially to separate methane from ethane.

4.2.3 Polar solid stationary phases

Silica

Silica, or silica gel, as normally encountered, is a powerful adsorbent of large specific surface area. Ordinary silica gel consists of an agglomeration of small globules of silica (with a specific surface area of 800–900 m^2/g), between which are pores some tens of μm wide. It has been used almost exclusively in permanent gas and low boiling hydrocarbon analysis.[737-740] Separations of the oxides of nitrogen and sulphur have also been reported.[741-743]

Ordinary silica gel, however, is ill defined, and as an adsorbent it is extremely heterogeneous. Adsorption forces are different at curved surfaces from those at flat surfaces, and, as the small pores necessarily contain many regions of curvature of random small radius, it is these that cause the heterogeneity. The heterogeneity can, therefore, be removed by increasing the average pore diameter, with consequent reduction in total surface area. Pore size can be increased by treatment of the gel with steam under pressure at high temperatures (800–900°C).[744]

The surface of non-porous and porous silica, as a rule, is covered by hydroxyl groups.[745, 746] The concentration of these groups on the surface and within the bulk of the particles depends on the nature and crystalline structure of the silica (for amorphous silica, on the nature of the sample). The concentration of hydroxyl groups also depends on whether the sample has been subjected to hydrothermal or simple thermal treatment or thermal treatment *in vacuo*.

Reproducible values for the concentration of hydroxyl groups on the silica surface (e.g. aerosil, aerosilogel,[747] Silochrom C 80,[748] Porasil, silica gel) can be obtained when reproducible conditions are used for the hydroxylation procedure or the dehydroxylation of the surface. The concentration of hydroxyl groups on the surface can be determined by means of the deuterium-exchange method[749] or by means of infrared spectroscopy. The infrared spectrum of silica has been the subject of intense investigation;[750-755] the strong adsorption bands at 1200, 1100, and 800 cm^{-1} have been assigned[753] to fundamental silicon–oxygen vibrations, while those in the 3000–4000 cm^{-1} region have been attributed to surface hydroxyl groups and molecular water.

Most workers agree that the 3743–3750 cm^{-1} band is due to the stretching of isolated hydroxyl groups, the 3660–3680 cm^{-1} band to weakly hydrogen-bonded surface hydroxyl groups, and the 3400–3500 cm^{-1} band to strongly hydrogen-bonded hydroxyl groups and/or adsorbed water.

Peri[751] concludes that the bands in the 3400–3500 cm^{-1} region arise from adsorbed water because they generally are associated with a band near 1630 cm^{-1} (which he presumes to be caused by the deformation of water molecules). In contrast, others[752] attribute a band at 3500 cm^{-1} to *gem*-hydroxyl groups.

The various types of surface hydroxyl groups generally have been assumed to be the only centres for the adsorption of water, alcohols, ammonia, and other molecules capable of forming hydrogen bonds. However, Kiselev[754] has carried out spectral studies of the adsorption of water and pyridine on silica gel and has obtained results which contradict such an assumption. He conditioned silica gel *in vacuo* at 550°C for 6 h to remove adsorbed water before recording the infrared spectra of the adsorbent in the presence of various amounts of water vapour. The resulting spectra were interpreted as showing that at low surface coverages, free hydroxyl groups are not adsorption centres, and water is adsorbed at more active sites. Similar experiments by Kiselev involving the adsorption of pyridine on silica gel in the presence of small amounts of water indicate that the pyridine is attached to co-ordinately bound water molecules rather than free hydroxyl groups. Silicon is known to form both five- and six-coordinate complexes (e.g., SiF_6) through sp^3d and sp^3d^2 hybridization. Hence Kiselev has postulated that the surface silicon atoms are unsaturated to a certain extent and that water may be adsorbed at these sites. For the orbitals to interact and this process to occur, the central silicon atom and the adsorbed molecules must have a correspondence between their symmetries. This is in accord with the adsorbance of water (C_{2v} symmetry) and ammonia (C_{3v} symmetry) and the absence of donor–acceptor bonds between silicon and pyridine. In the latter instance, steric factors also may be important.

Figure 29 illustrates the adsorption of water (B). This weakens the O–H bonds, which act as surface proton centres. The magnitude of the dipole set up in the O–H bond depends on how tightly the unshared electron pair is held by the inner d-orbital of the silicon. The latter is affected by the nature and number of the functional groups attached to the silicon. Thus dehydration and chemical modification of the surface should change the co-ordination unsaturation of the surface atoms and, consequently,

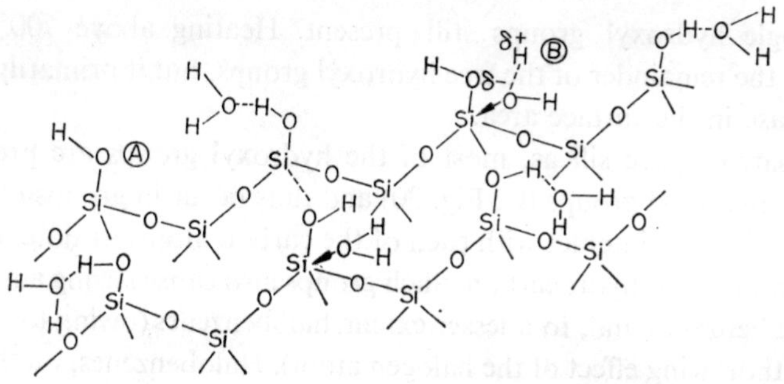

FIG. 29. Surface of silica gel before activation.[755]

their acid strength. Consideration of these arguments, the infrared data, and the adsorption data leads to the conclusion that the surface of the silica gel is covered initially by a considerable amount of water (B) bonded to surface silicon atoms and surface hydroxyl groups; some free hydroxyl groups (A) also are present.[755] Such a silica gel surface is illustrated in Fig. 29.

On raising the activation temperature to 200°C, most of the hydrogen-bonded water appears to be removed, with hydrogen bonds (C) forming between favourably placed single hydroxyl groups, as shown in Fig. 30.

Further heating to 500°C probably removes all hydrogen-bonded water and most of the co-ordinated water molecules, and increases hydrogen bonding between the hydroxyl groups. At 600°C, the hydrogen-bonded hydroxyls appear to condense to form surface siloxane linkages with a

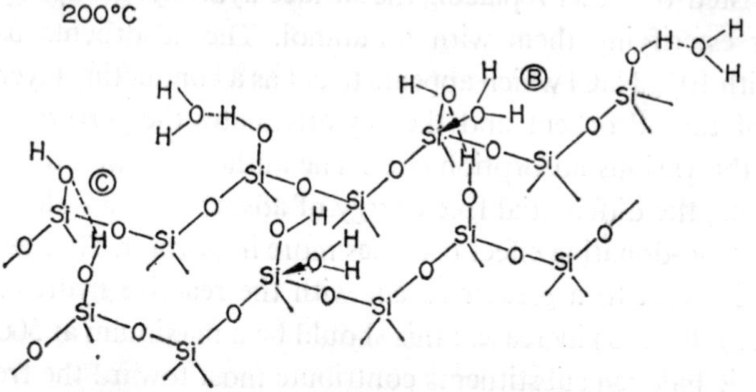

FIG. 30. Surface of silica gel after activation at 200°C.[755]

few single hydroxyl groups still present. Heating above 700°C may remove the remainder of the free hydroxyl groups, but it primarily causes a decrease in the surface area.

On narrow pore silicas, most of the hydroxyl groups are present as reactive hydroxyl groups (C, Fig. 30) and thus occur in groups of two or more, which can interact with each of the carbon atoms or double bonds of an unsaturated hydrocarbon. Such groups also cause strong adsorption of alkyl benzenes and, to a lesser extent, halobenzenes (owing to the electron withdrawing effect of the halogen atom). Halobenzenes, on the other hand, should be strongly adsorbed at the proton centres (B, Fig. 30) produced by the co-ordinated water molecules because of the higher electron density at the halogen atoms. Little, if any, adsorption will take place at the single hydroxyl groups (A, Fig. 29) if other more active proton centres are available. As the activation temperature of the silica gel is raised from 200° to 500°C, type (B) sites disappear and type (C) sites are formed. Thus, halobenzenes exhibit slightly reduced interactions at 500°C, whereas alkyl benzenes are adsorbed more strongly. Further increase in the activation temperature causes type (C) sites to disappear so that the interaction of both alkyl and halobenzenes with the surface is reduced drastically.

Gas chromatographic data obtained by Cadogan and Sawyer[755] using various thermally activated and chemically modified silicas (Silica gel Davison 62, 100–120 mesh, and Porasil C 100–150 mesh) are in accord with this model of the silica surface. Chemical modification of the adsorbents consisted of either replacing the surface hydroxyl groups by fluorine atoms or esterifying them with n-butanol. The adsorbents used were coated with 10% NaCl which appears to act as a conducting layer over the surface of the adsorbent and thereby attenuates the perturbations and forces at the various adsorption sites. The authors showed that for alkyl substituents, the differential free energy of adsorption $\Delta(-\Delta G)$ increases as the electron-donating effect becomes more important. Thus as the alkyl benzenes interact to a greater extent with the reactive hydroxyl groups (C, Fig. 30) $\Delta(-\Delta G)$ increases; this should be a maximum at 500°C.

Aromatic halogen substituents contribute most toward the free energy of adsorption when the halobenzenes are adsorbed at the co-ordinated water sites (B, Fig. 30). Their contribution thus is a minimum at 600°C when all type (B) and (C) sites have been removed. The increased contribution to the free energy of interaction above 600°C is considered to arise from the co-ordination of the halobenzenes to surface silicon atoms. The sizes of the maxima and minima also reflect the relative electron-donating or -withdrawing effects of the substituent atoms.

Comparison of the data obtained from Porasil C treated with ammonium fluoride, which leads to the substitution of the hydroxyl groups on the silica gel surface by fluorine atoms, with the data obtained from non-treated Porasil C, indicates that the contribution of aromatic halogen substituents to the heat of adsorption is increased by treatment with ammonium fluoride while that of aromatic alkyl substituents is reduced.

This phenomenon is explained by the authors by using the model of the silica surface described above. Activation at 230°C of Porasil C treated with ammonium fluoride should produce a surface very similar to that shown in Fig. 30 except that some of the hydroxyl groups are replaced by fluorine atoms. The effect of this replacement is to reduce the number of reactive hydroxyl groups (C, Fig. 30) and also to make the remaining surface hydroxyl groups and co-ordinated water molecules (B) more acidic because of the inductive effect of the electronegative fluorine atoms.

If the previous postulation, that alkyl-substituted aromatic compounds tend to adsorb at type (C) sites and halogen-substituted aromatics at type (B) sites, is correct, then the ammonium fluoride treatment should cause decreased alkyl benzene and increased halobenzene interactions. The esterification of the surface with n-butanol results in a slightly enhanced interaction with alkyl benzenes and a decreased interaction with halobenzenes. At an activation temperature of 700°C, the surface of silica gel that has not been treated with n-butanol is covered with siloxane linkages. Interaction between alkyl benzenes and such a surface probably occurs at these linkages, whereas halobenzenes may co-ordinate either with the freshly dehydroxylated silicon atoms or with impurity atoms that migrate to the surface at high temperatures. If the above assumptions are correct, esterification of the surface may well present similar sites (ether linkages) for alkyl benzene adsorption, but the halobenzene adsorption sites would be inaccessible.

17*

High-resolution NMR has been used by Pickett et al.[756] to study the adsorption of molecules at low surface coverages on wide-pore silicas with the aim of investigating specific interactions that are commonly encountered in gas–solid chromatography. Line widths were measured for cyclohexane, benzene, acetone, and methanol because the strength of interaction of these compounds with a hydroxylated silica surface is known to vary from weak to strong. Hydrogen bonding to the surface was evident for the last two compounds from the decrease in line width when the silica was silanized. Attempts were made to explain the magnitude of the changes by comparison with data from infrared spectroscopy and gas–solid chromatography. The capacity ratios, chemical shifts and line widths suggest that the silanized surface was much less active than the hydrated surface.

Applications of silica in gas chromatography is to a great extent determined by the chemical purity of the material. The presence on the silica surface of impurities (such as aluminium and boron) which form strong Lewis acid centres causes an increase in the energy of bonds formed with electron-donor molecules. The effect is especially noticeable upon dehydroxylation when strong Lewis acid centres become exposed. This may be illustrated by some calorimetric data obtained by Ash et al.[757] for the heat of adsorption of vapours of triethylamine, a strong organic base. After gradual monolayer coverage of the hydroxylated silica surface, the heat of adsorption of triethylamine was about 20 kcal/mole. This value corresponds to a specific molecular adsorption leading to the formation of a $SiOH \ldots N(C_2H_5)_3$ hydrogen bond.

Strong dehydroxylation of the surface, when a pure silica sample is heated at $1100°C$ (without altering the surface area), leads to the formation of siloxane groups which are incapable of specific molecular interaction. In this instance, the heat of adsorption of triethylamine decreases to 10 kcal/mole. This value corresponds to the energy for the non-specific interaction of triethylamine with a silica surface. If the silica contains an aluminium impurity (0.36% by wt.), dehydroxylation of the surface at $1100°C$ results in the formation of a large number of Lewis electron-acceptor centres besides the siloxane groups. The molecules of triethylamine are chemisorbed at these centres, with a heat of adsorption of about 50 kcal/mole. Only when all such centres accesible to the triethyl-amine molecules are filled does the heat of adsorption sharply decrease to a level corresponding to non-specific adsorption on the siloxane groups of the remaining sample surface.

The heat of adsorption of cyclopentane is not affected either by the conversion of silanol groups into siloxane groups or by the exposure of the aluminium centres. Thus a separation on columns made of silica gel containing Al_2O_3 as an impurity gives chromatograms with symmetrical peaks only for hydrocarbons (especially saturated hydrocarbons). Aroma-tic hydrocarbons, ethers, and ketones are eluted from the column only with difficulty, and the corresponding chromatograms are characterized by expanded peaks. On the other hand, such compounds as amines, pyri-dines, quinoline, and other strong organic bases remain on the column, i.e. they are held by chemisorption on strongly acidic centres distributed on the silica surface. For porous glass adsorbents, the boron atoms[758] act as such centres.

In addition, the hydroxyl groups on pure silica are neither too active energetically nor heterogeneous. The hydroxylated surface of chemically pure silica does not behave like a chemisorbent toward various electron-do-nor molecules, including aniline, but merely like a polar adsorbent which forms hydrogen bonds of different strengths with such molecules.[759, 760] The dehydroxylation of a pure silica surface, which decreases its specificity, only decreases the retention of such molecules and has little effect on the retention of non-polar molecules.

For molecules containing several functional groups, the geometrical structure of the molecule is of special significance. With dioxane, the presence of two ether groups isolated from each other by methylene groups does not result in a doubling of the value for the energy of inter-action with the silanol groups on the silica surface.[761] This is because the favourable orientation of one ether group of the chair-like molecule of dioxane precludes the other ether group approaching close enough to the silanol group to form a strong hydrogen bond with the surface.

All the factors discussed above affect the retention volumes of chroma-tographic columns filled with Silochrom C, a high purity, wide-pore silica prepared from a non-porous, highly dispersed aerosil.[748, 759, 762]

Macroporous silica (pore volume = 1.29 cm^3/g) with a specific surface area of about 80 m^2/g and pore diameter of about 55 nm, has been named Silochrom C 80.[748] The specific surface of the liquid benzene film formed on the surface of pores at the beginning of capillary condensation is 75 m^2/g for Silochrom C 80. This is, for this adsorbent, close to the specific surface area of the Silochrom itself, indicating the absence of fine pores. This shows the high geometrical homogeneity of the adsorbent.

A gas chromatographic study of Silochrom C 80 supports this conclusion about the pore structure. The chromatograms of mixtures of non-specifically adsorbed n-alkanes show symmetrical peaks. This indicates the absence of fine pores, fractures, and other geometrical inhomogeneities which would result in a local increase in non-specific adsorption of n-alkanes and peak asymmetry. Aldehydes and alcohols, which are strongly adsorbed on a hydroxylated surface and are capable of forming hydrogen bonds, also emerge rapidly from a column of Silochrom C 80 at 140°C with narrow peaks.

Ethers, esters, and ketones having lone electron pairs on the oxygen atoms which strongly interact with the hydroxyl groups of the silica surface, likewise emerge from the column with symmetrical peaks.

Figure 31 shows the separation of a mixture of C$_5$–C$_{17}$ n-alkanes and some esters on Silochrom C 80 (0.25–0.5 mm) by a column 100 cm × 3 mm, with a nitrogen flow rate of 40 ml/min.

When macroporous silica gel with the same specific surface area (80 m^2/g) but containing impurities is used, the peaks for aniline, nitrobenzene, and acetophenone are quite asymmetrical. These data show that structural hydroxyl groups on the silica surface do not by themselves cause the surface to be inhomogeneous because the properties of the silanol hydroxyl groups are all the same. Inhomogeneity is mainly the result of the different properties of the hydroxyl groups bound to the impurity centres (aluminium, iron, etc.) and especially of the dehydroxylation of these centres.

Chromatographic data show that Silochrom C 80 is homogeneous not only geometrically but also chemically. This adsorbent is especially suitable for the separation of high boiling substances, for analyses under

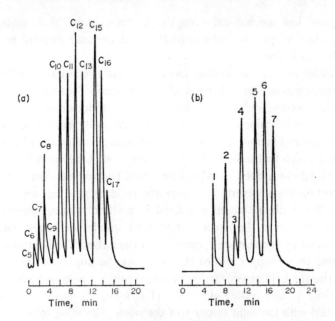

FIG. 31. Chromatograms on Silochrom C 80. (a) Mixture of C_5–C_{17} n-alkanes: column temperature programmed from 100 to 275°C at 10°C/min; sample size, 0.8 μl. (b) Esters: (1) ethyl formate, (2) ethyl acetate, (3) amyl formate, (4) ethyl butyrate, (5) ethyl valerate, (6) ethyl caproate, (7) amyl butyrate: column temperature programmed from 125 to 275 °C at 6°C/min; sample size: 0.4 μl.[748]

temperature programmed conditions and for repetitive determinations, i.e. in those instances where gas–solid chromatography is preferable and an adsorbent with a low non-specific adsorption energy is required.

Halász and Gerlach[763] used for the first time in gas chromatography columns produced by drawing out to 0.4 mm i.d., glass tubes loosely packed with highly dispersed silica (Aerosil) of 1 μm particle diameter. The powder, which proved to be a suitable stationary phase, was made by flame hydrolysis of $SiCl_4$. It is a loose, bluish-white, flaky, X-ray amorphous powder, which forms a colloidal solution with water.[764] The primary particles (3–15 nm diam.) are spherical, free of pores, and have a specific surface area of 380 m²/g. On account of its loose consist-en^y and low mechanical strength, the product is hardly suitable for filling classical packed colums and cannot be used directly in packed capillary columns.

In order to give the product more mechanical strength, the following procedure was adopted. Aerosil, which very easily forms gels with apolar organic solvents, was treated with carbon tetrachloride until a gelatinous product was obtained. The solvent was carefully evaporated and the residue ground in a mortar. A very voluminous powder was obtained as a product which had to be freed of agglomerated particles. Aerosil, as well as the treated product, could not be sieved because the particles became electrostatically charged. Therefore the product was put into a plastic bowl, 20 cm diam., which was fixed in a vibrator at an inclination of 30°. On vibration the particles described smaller or larger semicircles determined by the size and "density" of the flakes. The lightest particles travelled to the upper rim of the bowl. At the uppermost point of the bowl rim the light sieve fraction passed through a hole and was collected in a dish. A thick-walled glass tube (1 m long, 2.2 mm i.d., 6 mm o.d.) was filled with the light fraction of the silica. The filled tube was drawn out to a capillary of 0.4 mm i.d. according the method described by Desty.[81]

On account of their particularly loose packing (only 2.5% of the column volume was occupied by the stationary phase), columns filled with highly dispersed Aerosil are known as aerogel columns. Their main advantage is the short analysis times that can be obtained. Such columns were used for the separation of C_1–C_6 alkanes and olefins. Figure 32 shows the separation of saturated and unsaturated C_1–C_6 hydrocarbons at room temperature within a minute. The use of moist carrier gas proved very advantageous (the carrier gas was moistened over $Na_2SO_4 \cdot 10\,H_2O$ at room temperature.

High-resolution capillary adsorption columns have been prepared by Schwartz et al.[765] by coating stainless steel tubing with colloidal sols containing NALCO CD-100, a hydrophobic colloidal silica. It is interesting that such a column (200 ft × 0.01 in. i.d. (31 m × 0.4 mm) stainless steel coated with CD-100 from a solution containing 10% CD-100 in cyclohexane) permits the separation of the C_6–C_8 aromatic hydrocarbons

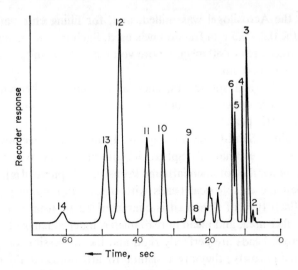

FIG. 32. Separation of C_1–C_6 hydrocarbons on an aerogel column at 27°C.[763] (1) Methane, (2) ethane, (3) propane, (4) propylene, (5) *iso*-butane, (6) n-butane, (7) 1-butene, (8) *iso*pentane, (9) n-pentane, (10) n-1-pentene, (11) 3-methyl-1-butene, (12) 2-methyl-1-butene, (13) 2-methyl-2-butene, (14) n-hexane. Column length: 1.6 m; carrier gas, hydrogen moistened over $Na_2SO_4 \cdot 10 H_2O$ at 20°C; $\Delta p = 4$ atm.; $p_0 = 1$ atm.; carrier gas flow: 22 cm/s.

at room temperature. The elution of hydrocarbons from such columns is more rapid than from partition columns of the same dimensions.

Symmetric elution peaks from various electron-donor molecules were obtained by Bebris *et al.*[759] using chromatographic columns (100 cm × 4 mm) of macroporous pure silica, Aerosilogel. This adsorbent was prepared as follows: 280 g of Aerosil (specific surface area 175 m^2/g) was mixed with 1 litre of water and strongly stirred for half an hour with a little metal shovel until a uniform, viscous mixture was obtained. The mixture was introduced into a crystallizer and left at room temperature until a solid mass formed. This mass was broken up, dried at 140°C to constant weight, and treated with steam for 6 h at 850°C. This treatment allows the formation of a more uniform pore structure and gives a product more suitable for gas chromatographic applications. After this treatment the Aerosilogel was milled, and, for filling chromatographic columns, the 0.25–0.5 mm fraction was used. Such treated material has a specific surface area of 80 m²/g, a pore volume of 1.24 cm³/g and a mean pore diameter of 50 nm.

Strongly adsorbed substances, such as aniline, nitrobenzene, cyclohexanol, and acetophenone, elute from an Aerosilogel column at 200°C giving symmetric peaks.

The Péchiney–Saint-Gobain company, France, has developed a type of porous silica bead named "Spherosil" (distributed by Waters Associates, USA, under the commercial name Porasil).[766] Spherosil is pure silica, and is available as spherical beads with diameters ranging from 20 to 300 μm. Each type of Spherosil is characterized by total pore volume, specific surface area, and mean pore diameter, independent of the particle size. Spherosil beads are perfectly rigid and incompressible despite their considerable porosity; they are resistant to attrition and do not swell

TABLE 20. PHYSICAL CHARACTERISTICS OF
SPHEROSILS[767]

Type of Spherosil	Nominal specific surface area (m²/g)	Average pore diameter (nm)
XOA 400	400	8
XOA 200	200	15
XOB 075	100	30
XOB 030	50	60
XOB 015	25	125
XOC 005	10	300

in any liquid. The different grades of Spherosil may be subjected to temperatures up to 600°C without any change in texture. Table 20 gives the commercially available grades in the particle size ranges 40–100 μm and 100–200 μm.[767] The characteristics indicated are nominal and may vary slightly from one batch to another. Their specific pore volumes are all 1 cm³/g.

Feltl and Smolkova[768] recently made a gas chromatographic study of the sorption properties of the Spherosils. For example, the shape of the adsorption isotherms for benzene is essentially identical for Spherosils with specific surface areas of 8, 25, and 66 m^2/g, whereas the Spherosil with a specific surface area of 428 m^2/g can be characterized as a narrow-pore material. Because only the surface concentration of hydroxyl groups is decisive for the chromatographic process on wide-pore Spherosils whereas the adsorption on the narrow-pore Spherosil is affected by capillary condensation which can take place at low relative pressures,[769] it may be assumed that capillary condensation is the main cause of the increased surface concentration of benzene observed.

In view of the characteristic properties of these silica beads (chemical inertness, excellent heat resistance, spherical shape, and controlled texture) this adsorbent appears to offer many advantages for analytical purposes. Chromatographic columns (1–3 m $\times$ 4 mm i.d.) filled with 60–80 mesh silica beads have enabled saturated, unsaturated, halogenated, aliphatic, and aromatic hydrocarbons to be separated.[766, 770]

Adsorbents similar to silica gel have also found applications as stationary phases. These include Fluorisil, a synthetic magnesia–silica gel, and Florex, a natural fuller's earth.[771, 772] These adsorbents were used to separate C_1–C_4 hydrocarbons.

Porous glass was also effective for the separation of both low- and high-boiling substances. Such glass contains the products of treatment of alkaline borosilicate glasses with acids and water.[773] If the composition of the original glass is changed by changing the conditions of hydroxylation and subsequent thermal treatment, glasses are obtained with pore dimensions of from < 1 nm to 1 μm with a very limited pore distribution.

Porous glass with fine pores ($\sim$ 1 nm) shows properties similar to molecular sieves. It can be used to separate gases with very low boiling points.[774] Helium, nitrogen and oxygen, carbon monoxide, and methane elute in this order from a column filled with such an adsorbent. Porous glass with pore diameter of 3–10 nm was used to separate light hydrocarbons, and that having a pore diameter of 30–50 nm allows the separation of some hydrocarbons with boiling points up to 200°C.[775]

The retention of unsaturated and aromatic hydrocarbons on such columns is greater than that of saturated hydrocarbons with the same number of carbon atoms. This is explained by the formation of π-complexes between the unsaturated hydrocarbons and the hydroxyl groups on the porous glass surface. A stainless steel column (6 ft $\times$ $\frac{1}{8}$ in. o.d.) (1.8 m$\times$3 mm) packed with porous glass (50–80 mesh, surface area 173.2 m²/g, pore volume 0.109 ml/g, average pore size 2.52 nm) allows the separation of C_5–C_{16} alkanes.[776] The highest boiling compound of the series, hexadecane (b.p. 287.5°C), was eluted at about 250°C in 30 min. The experiment indicated that temperatures as high as 850°C do not affect the adsorbent characteristics of porous glass and that the column can be operated up to temperatures at which pyrolytic decomposition of analytes begins.

Alumina

Alumina is an adsorbent with a relatively high specific surface which, together with the pore diameter and volume, depends on the treatment temperature. Alumina A (Péchiney–Saint-Gobain), with a particle size of 250–315 μm, has a specific surface area of 400 m²/g. The same alumina after heating for 6 h at 520°C has a specific surface area of 170 m²/g and after heating to 1000°C, a specific surface area of 85 m²/g.[777] The pore radius and volume of alumina A are 0.2 μm and 0.53 ml/g, respectively, after calcination at 1350°C and 0.4 μm and 0.1 ml/g after calcination at 1800°C.

Numerous studies of the properties of alumina surfaces have been carried out, particularly by infrared spectroscopy.[778]

The active sites on an alumina surface are:

(a) Three or five types of hydroxyl group, the type being determined by the number of nearest oxide neighbours. Three of these types are distinguishable spectroscopically; they show slightly different reactivities.

(b) The Lewis acid (electron-accepting) sites which are aluminium ions attached to three oxygen atoms.

(c) Oxide ions.

From catalytic studies, it is generally agreed that the Lewis acid sites are the most active surface sites on alumina. These sites are thought to arise when the hydroxyl groups condense and eliminate water at temperatures above 400°C, leaving oxide ions and vacancies on the surface; if two vacancies are adjacent, an aluminium ion is exposed and this is a Lewis acid. These Lewis acid sites can adsorb molecules in several ways. Ketones, for example, form stable complexes with the Lewis acid sites via the oxygen atom of the carbonyl group. The acceptance of the electrons of the oxygen atom into the vacant orbital of the Lewis acid causes a lowering in the infrared absorption frequency of the carbonyl group. An olefinic group adsorbs as a π-complex with the electron-deficient Lewis acid sites.[779]

Methyl alcohol reacts with an alumina surface at low temperatures to produce surface methoxide groups; on heating above 170°C, the methoxide group is oxidized to a surface formate group, probably by the Lewis acids.[780] Many other reactions between various adsorbates and alumina are reported in the literature, but these few examples illustrate the possibility of selective chemisorption on the known undesirable surface sites in order to obtain better chromatographic separations.

A parallel infrared spectroscopic-gas chromatographic study was undertaken by Neumann and Hertl[781] in which three representative compounds, n-butyl alcohol, pyridine, and acetic anhydride, were used to partially deactivate an alumina column (ALCOA type F-1, chromatography grade, 60–80 mesh). The treated alumina columns were evaluated by obtaining chromatograms of a series of standard hydrocarbons and also of a sample of commercial gasoline. The spectrum of the alumina before and after exposure to n-butanol at 250°C shows that, after exposure there is a small decrease in the intensity of the Al—OH band(s) near 3720 cm^{-1}; C—H stretching bands from the chemisorbed butanol appear in the region 2800–3000 cm^{-1}, and bands appear at 1460 and 1560 cm^{-1} as a result of the oxidation of the alcohol to a carboxylate bonded to the surface.[780] The same spectra were obtained from another alumina sample when the butanol was added at 400°C. Exposure of this treated alumina to water vapour at 200°C for 1 h resulted in no significant changes in the intensities of the C—H or C=O bands, although water did readsorb on the surface. Thus the bonded species is stable with respect to hydrolysis. On exposure to air at 250°C, there was a slow but steady decrease in the intensities of the C—H bands owing to the oxidation of the hydrocarbon chain; the bonded C=O groups showed no change.

When pyridine binds by co-ordination to the Lewis acid sites on alumina, characteristic bands appear at 1453 and 1457 cm^{-1}. When pyridine was added to the butyl alcohol-treated surface at 400°C, strong bands appeared at 1445 and 1570–1600 cm^{-1}. Concurrent with the addition of the pyridine there is a frequency shift of the C=O band at 1460 cm^{-1}, showing that there is an interaction between the pyridine and the surface C=O groups.

When acetic anhydride reacts with a fresh alumina surface at 400°C, a large decrease in the amount of adsorbed water is observed and two strong bands appear at 1465 and 1585 cm^{-1}. These two bands arise from the symmetric and asymmetric stretching of the carboxylate C=O group, and are similar, but not identical, in frequency to the bands observed when butyl alcohol has reacted with the alumina. Although acid anhydrides show a typical doublet C=O band (at 1824 and 1784 cm^{-1} for acetic anhydride), it is more likely that the acetic anhydride was hydrolysed before its reaction with the surface because a decrease in the quantity of adsorbed water was observed and the separation of the C=O bands was about 120 cm^{-1} after the acid had reacted. On addition of pyridine to the acid-treated surface, bands appeared near 1450 and 1600 cm^{-1} owing to the presence of a pyridine-C=O complex.

The chromatograms obtained for a commercial gasoline sample show that the butyl alcohol-treated surface is not significantly different in its separation properties from untreated alumina. Pyridine or acetic anhydride-treated alumina gives improved separations and shorter retention times with improved peak shapes, showing that the surface bonding sites for pyridine and acetic anhydride are different to those for butyl-alcohol.

Pyridine was applied to the column at 250°C, with beneficial results. If the column temperature was raised to 400°C for 2 h and then returned to 200°C, the chromatograms obtained were very poor. The degeneration of the column probably arises from traces of oxygen in the carrier gas, leading to oxidation of the pyridine, because the spectroscopic experiments showed that the bonded pyridine is stable at 400°C *in vacuo*.

Using spectroscopic arguments in conjunction with chromatographic data, Neumann and Hertl[781] state that, for the butyl alcohol reaction followed by oxidation, the C=O groups are bonded principally to the oxygen atoms or the alumina surface, but do not react with the Lewis acid sites, whereas for acetic anhydride and pyridine, the bonding is principally to the surface aluminium atoms, thus effectively removing most of the Lewis acid sites.

Klemm and Airee,[782] in an effort to gain further insight into the geometry of adsorption on alumina which was either untreated or pre-treated with non-eluting quinuclidine, determined the retention times and heats of adsorption of a series of compounds such as benzene, eight monoalkyl-benzenes, *m*-xylene, naphthalene, and 2,6-lutidine. The peaks of these solutes showed a strong asymmetry when the compounds were eluted from a spiral copper column (122 cm×6.4 mm i.d.) packed with F and M 30–60 mesh alumina, using helium (40 ml/min) as eluting carrier gas. The measurements were carried out isothermally, between 173° and 367°C.

For the series pyridine, 2-picoline, 2,6-lutidine, the order of increasing retention times is a marked inversion of the order of increasing boiling points. Also, 8-methylquinoline and the dimethylquinolines have retention times less than quinoline, whereas all the dimethylnaphthalenes are more strongly retained than naphthalene. Quinuclidine, moreover, is strongly retained, although its molecular weight approximates to that of ethyl-benzene and it sublimes readily when heated in air.

These results have been interpreted on the basis of preferential adsorption of azacyclic compounds by means of the n-electrons on the nitrogen atom. In the pyridine molecule steric hindrance to adsorption is provided by alkyl groups in an α-position but not in a β-position. For quinoline, steric hindrance to adsorption is largest for an alkyl group in the 8-position (where the C_{Ar}–C_{Al} bond is parallel to the spatial direction of the n-electrons) and smaller in the 2-position (where corresponding bonds are at an angle of about 120°C).

In general the retention times for alkyl benzenes increase with increasing boiling point. Isobutyl benzene and s-butyl benzene, however, provide an exception to this trend. The inverted order for this pair may be ascribed to a larger steric hindrance by the s-butyl group ($—CH(CH_3)C_2H_5$) as compared to the isobutyl group ($—CH_2CH(CH_3)_2$) towards flatwise adsorption of the benzene ring. Generally, both retention times and heats of adsorption decreased on impregnating the alumina with quinuclidine.

The chromatographic peaks obtained, even for lighter hydrocarbons, when using perfectly dry alumina, tail very badly and have high retention volumes. To reduce the retention volumes and to prevent tailing effects, the active centres on the alumina must be blocked. This is most simply done with water (commercial alumina is never perfectly dry; it is always covered with an undefined amount of sorbed water). There are two ways of coating alumina reproducibly with water:

(1) The alumina has a small amount of water, irreversibly adsorbed, and therefore a constant amount, at the temperature of the analysis. It is prepared by drying moist alumina by flushing with dry carrier gas at a temperature higher than that of the analysis but lower than that of calcination. Reproducible results can only be obtained by working with well-dried carrier gas at the temperature of the analysis.

(2) The water content of the alumina can also be controlled by using a carrier gas with a controlled partial pressure of water. The carrier gas is moistened by passing it over $CuSO_4 \cdot 5 H_2O$. In this way the partial pressure of the water in the carrier gas is not too sensitive to slight changes in the temperature of the moistening agent. Using moist carrier gas, the relative retentions remain practically constant.[783]

Scott[784] showed that the amount of water required to reach minimum polarity is equivalent to that required to form a monolayer on the surface of the alumina.

Hoffmann and Evans[785] have shown that the retention characteristics of the hydrocarbons on activated alumina and the efficiency with which hydrocarbons are separated are strongly influenced by the carrier gas employed. Their investigation included eight carrier gases (hydrogen, deuterium, helium, neon, nitrogen, carbon dioxide, argon, and krypton) for elution of n-butane from a 4 ft $\times \frac{1}{4}$ in. o.d. (1.2 m $\times$ 6.3 mm) aluminium column of 60–80 mesh Microtek activated alumina. Though carbon dioxide is strongly adsorbed on alumina and should be an effective competitor for active sites on the adsorbent, both deuterium and hydrogen provide shorter retention times at all temperatures. Helium had an average efficiency between the extremes exhibited by deuterium and hydrogen. Carbon dioxide was the most efficient eluant and, except for the hydrogen isotopes, gave the shortest retention times. The butane retention times observed using deuterium, hydrogen, and helium were explained by Hoffmann and Evans on the basis that deuterium is smaller and more polarizable than helium but heavier than hydrogen. These differences should make deuterium the most efficient competitor against butane for active sites on the alumina surface. In comparison to hydrogen, the increased mass of deuterium should account for its ability to elute butane more quickly. The shorter retention times obtained by eluting with carbon dioxide undoubtedly result from its high affinity for active sites on the alumina. Flow rate affects column efficiency and, with the exception of hydrogen, all gases tested showed maximum efficiency at low flow rates.

Retention times for n-butane were longer with the rare gases than the other gases, and increased from helium to krypton. In this instance, the correlation between retention time and atomic cross-section of the rare gases indicates the possibility of a desorption hindrance effect.

When a flame ionization detector is used, nitrogen appears to be a good compromise choice for maximum resolution and elution time. At elevated temperatures, column efficiencies for the different carrier gases became similar.

Alumina columns are very little used for permanent gas analysis. Hoffmann et al.[786] obtained a good separation of oxygen, nitrogen, and carbon dioxide on a single 4 ft $\times \frac{1}{8}$ in. (1.2 m $\times$ 3 mm) aluminium column packed with 60–80 mesh F-20 Microtek alumina (activated by passing helium for 1 h at 350°C). The separation was done isothermally at 0°, at a carrier gas flow of 30 ml/min, and took 30 s. Because alumina at low temperatures tends to adsorb carbon dioxide irreversibly, the column was first purged with that gas for 1 h at 0°C to reduce adsorbent activity.

The most numerous applications of this adsorbent in gas chromatography are separations of hydrocarbons. List *et al.*,[787] using an 8 ft × $\frac{1}{4}$ in. aluminium tube packed with 60–80 mesh activated alumina (F-20), temperature programmed from 75° to 300°C at 21°C/min, have separated the C_1–C_8 n-alkanes and α-olefins. The saturated hydrocarbons are eluted before the olefins with the same number of carbon atoms because the electron-rich unsaturated bonds of the olefins are more firmly bound to the electropositive sites on the alumina lattice.

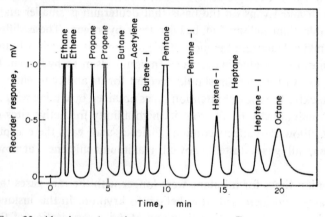

Fig. 33. Alumina adsorption gas chromatogram of a hydrocarbon mixture.[788] Column: 6 ft×$\frac{1}{4}$ in. (1.8 m×6.3 mm) diam. aluminium; support: 60–80 mesh activated alumina; no stationary phase; carrier gas: nitrogen at 65 ml/min; injector temperature: 150°C; flame ionization detector; block temperature: 150°C; programme rate: 21°C/min from 75 to 300°C; analysis time: 27 min.

Figure 33 illustrates a separation, achieved with uncoated activated alumina in a column 6 ft×$\frac{1}{4}$ in. diam. (1.8 m×6.3 mm),[788] of compounds having a boiling point range of 215°C, from ethane (boiling point −88.3°C) to n-octane (boiling point 125.8°C), and including alkanes, olefins and acetylene. By using temperature programming, the entire 13-member series was cleanly resolved in less than 30 min. The components

were eluted according to their molecular weight and degree of unsaturation. In spite of the large difference in molecular weight (26 vs. 58) and
boiling point ($-84°C$ vs. $0.5°C$), acetylene was retained longer than n-
butane, indicating the strong degree of association between its π-electrons
and the alumina lattice.

The separation and analysis of some *cis* and *trans* olefins on activated
alumina was reported by List *et al.*[789] On a 3 ft$\times \frac{1}{16}$ in. aluminium column
packed with 60–80 mesh activated alumina and with nitrogen (40 ml/min),
as carrier gas, *cis*- and *trans*-butene-2 (at 75°C), pentene-2 (at 100°C),
hexene-2 (at 115°C), heptene-3 (at 150°C), and octene-2 (at 175°C) were
separated. In all instances the *trans* isomer was eluted before the *cis* isomer
because the electron clouds surrounding a *cis* double bond can lie closer
to and thus bind more strongly with the alumina surface than those of
the *trans* isomer.

The analysis of a complex mixture of saturated (up to C_{10}) and unsaturated hydrocarbons (olefins, dienes, acetylenes) is possible if a pre-
column of 17% silver nitrate on alumina is utilized. Solid silver nitrate
coated on alumina removes alkenes and alkynes by oxidative degradation.[790]

It may be stated that under the appropriate chromatographic conditions
(column length and diameter, sample size, column temperature, carrier
gas), gas–solid chromatography on alumina lends itself well to the analysis
of volatile hydrocarbons. It can separate saturated and unsaturated
compounds as well as linear and cyclic hydrocarbons, and *cis* and *trans*
isomers.

Chromium(III) oxide gel

Van der Vlist and de Jong,[791] during studies on the reversible adsorption of oxygen and nitrogen on various materials, discovered the use of
black chromium(III) oxide gel as a solid adsorbent for the gas chromatographic separation of oxygen and nitrogen. The gel was prepared by
reduction of an aqueous solution of chromium(VI) oxide in ethanol, and
dried at 150°C for 24 h. For column preparation the 420–600 μm sieved
fraction was used. The column dimensions were length 1.50 m$\times$4 mm
i.d. The column was activated by heating at 350°C for 4 h in a stream of

helium (30 ml/min). At 100°C, the separation of oxygen from nitrogen takes place in a minute. This adsorbent has no advantages over the normally used molecular sieve except that it may be a less expensive alternative.

Hydrated iron(III) oxide

Investigations concerning the use of hydrated iron(III) oxide as an adsorbent in gas chromatography were made by Kilarska and Zielinski.[792] They filled chromatographic columns 100 and 200 cm long and 4 mm i.d., with the 0.3–0.6 mm fraction. These columns were tested for the analysis of natural gas and of the components of furnace gases. Air, as well as methane and ethane from natural gas, was analysed using a 200 cm column where the adsorbent had been activated by heating under vacuum for 5–6 h at 130°C. The determinations were carried out at room temperature using hydrogen as carrier gas. Under these conditions, carbon dioxide was completely adsorbed on the column, as were the higher hydrocarbons.

A column of hydrated iron(III) oxide does not separate the components of furnace gases (especially nitrogen, carbon monoxide, and carbon dioxide; the first two gases elute almost together). Thus it seems that such a column can be applied for the analysis of natural gas but is unsuitable for the determination of the components of furnace gases.

Barium sulphate

Barium sulphate was proposed as an adsorbent in gas chromatography by Belyakova et al.[793, 794] The possibility of separating a mixture of unsaturated and aromatic hydrocarbons on barium sulphate with a specific surface area of 2.5 m²/g was investigated.

Figure 34 shows a chromatogram obtained in the separation of a mixture of cyclic unsaturated and aromatic hydrocarbons on barium sulphate. All the xylene isomers are well separated. The order of emergence of aromatic hydrocarbons accords with the values characterizing the change in electron density on the benzene ring, depending on the nature and position of substituted groups.

On a 1 m column, using temperature programming from 210° to 350°C

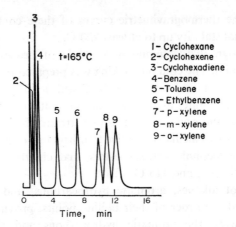

FIG. 34. Chromatogram of unsaturated cyclic and aromatic hydro-carbons on barium sulphate on a 4 m×0.6 cm diam. column; flame-ionization detector; nitrogen carrier gas flow rate: 50 ml/min.[793]

at 20°C/min, naphthalene, diphenyl, *o*-terphenyl, phenanthrene, *m*-terphenyl, and *p*-terphenyl eluted in this order and were separated.[794]

In the series cyclohexane, cyclohexadiene-1,3, and benzene, the values of the heat of adsorption on BaSO$_4$ increase, and the difference, as com-pared with n-hexane, per double bond in cyclohexene is 2.2 kcal/mole, per two double bonds in cyclohexadiene-1,3 is 2.9 kcal/mole, and for benzene is 4 kcal/mole. The heat of adsorption of xylene isomers con-siderably differ from each other, and decreases in the order *p*- < *m*- < *o*-. This order indicates that the energy of interaction increases with the in-crease in the asymmetry in the distribution of electron density in the aromatic molecule.

The heat of adsorption depends on the way in which the barium sulphate is prepared, which influences its specific surface area, its size and particle shape, surface irregularities, the kinds of coprecipitated ions, and the sign and value of the electrokinetic potential.

Alkali metal tetraphenylborates

A study of the behaviour of sodium, potassium, rubidium, and caesium tetraphenylborates as adsorbents in gas chromatography, has been reported.[197] The thermogravimetric curves of these compounds show very good thermal stability up to at least 220°C.

The chromatographic columns were made of heat-resistant glass, 1.40 m × 3.6 mm i.d. The column packing was prepared by mixing 60–80 mesh hexamethyl silanized Chromosorb W with a 15% solution of the various tetraphenylborates in acetone. The chromatographic column was filled by tamping, and was conditioned in a stream of nitrogen for 12 h at 180°C. Retentions, calculated relative to n-heptane, of alkanes, aromatic hydrocarbons, alcohols, ketones, esters, and chlorocompounds were determined at 75°, 95°, and 115°C.

The elution of alkanes, aromatic hydrocarbons, and monochlorobenzene followed the order of their boiling points, proving the lack of interaction between the aromatic hydrocarbons and the stationary phases. The alcohols, however, were strongly retained. The retention diminished from the primary to the tertiary alcohols. The elution of the primary alcohols and ketones was in the reverse order of their boiling points. Similarly, the elution of esters was generally in the order methyl butyrate to methyl acetate. There is a strong interaction between the alkali metal tetraphenylborates and the first members of a homologous series of alcohols, ketones, and esters. In some instances, this interaction overcomes the effect of the boiling points of a given series. The relative retention of the chloromethanes does not vary much from one column to another, except for methylene chloride which is retained longer on sodium tetraphenylborate.

4.2.4 Modified adsorbents

Gas–solid and gas–liquid chromatography are the techniques traditionally used in gas chromatography, and though the attractive features of the former have been recognized, the latter has been applied far more extensively. The factors limiting the use of gas–solid chromatography have already been mentioned, the main drawback being surface heterogeneity, which is responsible to a greater or lesser extent for peak asymmetry and, in some instances, for irreversible adsorption.

The curvature of the adsorption isotherm and the shape of the peak depends on the presence of "active sites" on the adsorption surface. This generic term is used to indicate every part of the adsorbent surface where the heat of adsorption will be greater than on a plane surface. Such sites can arise from irregularities on the surface such as pores and cracks, or from the presence of chemically active groups such as ions, electron donors, or acceptors.

Many attempts have been made to deactivate an adsorptive surface, and various types of liquid or solid "tailing reducers" have been used. Basically, the technique consists in modifying the properties of an adsorbent by adding a limited amount of stationary phase resulting in an important decrease in analysis time, suppression of peak tailing, and increase in selectivity. The use of these compounds to obtain symmetrical peaks is a well-known practice in gas chromatography.

(a) *Liquid modifiers*

If an adsorbent is treated with a small proportion of a non-volatile liquid such as any of those previously described as stationary liquid phases, the liquid is adsorbed and forms a stable layer on the surface. If the adsorbent is heterogeneous, the liquid will tend to occupy the sites of greatest activity, leaving empty those of lesser activity, with the result that the adsorbent becomes more homogeneous. In practice, this renders chromatographic peaks more symmetrical.

The use of small proportions of a non-polar stationary phase to modify an active charcoal in order to enable it to produce symmetrical peaks was first described by Eggertsen et al.[795] They succeeded in separating hydrocarbon mixtures by adding squalane to a Peletex Carbon column and, in the sub-monolayer region of liquid concentration, observed a sharp decrease in the retention time as the liquid to solid ratio was increased.[796] In this instance, the most active surface sites are preferentially coated by the liquid phase and are prevented from playing any dominant role in the separation process.

Di Corcia et al.[401] found that Graphon, a partially graphitized carbon black with a surface area of about 100 m^2/g (which shows good mechanical resistance in contrast with more graphitized carbon black), could be satisfactorily used as a versatile medium for modified gas–solid or gas–

liquid–solid chromatography.[797] A typical application of Graphon deactivated with very small amounts of liquid phase is the separation of light hydrocarbons from air, hydrogen sulphide, and carbon dioxide at a relatively high temperature (0°C). In this instance, Graphon (40–60 mesh) was modified with 0.3% w/w glycerol. Such small amounts of liquid phase serve only to deactivate the few active sites of the adsorbent so that acetylene and hydrogen sulphide also yield symmetrical peaks. In this way, the selectivity of graphitized carbon black is completely utilized. The elution order follows approximately that of the polarizability of the molecules, the most polarizable being preferably retained. In 5 min, a good separation of air, methane, carbon dioxide, acetylene, hydrogen sulphide, ethylene, and ethane is obtained at 0°C with a 2 m×3 mm i.d. column with hydrogen as carrier gas flowing at 60 ml/min.

A column (2 m×3 mm) packed with Graphon coated with an equimolar mixture of glycerol and phosphoric acid, 0.3% w/w, operating at room temperature, separates air, sulphur dioxide, and water (the gas chromatographic analysis of sulphur dioxide is one of the most pressing in air pollution control). No tailing is observed for the sulphur dioxide peak even at very low concentrations, and with an electron capture detector a detectable peak is obtained even when the quantity is about 0.1 μg. Water is eluted much later than sulphur dioxide, so that the amount of water does not affect the analysis of air, sulphur dioxide, and water.

When coated with various amounts (between 0.5 and 10% w/w) of a basic liquid phase such as tetraethylenepentamine, Graphon was successfully used for the linear elution of aliphatic amines. Linear elution of nanogram amounts of acids and phenols has been achieved by coating the surface of the graphitized carbon black with a suitable amount of an acidic stationary phase such as FFAP.[798, 799]

These results show clearly that by modified gas–solid chromatography, linear elution of very polar compounds is made possible. Mainly by varying the nature and the surface concentration of the liquid, an extremely wide range of selectivity is made available.

An extensive study of adsorptive modifications obtained by the addition of various amounts of a liquid phase to graphitized carbon black was made by Di Corcia et al.[800] They used Graphon and graphitized

Sterling FT-G (60–80 mesh) with specific surface areas of about 110 and 13 m²/g, respectively, coated with various amounts of poly(ethylene glycol) 1500 (PEG 1500). All the chromatographic columns used were 2 mm i.d. glass tubing, 1.4 m long. In order to obtain information on modifications to the adsorptive process, the values of heats of adsorption were calculated for an alkane (pentane) and for a set of alcohols (methyl, ethyl, and propyl alcohols) as a function of the percentage of the liquid phase (PEG 1500) added. In the Sterling FT-G+PEG 1500 system, at low percentages of PEG 1500, the heat of adsorption of pentane remained unchanged, whereas it increased for the alcohols. By increasing the amount of PEG 1500, a rise in the heat of adsorption was observed for all the eluates, and a maximum appears. After the first maximum, the heats of adsorption show, to a greater or lesser extent, a steep decrease. At relatively high percentages of PEG 1500, small but real second and third maxima appear.

This behaviour can be explained by noting that the general shape of these curves is analogous to those obtained for the multimolecular adsorption of a single component on a homogeneous surface. The thermodynamic evaluation has been also made in terms of surface coverage θ, which is the ratio between the number of adsorbed macromolecules and the ones needed to form one complete monolayer. The formation of one complete monolayer is assumed just after the first maximum is observed. Assuming that the adsorbing surface is very homogeneous, at low surface coverage of PEG 1500 it is hardly possibile for a single molecule of eluate to be adsorbed on pre-adsorbed macromolecules of PEG 1500. The heat of adsorption thus remains unchanged up to a certain surface coverage.

The progressive rise in the heat of adsorption with the increase in surface coverage is considered to arise from lateral interactions which can be established on the adsorbing surface between molecules of PEG 1500 and those of the eluate. As soon as macromolecules of PEG 1500 of the first layer are more or less closely packed, a two-dimensional condensation can take place. Adsorption heats for eluate molecules on the top of this monomolecular layer are generally smaller than the heats for adsorption on the last portions of the graphitized carbon black surface being occupied. As a consequence, a steep drop in the heat of adsorption–coverage dependence is observed near the completion of the first layer. On top of this layer, other macromolecules of PEG 1500 can be adsorbed, and a second layer starts to form. The completion of the second layer is again indicated by the appearance of a more or less pronounced maximum in the value of the heat of adsorption. The subsequent maxima and minima are less pronounced as the adsorption on the second and succeeding layers is not so well defined as on the first.

Evaluating the differences in the adsorption process between polar and non-polar eluates in terms of the surface coverage of liquid phase, it has been shown that at a low surface concentration (0.2% PEG 1500), heats of adsorption for alcohols show a rise which is more rapid for the more polar compounds, whereas adsorption of pentane does not show any change in the heat of adsorption. The hydroxyl groups of adsorbed alcohol molecules can establish long-range, specific, attractive forces with the oxygen atom of the $-CH_2OCH_2-$ links of poly(ethylene glycol) molecules, whereas for pentane, only weak, non-specific, van der Waals forces are operative. The completion of a more or less closely packed monolayer of PEG 1500 is apparent at about 0.65% w/w. At the completion of this monolayer, a new adsorbing surface containing functional groups alternating regularly with hydrocarbon groups is achieved. This layer is thus able to adsorb polar molecules more strongly than non-polar ones. Pentane shows, therefore, a much sharper decrease in heat of adsorption than alcohols. Only about one-half of the number of molecules required to fill the first monolayer is needed to fill the second layer. This decrease can presumably be accounted for by the large macromolecules bridging depressions on the surface of the solid. It follows that the surface area of the first layer is less than that of the adsorbing solid. In the filling up of the second layer, lateral interaction contributions once more cause an increase in the heat of adsorption for alcohols, and a second maximum appears when the second layer of PEG 1500 is nearly complete.

For pentane, weak, lateral interactions with macromolecules of the second layer result in a distinct step in the heat of adsorption dependence which rapidly tends to approach the heat of solution.

Surface heterogeneities can hinder to a greater or lesser extent the multi-layer deposition of the liquid phase. Graphon has a higher degree of surface heterogeneity than Sterling FT-G. Thus the behaviour of heats of adsorption on a Graphon-PEG 1500 system presents some differences, though the general shape of the dependence of the heats of adsorption is analogous to that already discussed. Thus the initial decrease in the heat of adsorption for pentane and propanol is considered as resulting from the preferential adsorption of PEG 1500 on high-energy sites on the Graphon surface, which thus become unavailable for adsorption of eluates. The appearance of maxima and minima in the heat of adsorption–coverage dependences is not so well defined as for adsorption on Sterling FT-G + PEG 1500. The smoothness in the heat of adsorption dependences indicates that multimolecular deposition of the liquid phase does not occur in a regular way and adsorption of molecules of PEG 1500 can take place on the top of other macromolecules before the underlayer is complete. The formation of one layer of PEG 1500 in this instane is approximately established at about a 4.5% w/w loading.

The influence of the molecular weight of the poly(ethylene glycol) on the density and structure of the monolayer on graphitized carbon black was investigated by Kiselev et al.[801] Polyethylene glycols with average molecular weights of 300, 3000, and 15,000 were used. The authors concluded that among the PEG samples investigated, only PEG 300 molecules, with the lowest molecular weight, straighten out almost completely and pack densely on the surface of the carbon black under the influence of the adsorption forces. With PEG 3000, and particularly with PEG 15,000, complete stretching out of the macromolecules on the surface of carbon blacks does not occur. It is considered that these macromolecules are not completely linear and that they do not contact the surface of the carbon black with all their units. This is supported by the fact that the concentration of ether groups and terminal hydroxyl groups on PEG which are capable of specific interaction with adsorbate molecules on the surface of a dense monolayer, is greater for PEG 300 than for PEG 3000 and PEG 15,000. Accordingly, on dense monolayers of PEG with different molecular weights, the retention volumes of the n-alkanes which are adsorbed non-specifically are the same, but the retention volumes of compounds capable of specific interaction with the functional groups of PEG decrease with increasing molecular weight of PEG.

Bruner et al.[802] report a detailed study of the changes in the chromatographic parameters and thermodynamic functions induced by modifying the adsorptive surface of Sterling FT (specific surface area, 15 m²/g) with various amounts of different liquid phases. Squalane, phenanthrene, and glycerol were chosen as modifiers because of the well-defined chemical structure of these compounds with respect to more common liquid phases for gas–liquid chromatography and because the decisive difference between squalane (non-polar) and glycerol (very polar) could make the interpretation of the results easier. Isotopic molecules and isomers were taken as the solutes because the separation of the former is essentially

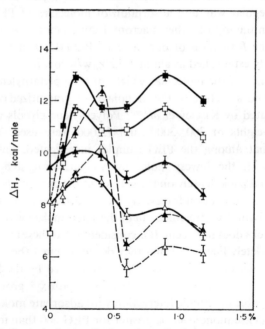

FIG. 35. Isosteric heats of adsorption of n-butane, n-pentane, ethyl alcohol, and propyl alcohol at various loadings of liquid phase on Sterling FT.[802] (△) butane; (□) ethyl alcohol; (▲) pentane; (■) propyl alcohol; (—) glycerol; (- - -) squalane.

affected by differences in molecular polarizability and of the latter by differences in geometrical structure.

In Fig. 35 the heats of adsorption of n-butane, ethyl alcohol, n-pentane, and n-propyl alcohol are plotted against the percentage of glycerol and squalane. A sharp increase in the heat of adsorption is observed for both ethyl alcohol and n-propyl alcohol in the first part of the curves. The maximum heat of adsorption is reached at 0.2%. After the first maximum the formation of a monolayer of glycerol produces an inflection in the curve. The monolayer hinders the interaction with the surface. A second maximum also occurs, because of the lateral interactions of the alcohol molecule during the formation of a second glycerol mono-layer.

Lateral interactions also occur between hydrocarbons and glycerol, as shown by the two maxima in the heat curves, but the energy involved is much less because of the absence of the hydrogen bond and the "squeezing" effect of the glycerol molecules. The smooth decrease in the mono-layer region (0.4–0.6%) is explained by the irregularity of the carbon black–glycerol surface. The effect of lateral interactions is very important for hydrocarbons when the liquid phase is squalane.

It is interesting to note that the difference in the heat of adsorption at zero loading and at the maximum of the lateral interactions increases with the number of carbon atoms at about 1 kcal/mole per carbon atom. This has been verified experimentally from propane to n-hexane and is considered as the contribution of lateral interactions per —CH_2 group.

A very sharp fall in ΔH is found for hydrocarbons when using $> 0.4\%$ of squalane. In this region, the value found for squalane is lower than for glycerol at the same coverage. This is explained by a strong "shielding effect" by the monolayer on the adsorption forces. The hydrocarbon molecule is surrounded by a liquid of similar structure and a process approaching gas–liquid chromatography takes place. This is confirmed by the separability of isomers. Thus after a monolayer of liquid phase has been formed, the interactions with compounds of similar structure (alcohols on glycerol and hydrocarbons on squalane) act so as to diminish the differences in heats of adsorption. The adsorbed molecule interacts preferentially with the liquid phase of the analogous molecular structure, neglecting the adsorbent because of the "shielding effect". This makes the differentiation between the two isomers problematic. If the liquid phase and the molecules eluted are insoluble in each other (as for hydrocarbons on the glycerol), adsorption on the two-dimensional liquid layer takes place after the monolayer is formed. This explains the high values for the differences in the heats of adsorption of hydrocarbon isomers at high glycerol percentages.

The effect of the addition of modifiers on the separation of isotopic molecules has been studied making use of squalane and phenanthrene

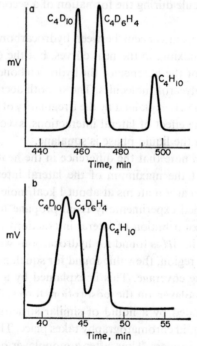

Fig. 36. Chromatogram showing the separation of C_4H_{10}, $C_4H_4D_6$, and C_4D_{10}. (a) Column: 105 m×4 mm containing Sterling FT+0.2% squalane; temperature: 45°C; hydrogen carrier gas flow rate: 300 ml/min; inlet pressure: 20 kg/cm². (b) Column: 5 m×2 mm, containing Sterling FT+4% squalane; temperature: 0°C; hydrogen flow rate: 100 ml/min; inlet pressure: 7 kg/cm².[802]

as modifiers, with C_4H_{10}–C_4D_{10} and C_6H_6–C_6D_6 as the isotopic systems. This allows a comparison to be made between the effects on the isotopic separation of interactions with molecules of analogous and different chemical structure.

It has been established that lateral interactions with squalane enhance the separation of both the isotopic pairs. This is considered to arise from the differences in molecular polarizability which are mostly responsible for the separation of deuterated and hydrogenated species in gas chromatography. Squalane, because of its hydrocarbon structure, is particularly suitable, as has also been shown in a study using capillary columns for gas–liquid chromatography by Bruner et al.[803] The ratio of retention times found for the pair C_6H_6—C_6D_6 was, for gas–liquid chromatography using squalane, 1.04 at 0°C. The maximum value found in modified gas–solid chromatography was 1.08, showing again the analytical importance of this technique. Lower values for both C_6H_6—C_6D_6 and C_4H_{10}—C_4D_{10} were found on phenanthrene. In this instance the π-electron system is less useful for exploiting the differences in polarizability. As an example, Fig. 36 shows a comparison between two separations obtained on different columns for an isotopic system, C_4H_{10}—$C_4D_6H_4$—C_4D_{10}.

As a second example of the effectiveness of modified gas–solid chromatography, Fig. 37 shows a chromatogram illustrating the elution of an aqueous solution containing about 30 ppm each of C_2–C_5 aldehydes on Sterling FT-G $+0.2\%$ PEG 1500.

It is interesting to note the separation of 2-methyl butyl alcohol and 3-methyl butyl alcohol. The difficulty in separating this isomeric pair by gas–liquid chromatography arises from the slight difference ($\sim 0.2°C$) in the boiling points of the two components. The non-polar, flat surface of the graphitized carbon black is, however, particularly suitable for separating molecules according to differences in their geometrical structure, disregarding the existence of functional groups in the molecule. Thus the action of a liquid phase added to homogeneous adsorbing media such as graphitized carbon black causes a deactivation of the surface at very low percentage loadings. It should be pointed out, however, that partial orientation of the liquid layers nearest to the solid can affect the

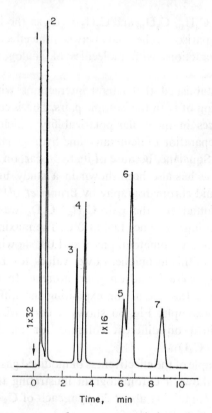

FIG. 37. Chromatogram of an aqueous solution of C_2–C_5 aliphatic aldehydes. Column: 1.4m×2 mm containing Sterling FT-G+0.2% PEG 1500; column temperature: 101°C; linear carrier gas velocity: 11 cm/s; sample size of each aldehyde, ~30 ng. (1) Acetaldehyde, (2) propionaldehyde, (3) 2-methyl propionaldehyde, (4) butyraldehyde, (5) 2-methyl-butyraldehyde, (6) 3-methyl-butyraldehyde, (7) pentanal.[800]

partitioning properties of the liquid phase to a greater or lesser extent, depending upon the surface activity of the adsorbing medium and the nature and amount of the liquid phase.

Guillemin et al.[767] have systematically studied the relation between the physical characteristics and chromatographic properties of modified Spherosil coated with various stationary phases. Coating Spherosil with liquid phases is similar to coating classical supports. It may be useful to distinguish between hydrophilic and hydrophobic phases. For hydrophilic coatings like β, β'-oxydipropionitrile or Carbowax 20 M, no special precautions have to be taken and, in particular, preliminary dehydration of Spherosil is not necessary. Dehydration is required, however, if hydrophobic coatings, such as silicone oils, are used. The experimental study was carried out with Spherosil of specific surface areas 3, 7, 20 and 64 m²/g coated with β, β'-oxydipropionitrile at 1, 2, 5, 10, 20, and 30% w/w loadings. A synthetic mixture of ethyl ether, acetone, methyl ethyl ketone, and methyl *iso*butyl ketone was used under the following conditions: column, 1 m×4 mm i.d.; particle size, 150–200 μm; carrier gas, nitrogen, 3 l/h; column temperature, 110 °C.

It was found that the asymmetry of the peaks decreased rapidly with increasing thickness of β, β'-oxydipropionitrile layer on the various types of Spherosil. The specific retention volumes for these solutes first decreases with increasing layer thickness of stationary phase, passes through a relatively well-defined minimum and finally increases upon further addition of liquid phase. This dependence can be explained by considering that for very small values of average layer thickness, gas–solid adsorption is still the dominating process, and gives rise to broad asymmetrical peaks. The presence of a minimum, however, indicates that the separation mechanism gradually changes. The authors hold that the intermediate transition region may serve as a definition for modified gas–solid chromatography and is characterized by the occurrence of the smallest retention time and of narrow and nearly symmetrical peaks.

The value of the average layer thickness for which the specific retention volume is minimal depends on the nature of the liquid phase. For β, β'-oxydipropionitrile, it is about 1.2–1.5 nm for the different types of Spherosil investigated, except for that with the smallest surface area (3 m²/g), which behaves differently and for which, apparently, the average layer thickness is much greater. This discrepancy is believed to arise from non-uniform coating.

To a first approximation the practical rules which govern separations using Spherosil for modified gas–solid chromatography are the following: at constant surface area and with variable average layer thickness, selectivity is variable; with variable surface area but constant average layer thickness, selectivity is constant.

The specific surface area of the chosen Spherosil should be smaller, the larger the polarizability of the molecules to be separated.

When Spherosil is used in modified gas–solid chromatography, the analytical applications of this adsorbent are extended to the separation of polar compounds such as alcohols. A separation of twelve alcohols (from methyl alcohol to 1-decanol) may be carried out in 20 min on Spherosil (5 m^2/g)+10% Carbowax 20 M (column, 2 m×3 mm i.d.; nitrogen flow, 3 l/h; temperature programmed, 70–230°C at 20°C/min). Spherosil used in adsorption chromatography gives, for instance, only an incomplete separation of butenes.[770] In the modified gas–solid technique, however, this separation is complete. A mixture of C_2–C_4 hydrocarbons is entirely separated on a 4 m column of Spherosil (85 m^2/g)+10% squalane at 60°C, using a nitrogen flow of 3.6 l/h. Also, aromatic hydrocarbons, high boiling aliphatic hydrocarbons, aliphatic halogenated hydrocarbons, aldehydes, thiols, various oxygenated compounds, and pesticides have been separated in this way.[767]

MacDonell et al.[804] studied the potential usefulness of pretreated porous glass in gas chromatography. All the porous glass was initially heated to 500°C and held at this temperature for 2 h. Isobutyl alcohol, methyl isobutyl ketone, and dibutyl ether were reacted with the porous glass (50–60 mesh) by simply soaking the glass in one of the solvents or by refluxing the solvent containing porous glass for several hours. After this treatment the porous glass was filtered and dried at 140°C for 4 hours. Chromatographic columns prepared from 6 ft×¼ in. (1.8 m×6.3 mm) copper tubing packed with the porous glass, and which received no additional treatment, were unsatisfactory in that they retained all samples most tenaciously, resulting in long retention times and low, wide peaks with considerable tailing.

Considerable shortening of retention times as well as an improvement in peak response was noted when the porous glass was pretreated with organic compounds. Simply soaking the glass in a solvent reduced its adsorptive character considerably, and refluxing for several hours reduced it still further because of the blocking or deactivation of silanol groups. It is to be remembered that although the retention times of the aliphatic compounds were shortened with almost every treatment, oxygen-containing species were still very strongly adsorbed.

It is believed that the improvement of porous glass for chromatographic use is a result of a reaction or reactions which permanently destroy the hydroxyl component of some silanol groups. In addition, all silanol groups are not permanently deactivated during refluxing because of hydrogen bonding during or before refluxing, which blocks the more permanent reactions at some sites; on subsequent heating these temporarily blocked groups revert to active silanol groups. Finally, the ability of modified porous glass to retain oxygen-containing organic compounds arises from the presence of both residual and reverted silanol groups.

Rowan and Sorrell[805] also found that chemical modification of Davison's wide-pore silica gel by the reaction of an alcohol or a chlorosilane with the surface hydroxyl groups of the gel, increased the column efficiency and reduced peak tailing for the adsorbates studied (hexane, heptane, 1-hexene, octane, cyclohexane, and benzene). The modified gels which gave higher column efficiency and more symmetrical peak shapes were those which had a lower concentration of high-energy adsorption sites.

A fine alumina, fibrillar colloidal boehmite, has been found to be a useful adsorbent for carrying out certain separations by modified gas–solid chromatography by Kirkland.[806] For instance, the stearic acid-modified material permitted the separation of the Freon fluorocarbon mixture at 50°C, similar to that achieved by the unmodified column at 100°C. It is believed that the stearic acid is probably chemisorbed as an oriented monomolecular layer on the surface of the boehmite.

McKenna and Idleman[807] have separated isomeric 2-butenes on alumina modified with propylene carbonate. Urbach has reported the separation at 92°C of C_6–C_9 alkanoic acids on a column 200×0.4 cm packed with Dowex 50 (52–80 mesh) coated with 1.5% squalane.[808]

Generally, modification of non-porous and widepore adsorbents by adsorbed dense monolayers gives the possibility of obtaining quite homogeneous surfaces of different chemical composition with low adsorption energies. Furthermore, monolayer films held by adsorption forces on a large surface area of a strongly sorbing adsorbent are significantly more stable to heat than thick layers of these compounds on supports with small and weakly adsorbing surfaces such as are used in gas–liquid chromatography.[396] Also, exchange processes (adsorption and desorption) on the surfaces of monolayers take place much more rapidly than in the bulk of liquids.

Modification of adsorbents with different liquid stationary phases has some particular analytical features:

(i) it is of general applicability, the advantages of gas adsorption chromatography being combined with those of gas–liquid chromatography;

(ii) it gives high selectivity;

(iii) it gives great versatility (any kind of liquid or solid modifier may be used);

(iv) higher column efficiency is achieved than in gas–liquid chromatography.

Finally, it should be noted that the term gas–liquid–solid chromatography has a mainly historical meaning because the liquid phase adsorbed on the surface behaves completely differently from the bulk liquid and because solids can be used as modifiers with the same mechanism.

(b) *Solid modifiers*

Vidal-Madjar and Guiochon[809] propose the use of crystallized organic compounds coated on graphitized carbon black for preparing suitable adsorbents of various selectivities for gas–solid chromatography. Thus excellent chromatographic results can be obtained with anthraquinone or copper phthalocyanine coated on graphitized carbon black. With the latter modifier, high efficiency was obtained for various phenols, aniline, and nitrobenzene. The high retention of anthracene relative to phenanthrene (1.16 at 265°C) is unusual as is the high selectivity to xylene isomers, which is much greater than the selectivity of uncoated carbon black.

A column, 1 m×2 mm i.d., packed with 1% copper phthalocyanine on graphitized carbon black Sterling MT, separates isothermally at 265°C a mixture of polynuclear aromatic hydrocarbons (naphthalene, diphenyl, 1-methylnaphthalene, acenaphthene, fluorene, phenanthrene, and anthracene) in 10 min. The column is stable at 300°C for at least several weeks.

Studies on copper, nickel, cobalt, zinc, and iron phthalocyanines coated on graphitized carbon black showed the great affinity of these metal phthalocyanines for aromatic rings.[810, 811] They also showed that cobalt phthalocyanine, because of its specificity toward π-electrons, has very useful properties for the separation of aromatic amines, pyridine and methyl-substituted pyridines. Franken *et al.*[812] compared the behaviour of cobalt phthalocyanine toward these nitrogen derivatives when used as a packing in conventional columns and as a porous coating on the walls of open-tube columns.

The chromatogram in Fig. 38a shows the analysis of a mixture of aniline, *N*-methylaniline, *N,N*-dimethylaniline, and the three toluidines, using

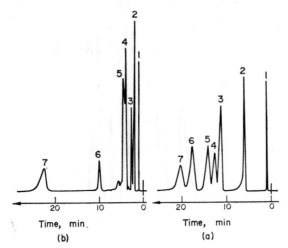

FIG. 38. Analysis of nitrogen compounds on a conventional packed column: 5% cobalt phthalocyanine on Sterling MTG. Column length: 4 m×0.2 cm i.d.; carrier gas: hydrogen; temperature: 178°C; $\Delta P = 4$ atm. (a) (1) Methane, (2) aniline, (3) *o*-toluidine, (4) *N,N*-dimethylaniline, (5) *N*-methylaniline, (6) *p*-toluidine, (7) *m*-toluidine. (b) (1) Methane, (2) 2-picoline, (3) 2,6-lutidine, (4) 2,3-lutidine, (5) 2,4-lutidine+ 2,5-lutidine, (6) pyridine, (7) 3-picoline+4-picoline.[812]

the conventional packed column at 178°C. The chromatogram in Fig. 38b shows the separation of a mixture of pyridine, picolines, and lutidines on the same conventional packed column, under the same experimental conditions. As may be seen, the selectivity toward nitrogen compounds is large and the eluting peaks are reasonably symmetrical. This is illustrated by the retention behaviour of pyridine which is retained more than aniline, 2-picoline, and the 2-substituted lutidines, although pyridine has a lower molecular weight.

The chromatogram in Fig. 39 shows the separation at the same temperature of a similar mixture using a porous-layer open-tube glass capillary column prepared with the same adsorbent. The two mixtures which were analysed separately on the packed column, are analysed together in only one run in a very short time (150 s). However, when using the capillary

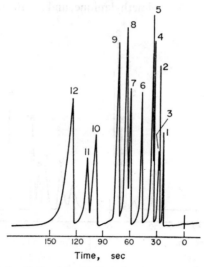

FIG. 39. Analysis of nitrogen compounds on a porous-layer open-tube capillary column: 5% cobalt phthalocyanine on Sterling FTG. Column: 10 m×0.5 mm i.d.; carrier gas: hydrogen+ammonia; temperature: 178°C; $\Delta P = 0.35$ atm. (1) Methane, (2) 2-picoline, (3) 2,6-lutidine, (4) 2,3-lutidine, (5) 2,4-lutidine+2,5-lutidine, (6) aniline, (7) pyridine, (8) o-toluidine+N,N-dimethylaniline, (9) N-methylaniline, (10) p-toluidine, (11) m-toluidine, (12) 3-picoline+4-picoline.[812]

column, ammonia must be added in small concentration to the carrier gas to deactivate the acidic sites on the glass surface.

The selectivity of cobalt phthalocyanine towards nitrogen-containing molecules may be shown by the influence of steric hindrance on the retention behaviour. Thus, 2-picoline and the lutidines, which have a methyl group in the 2-position, elute before pyridine with an unhindered nitrogen atom, although they have higher molecular weights. This effect is also demonstrated by the fact that 2,6-lutidine and 2-picoline have almost the same retention time. As might be expected from the increased shielding of the nitrogen atom in 2,6-lutidine, this compound elutes before 2, 3-, 2, 4-, and 2, 5-lutidine which have only one methyl group adjacent to the nitrogen atom.

This explanation also appears to be valid for substituted anilines; N, N-dimethylaniline is less retained than N-methylaniline. From the shorter retention time of o-toluidine compared to N-methylaniline, it may be concluded that the shielding by the methyl group in the o-position is greater than the shielding by a N-methyl group in substituted anilines.

The selectivity toward nitrogen compounds arises from the cobalt phthalocyanine, as the elution order on graphitized carbon black or phthalocyanine itself is completely different and follows the molecular weight order.

Cadogan and Sawyer[120] show that coating Graphon with lanthanum chloride which contains partly filled d- or f-shells, also may have analytical advantages. Because such salts are strong Lewis acids, they should form complexes of varying strengths with electron-donor molecules. Thus the elution of n-hexane is essentially unaffected by substituting lanthanum chloride for sodium chloride on the surface. In contrast, aromatic adsorbates interact much more strongly with lanthanum chloride. This implies some type of specific interaction between the aromatic ring and the lanthanum chloride. Electron-donating substituents (alkyl groups) enhance the interaction of the aromatic ring. Also 1, 4-cyclohexadiene appears to complex somewhat with lathanum chloride on Graphon, an effect that is less pronounced for 1-hexene. Similar behaviour occurs at the lanthanum chloride-silica gel surface.

An industrial macroporous silica gel of specific surface area 17 m²/g and mean pore diameter about 50 nm was used in modified gas–solid chromatography by Kiselev *et al.*[801] The modifying compound was polyarylate Ph-I of average molecular weight about 30,000. The macromolecule is composed of the monomer units shown below. It is a specific adsorbent carrying on its surface negative charges concentrated mainly on the oxygen atoms of the carbonyl and ether groups.

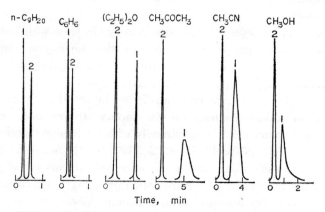

The average concentration of the polymer on the surface of the silica gel, as determined by the loss of weight during calcination of the modified adsorbent at 800°C in air, was 2.3 mg/m². The modified silica gel was sieved and the fraction between 0.25–0.5 mm taken, dried *in vacuo* at 200°C, loaded into the chromatographic column, 100×0.4 cm, and given

FIG. 40. Examples of chromatograms of a number of organic compounds on (1) initial wide-pore silica gel, (2) wide-pore silica gel modified with a thin layer of polyarylate. Temperature: 150°C; column: 100 cm×0.4 cm; carrier gas (nitrogen): flow rate 40 ml/min.[801]

a preliminary heating in a current of nitrogen at 200°C until a stable base line was obtained.

Figure 40 shows that a deposit of a thin layer of Ph-I on the surface of the silica gel leads to increased retention times for n-alkanes and aromatic hydrocarbons. However, for ethers, ketones, acetonitrile, and alcohols, which are all capable of forming strong hydrogen bonds with the silanol groups of the silica gel, a sharp decrease in retention times is observed after modification of the silica gel surface with the Ph-I.

The adsorption properties of the polymer film deposited on the silica gel are substantially different from the properties of the porous Ph-I polymer. Thus naphthalene emerges from a column of silica gel, modified by a film of Ph-I, with a very narrow peak and much more rapidly than from a column of Ph-I. In order to explain this behaviour, it was assumed that the macromolecules of the polymer are probably arranged in a more orderly fashion on the surface of silica gel than in the network of the same porous polymer, being oriented to the surface of the support by their polar groups. This is demonstrated in particular by the smaller heats of adsorption (smaller by 2–3 kcal/mole) of molecules such as diethyl ether, acetone, and acetonitrile on silica gel modified by a film of Ph-I, than on the porous Ph-I.

Modification of silica gel surfaces by polyarylates sharply diminishes their selectivity and decreases the retention times and asymmetry of the chromatographic peaks of polar compounds such as ethers, esters, ketones, nitriles, and nitro compounds. Also this use of polyarylates sharply reduces the strong retention of n-alkanes, and in particular of aromatic hydrocarbons, characteristic of porous polyarylates and industrial porous polymers based on styrene and divinylbenzene.

The effect of water deactivation on retention by powdered etched glass containing 10% NaOH was discussed by Alberini et al.[813] Tailed peaks which are always obtained from when the dry powders are used, are completely eliminated in the alkali-treated material deactivated by the water vapour in the carrier gas. The possibility of changing the gas chromatographic properties of etched glass by controlling the moisture content makes this material a versatile adsorption medium. The column can be used over a wide range of temperatures (0–80°C) and for various analyses, the only requirement being that the deactivating water trap temperature should be variable. A rapid determination of the isotopic molecules HD and D_2 with a conventional column 2 m$\times$4 mm i.d. filled with alkaline-etched and iron(III) oxide impregnated glass powder or beads, 0.09–0.12 mm diam. was carried out by Gaumann et al.[814] After activation at 300°C, a complete separation of HD and D_2 was achieved in 2 min at -196°C using H_2 as carrier gas, flowing at 75 ml/min.

Various coating salts are compared and data are presented of the functional group contributions to adsorption on Porasils modified with Na_2SO_4, NaCl, LiBr, Na_3PO_4, Na_2MoO_4, $NiSO_4$, $CoSO_4$, $Al_2(SO_4)_3$, and $Cr_2(SO_4)_3$ in a paper by Isbell and Sawyer.[815] All salt coatings on the porous glass bead adsorbents were 10% by weight. It was established that for the practical separation of hydrocarbons and their halogenated derivatives, salt-coated Porasil C offers the best compromise in terms of speed of analysis, separation efficiency, and column temperature. Separations of aliphatic compounds are most effective with either sodium sulphate or sodium phosphate as the coating salt. For substituted benzenes cobalt(II) sulphate is superior to all the other salts studied in terms of separation efficiency. A comparison of the data obtained for salt-modified Porasil with that obtained for salt-modified aluminas indicates that many differences exist.[190, 192, 816] Thus, salt-coated aluminas are more selective for mixtures composed entirely of alkyl-substituted benzenes, halogenated benzenes or a homologous series of saturated hydrocarbons. Salt-coated Porasils are more selective for the separation of positional isomers of olefinic hydrocarbons and for the separation of alkyl-substituted benzenes from halogenated benzenes. The slight superiority in the selectivity of the aluminas is offset somewhat by the higher operating temperatures required and the slower analysis times. The main advantage of the aluminas is their ability to separate the cis and trans isomers of olefins. Salt-coated Porasil columns do not resolve such cis and trans isomers unless the operating temperature is lowered substantially below the optimum conditions for rapid separation. The Porasils have the advantage of being spherical and easy to pack, and their chemical inertness allows a wide range of operating conditions without catalytic decomposition of most adsorbates.

The resolution and analysis of the components of hydrogen–deuterium mixtures by a chromium(VI) oxide–alumina column operated at $-196°C$, with neon as the carrier gas, was reported by Hunt and Smith.[817] The column packing was prepared from activated alumina (20–40 mesh) to which 6.7% by weight of chromium(VI) oxide was added, packed into copper tubing 12 ft$\times\frac{5}{16}$ in. o.d. (3.6 m$\times$8 mm).

Materials similar to chromium(III) oxide may also be deposited on the alumina for accomplishing these separations. For example, similar separations were carried out using iron(III) oxide as the coating for the alumina.[818–820] Complete separation of the six isotopic compounds of hydrogen (H_2, HD, HT, D_2, DT, and T_2) in 30 min at $-196°C$ using a 3 m$\times$4 mm column filled with alumina (125–150 μm) covered by hydrated iron(III) oxide was reported by Genty and Schott.[821] Neon was used as the carrier gas flowing at 200 ml/min. For a good separation, the column had to be deactivated with carbon dioxide.

Chromium trioxide and anhydrous or hydrated iron(III) oxides, produce rapid *para* and *ortho* hydrogen interconversion and therefore give a unified peak for these isomers.

The use of sodium chloride-impregnated, sodium hydroxide-modified alumina for the separation of polycyclic aromatic hydrocarbons and of a sodium hydroxide-modified molecular sieve for the separation of naphthenes and alkanes have been reported.[822, 823]

An excellent review of the use of alumina, silica, or silica–alumina modified with inorganic salts, such as the alkali metal halides or hydroxides, was written by Phillips and Scott.[395] Modified adsorbents which have been found to give elution chromatograms with symmetrical peaks were classified as follows:

(a) Alumina modified with: H_2O, I_2, LiOH, LiCl, LiBr, NaOH, NaCl, NaBr, NaI, Na_2CrO_4, Na_2MoO_4, Na_2WO_4, KOH, KF, KCl, KBr, KI, KNO_3, K_2CO_3, K_2SO_4, K_2CrO_4, $MgSO_4$, CuCl, AgCl, AgI, $AgNO_3$, Ag_2SO_4, CdI_2, Na salts of fatty acids, urea.

(b) Silica modified with: LiCl, LiBr, LiI, NaCl, NaBr, NaI, KCl, KBr, KI, CuCl, CuBr, Cu(II) alanine, $Cu(II)(en)_xSO_4$, Cu(II) tartrate, Cu(II) phenylalanine (en = ethylene diamine).

(c) Silica–alumina modified with: LiOH, LiCl, LiBr, LiI, NaOH, NaF, NaCl, NaBr, NaI, KOH, KCl, KBr, KI, CsCl.

Symmetrical elution peaks are obtained only when the weight of modifier used is sufficient to give at least monolayer coverage.

In certain instances, e.g. alumina modified with lithium chloride, or bromide, considerable hydrolysis can take place, so that the final column has the properties expected from a mixed salt–hydroxide modifier. Hydroxide-modified columns tend to be extremely hygroscopic and should be stored with their ends sealed. A sodium hydroxide–alumina column left with open ends for more than a month will have formed a hard deposit of material at the column ends making it impermeable to the carrier gas.

Many columns are also changed by interaction with organic vapours. Thus silver nitrate on alumina is reduced to silver on alumina by olefins and amines at temperatures which vary with the exact mode of preparation but are generally above 100°C. Copper(II) chloride on alumina or silica gel is highly reactive and produces chlorination, aromatization, and cracking of hydrocarbons above 100°C.

It was observed that the effect of a modifier in increasing the retention of polar solutes generally increases with the polarizability of the ions. This is illustrated in Table 21 for alumina modified by sodium salts.[395]

TABLE 21. RETENTIONS OF HYDROCARBONS RELATIVE TO n-HEPTANE ON ALUMINA MODIFIED WITH SODIUM HYDROXIDE AND HALIDES AT 100°C[395]

Modifier	Cyclohexane	2,2,4-trimethyl pentane	1-heptene	benzene
NaOH	0.36	1.21	1.34	1.05
NaCl	0.28	1.28	1.80	1.97
NaBr	0.27	1.32	2.00	3.00
NaI	0.30	1.37	2.31	3.90

Phillips and Scott found that salt-modified columns have, in general, a very long life, and retention volumes are normally constant to within 1 or 2% with the same column either before and after continued use or before and after storage of the column.

Comparison between different columns prepared in the same way shows that retention volumes are generally reproducible to within about

5%, but to this must be added the variability of the surface area resulting from modification which, even under apparently similar conditions, can be ±10%.

4.2.5 Porous polymers

In recent years, macroporous copolymers have been used frequently as packings for gas chromatographic columns. Copolymers of styrene with divinylbenzene or of ethylvinylbenzene and divinylbenzene, have been especially popular. They are usually prepared by pearl copolymerization; the polymerization and cross-linking is carried out in the presence of a diluent which is a good solvent for the monomers but in which the polymer is not soluble.

The preparation and some of the properties of the porous polymer beads have been described by Lloyd and Alfrey.[824]

The structure and properties of porous polymers, i.e. the swelling capacity, particle size, specific surface, and the distribution and diameter of pores, are characterized by:

(a) the type and quantity of the inert diluent;
(b) the ratio of the initial monomer concentrations;
(c) the conditions of the polymerization (temperature, polymerization time, suspension stabilizer, mixer revolutions, etc).

The specific surface area measured by the BET method varies, according to the preparation conditions, between 20 and 700 m^2/g.

The use of these materials for gel permeation chromatography was described by Moore.[825] Later Hollis[386] first described the use of the macroporous copolymers of the styrene-divinylbenzene type in gas chromatography. At present various porous polymer beads are available and are designated by different trade marks, such as Porapaks (Waters Associates, Inc.), Polypak (F & M Scientific), and Chromosorb 100 series (Johns–Manville). Some physical properties of porous polymers are given in Table 14 (p. 108).

It is considered that the separation processes on these sorbents are different from those encountered in gas–liquid or gas–solid chromatography. Most analysts agree that in the separation process on porous polymers, adsorption and absorption take place simultaneously.[386, 826-830] The adsorption effect can be distinguished quite easily in the shape of the peaks especially of solutes which, because of their polarity, do not diffuse into the skeleton, and are separated primarily by adsorption. The slight asymmetry in their peaks is an indication of a non-linear separation isotherm. The working temperature of the chromatographic column has to be about 50°C higher than that of a conventional gas chromatographic arrangement in order to obtain acceptable retention times. Moreover, the shape of the chromatographic peak and the position of the peak maximum depends upon the sample size.

On the other hand, the molecules of the sorbate penetrate into the copolymer structure during the absorption process, so that the whole of the mass of the porous polymer can behave like a polyaromatic liquid phase. A number of authors[831-834] have studied the retention of molecules with different electronic structures on porous polymers of styrene, ethyl styrene, and divinylbenzene (Porapak Q, Synachrom, polysorbs). It has been found that a linear dependence exists between the logarithm of the corrected retention volume and the number of carbon atoms in the molecule. The above dependence has been observed in particular in homologous series of alkanes, aromatic hydrocarbons, alcohols, ketones, and fatty acids. The logarithm of the corrected retention volume is also a linear function of polarizability, boiling point, molecular weight, and standard entropy of the molecules of the homologous series.

These porous polymers behave as weakly specific sorbents in which the contribution of non-specific dispersion interactions to the total energy of the molecules of different compounds on these polymer surfaces predominates. This can be demonstrated if the retention volumes of different organic compounds are compared with the retention volumes of their fluorinated analogues.[391] Thus, when the dispersion interaction energy is evaluated, the values of the electron polarizability and the van der Waals radii of groups of interacting molecules are very important. The polarizabilities of hydrogen and fluorine are slightly different, but the van der Waals radius of hydrogen is less than that of fluorine. Therefore the value of α/r^6 characterizing the dispersion interaction is less for fluorine than for hydrogen, and this must result in a decrease in the retention volumes and heats of adsorption (especially for the predominant appearence of non-specific dispersion interactions).

For mixtures of methyl acetate and trifluoromethyl acetate, propyl alcohol and perfluoropropyl alcohol, n-amyl alcohol and n-perfluoro-amyl alcohol, acetophenone and perfluoroacetophenone, the substitution of hydrogen atoms by fluorine atoms decreases the retention volumes (even for one fluorine atom in acetophenone). The heats of adsorption of the fluorinated compounds on weakly selective porous polymer are generally less than those of the non-fluorinated compounds. This shows that the predominant contribution to the total energy of interaction is made by the non-specific dispersion interaction.

The qualitative analogy established in the retention of molecules from different classes on Porapak Q, Polysorbs, and graphitized carbon black prove that the interactions are qualitatively similar.[833] The differences between the retentions of aromatic and cyclic hydrocarbons and normal hydrocarbons with the same number of carbon atoms in the molecule on porous polymers and graphitized carbon black are considered to arise from the particular geometrical structures of these sorbents.

Styrene, ethyl styrene, and divinylbenzene-based non-polar porous polymers can be obtained under the commercial names: Porapak P, Porapak Q, Chromosorb 101, 102, 105, Synachrom, Polysorbs 1–9, PAR-1 and PAR-2.

The non-specificity of molecular adsorption on these sorbents allows their use for the analysis of mixtures of strongly polar compounds. Water is rapidly eluted and its peak is symmetric. The extreme tailing of amine peaks eluted from a column of Chromosorb 102 has been ascribed to unreacted vinyl groups, which are the active sites, which can be removed by the addition of hydrogen fluoride to the double bond of the vinyl group.[392]

The introduction of monomers with different functional groups into the polymerizing mixture gives rise to porous polymers with specific sorption properties. Such porous polymers, known by commercial names such as Porapak N, R, S, T, Chromosorb 103, 104, and Polysorbate-2, differ from each other by the quantity and nature of polar centres. The use of polar porous polymers modifies the retention time of polar molecules. Thus, whereas on non-polar porous polymers water elutes between ethane and propane, it elutes between propane and *iso*butane from Porapak N, S (more polar polymers), and after n-butane from Porapak T.[835] General views of the separation possibilities using Porapak columns have been published.[836, 837]

The Porapak Q columns are considered to be universal, i.e. they may be used for the separation of gases, C_1–C_{10} hydrocarbons, alcohols, esters, and sulphur compounds. Porapak N is recommended for the determination of formaldehyde in aqueous solutions and for the separation of acetylene from other C_2 hydrocarbons. On Porapak R columns, a high efficiency has been observed for the separation of amines, nitriles and nitro compounds, for the separation of water and highly active organic compounds such as chlorine and hydrochloric acid. Weakly polar Porapak S is useful for the analysis of carbonyl compounds, alcohols, and halogenated compounds. Carboxylic acids, alcohols, and amines are more strongly retained on Porapak S than on Porapak Q. The most polar polymer, Porapak T, has high retention volumes and efficiency for the separation of acids and amines

Johns–Manville silanized Porapaks, Porapak P=S and Q=S, have increased efficiency, but if we look at the retention characteristics, they are identical with the respective non-silanized sorbents.

Gvosdovich *et al.*[838] investigated and compared the adsorptive and chromatographic properties of the porous polymers Chromosorb 101 and Chromosorb 104 with those of Chromosorb 102. Chromosorb 101 has the same chemical nature as Chromosorb 102 but has larger pores (average pore diameter of Chromosorb 101 is 300–400 nm and 102 only about 8 nm) and has approximately a fifth of the specific surface of Chromosorb 102. Chromosorb 104 (specific surface 100–200 m²/g, average pore diameter 60–80 nm) is a copolymer of acrylonitrile with divinylbenzene.[839] Benzene was chosen as a reference solute. The substitution of a methyl group or a halogen atom into a benzene ring causes approximately the same changes of free energy and heat and entropy of adsorption for both the sorbents. With the substitution of a nitro group or, especially, a hydroxyl group, these changes are considerable greater with Chromosorb 104. This indicates its greater selectivity as a stationary phase.

The retention of different compounds on Chromosorb 104 increased as the dipole moment of the compounds increased. The strong retention of water and alcohol may be explained by the possibility of hydrogen bond formation with the nitrile groups of the stationary phase. The elution order of alcohols and water from a Chromosorb 104 column is: butyl alcohol, ethyl alcohol, propyl alcohol, *iso*butyl alcohol, methyl alcohol, n-butyl alcohol, water, and amyl alcohol.[840] The poor retention of t- and *iso*butyl alcohols is likely to result from steric hindrance to the formation of hydrogen bonds with the nitrile groups on the surface of Chromosorb 104.

The retention of n-alkanes on Chromosorb 104 is less than on Chromosorb 101. The difference in the relative retention on Chromosorb 102, the average pore size of which is almost fifty times less than that of Chromosorb 101, becomes noticeable only for n-alkanes above C_8. On the other hand, the effective pore diameters of Chromosorb 104 and Chromosorb 101 differ only by about five times; thus geometrical factors should be less noticeable on comparison. The difference in the relative retention of n-alkanes on Chromosorbs 101 and 104 is connected with other peculiarities of the structure of these porous polymers, of a different chemical nature

In order to explain the retention behaviour of organic adsorbates on porous polymer columns in gas chromatography, an attempt was made to derive characteristic functional group incremental energies for non-polar Porapak Q and very polar Porapak T.[389] It has been demonstrated that Porapak T exhibits a characteristically strong interaction with the oxygen atom in alcohols, ethers and ketones and very likely reacts equally strongly with hydroxyl hydrogen atoms. Porapak Q, however, shows repulsion or only weak interaction with a hydroxyl, ether, or ketone oxygen but strongly interacts with hydroxyl hydrogen atoms. The strong interaction on Porapak Q indicates that certain specific interactions must take place, i.e. that the Porapak surface is slightly negative or polarizable as a result of its benzene rings. The endothermic incremental energy for the ether oxygen on Porapak Q seems to favour the assumption that there is a strong repulsion of either free electron pairs or negatively charged atoms or groups.

The particularly notable decrease in the adsorption energy of n-alkanes on Porapak T is less explicable in view of the unknown chemical structure of this adsorbent, but otherwise this behaviour completely corresponds to the adsorption behaviour of alkanes on other adsorbents. For example, the heat of adsorption of alkanes on silica gels (an adsorbent of the same general type as Porapak T, an electron acceptor) is also lower than on a non-specific adsorbent such as graphitized carbon black.

In order to provide a basis for the characterization of the different porous polymers which are commercially available, it was shown that retention indices can be used to compare the relative polarity of the various polymers, whereas the retention times for the n-paraffins are useful for estimating the relative speed of analysis.[310]

Among the porous polymers studied by Supina and Rose[310] (Porapaks N, P, R, S, T, Chromosorbs 101, 102, 103), it has been established that all of the polymers except Chromosorb 103 appeared to be suitable for the analysis of different compounds such as alcohols, acids, ketones, nitriles, glycols, nitro-alkanes, esters, ethers, and aromatic hydrocarbons. Chromosorb 103 apparently cannot be used for acids, nitriles, glycols, or nitro-alkanes, but is recommended by Johns–Manville for the analysis of amines, amides, aldehydes, and ketones. The Chromosorb "century series" produced by this company are now available in seven grades with the following specifications: Chromosorb 101 is recommended for the efficient separation of free fatty acids, glycols, alkanols, alkanes, esters, ketones, aldehydes, and ethers; Chromosorb 102 for light and permanent gases and low molecular weight compounds; Chromosorb 104 shows the highest polarity and is recommended for efficient separation of nitriles, nitro-alkanes, xylenols, and oxides of nitrogen and ammonia; Chromosorb 105 separates aqueous mixtures containing formaldehyde and acetylene from lower hydrocarbons and water from organic compounds; Chromosorb 106, a non-polar porous polymer, separates C_2–C_5 acids from C_2–C_5 alcohols; Chromosorb 107, a moderately polar porous polymer, separates formaldehyde and acetylene from lower hydrocarbons.

A review of the different separations made with these stationary phases shows that, although they are organic in nature, porous polymers have been found to separate most of the inorganic gases very efficiently and, in addition, all classes and types of organic compounds. Out of nearly 150 papers published in this field up to 1970, almost half deal with the elution of pure inorganic compounds or mixtures of organic and inorganic compounds. Components of air are well separated on the styrene–divinyl–benzene copolymer at $-78°C$ in the following order: H_2, He, N_2, O_2, Ar;[386, 841] CO_2 and N_2O can be separated at room temperature.[386, 842]

Light hydrocarbons are separated on porous polymers as a function of the number of carbon atoms in the molecule and the degree of chain branching.[843] The lighter hydrocarbons elute before their saturated analogues (ethylene before ethane, propylene before propane). Thus the retention of unsaturated compounds does not depend on the presence of double bonds in the molecule, and elution is in the order of boiling points.[386, 844]

Hydrogen, carbon monoxide, methane, nitrogen oxides, carbon dioxide, gaseous compounds of sulphur and fluorine, chlorinated hydrocarbons, hydrogen cyanide, phosgene, and sulphuryl chloride are readily separated on porous polymers.[386]

It often happens that in order to make certain separations, two-column systems and temperature programming must be used. In this way, Jones[845] separated a complex gaseous mixture (H_2, O_2, N_2, CO, CO_2, H_2S, NH_3, H_2O, C_1–C_5 hydrocarbons) on Porapak Q using a dual column chromatograph. One column was packed with Porapak Q on which N_2, O_2, CO were separated at the temperature of solid carbon dioxide. The second column separated CH_4 and CO_2 at room temperature. Temperature programming from room temperature to $125°C$ enabled the other components to be separated.

A similar system with two columns has been also used to analyse mixtures of N_2, O_2, Ar, CO, CO_2, H_2S, and SO_2.[846] An elaborate gas chromatographic method has also been devised to determine the possible components of the Martian atmosphere.[847] Porous polymers are ideal for such research because they are resistant to a vacuum, radiation, or temperatures up to $250°C$. The resulting composite chromatogram is shown in Fig. 41.

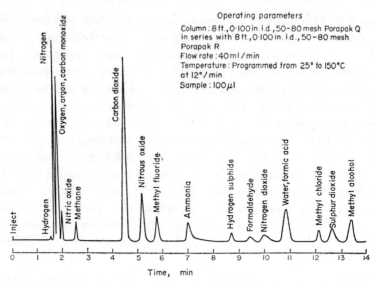

FIG. 41. Separation of possible constituents of a Martian atmosphere.[837]

Other separations of complex gas mixtures have been reported.[848-863] They are summarized in Table 22.

A wide field for the use of porous polymers in gas chromatography is the analysis of aqueous solutions. The rapid elution of water, and its symmetrical peak (the water elutes before organic solvents), were the main factors contributing to the use of porous polymer beads for water–alcohol analyses, studied by Gough and Simpson.[864] The quantitative analysis was found to be satisfactory using all types of porous polymer beads. Column wall material had no effect on the analyses. Column efficiencies and resolution, however, varied with the packing material and deteriorated after prolonged conditioning. The most satisfactory performance was achieved using Porapak Q-S.

The precision of water determination by gas chromatography using porous polymers as stationary phases is comparable to that of the Karl Fischer method. It allows the determination of 10 ppm of water with a thermal conductivity detector[835] and of 0.5–0.7 ppm with an electron capture detector.[865]

Hollis and Hayes[835] have determined water in hydrocarbons, in chlorinated hydrocarbons, and in alcohols, acids, nitriles, and atmospheric gases, using both polar and non-polar porous polymers. Analysis of aqueous solutions of formaldehyde[866, 867] and of traces of water in acrylonitrile,[868] nitrogen oxides,[869] and alcohols[870] has also been achieved. Aqueous solutions of volatile fatty acids were analysed quantitatively using Porapak Q, Q-S, N.[871, 872]

The possibility of separating hydroxylated compounds on polyaromatic adsorbents was reported by Hollis.[386] Such adsorbents have been used for the quantitative analysis of alcohols, diols, glycols, fatty acids, and ethers. The lower glycols are readily separated, ethylene glycol eluting before propylene glycol.[386, 849, 873, 874]

Modification of porous polymers with poly(ethylene imine), tetraethylene pentamine, and potassium hydroxide has been used to improve the peak shapes caused by the elution of amines and alkyl diamines.[386, 831, 875] Modification using a polar stationary phase (PEG 1500) has been used for the separation of alcohols with different structures but similar boiling points.[829]

Porous polymer stationary phases have also enabled the polar components of alcoholic drinks,[876] dairy products,[877] and blood[878–880] to be determined. Among other analyses which may be performed with porous polymers are the determination of ketones,[881] acetylcholine and its derivatives,[882–884] oxalic acid,[885] volatile carbonyl compounds,[886] nitrile isomers, and α-amino acids.[887]

The porous polymers based on styrene and divinylbenzene may be used to analyse substances having molecular weights below 200. The elution time of the larger molecules is longer, and their peaks show a marked assymetry.

Sie et al.[888] showed that the chromatographic combination of porous polymers with high pressure gases or supercritical fluids enlarges their range of application. For example, at 200–250°C, alkyl benzene derivatives with thirty-six carbon atoms can be separated very rapidly.

Another way to increase the applicability of porous polymers in gas chromatography is to synthesize new materials with better thermal

TABLE 22. GAS CHROMATOGRAPHIC ANALYSIS OF GAS MIXTURES USING POROUS POLYMERIC ADSORBENTS

Gas mixture	Adsorbent	Column dimensions	Analysis temp. (°C)	Carrier gas, flow (ml/min)	Reference
High purity ethylene:					
N_2, CH_4, CO_2, C_2H_4, C_2H_6	Porapak Q	1.8 m×6 mm	25	Helium, 60	849
H_2, N_2, O_2, CH_4, Kr, CO_2	Porapak Q+0.01% H_3PO_4	7 m×1.5 mm	Ambient	Helium, 15	849
CF_2O and CO_2	Porapak T+Porapak N	0.6 m×6 mm, 1.2 m×6 mm	23	Helium, 60	850
Pyrolysis products of town refuse:					
H_2, O_2, N_2, CH_4, CO, CO_2, C_2H_4, C_2H_6	Molecular sieve 5A + Polypak-2	2.4 m×6 mm, 1.2 m×6 mm	40	Helium, 83, Helium, 42	851
C_1–C_4 hydrocarbons	Porapak Q	3.1 m×4 mm	76	Helium, 100	852
Oxidation products of toluene:					
O_2, N_2, CO, CO_2, H_2O, C_6H_6, C_7H_8, C_7H_6O	Porapak R+molecular sieve 5A	0.9 m×3 mm, 0.9 m×3 mm	25; 25–250	Helium, 60	853
O_2, CO, CH_4, CO_2, C_2H_4, C_2H_6 in N_2	Porapak Q+molecular sieve+Porapak Q	1.5 m×6 mm, 1 m×6 mm, 1 m×6 mm	25, 20	Helium, 40	856
Burning products of gases:					
H_2, air, CO, CH_2, CO_2, C_2H_4, C_2H_2, C_2H_6	Porapak Q	3 m×5 mm	30	Helium, 40	855
O_2, N_2, CH_4, CO_2, N_2O	Column residue, Porapak Q+molecular sieve 13X	0.6 m×6 mm, 1.8 m×6 mm	Ambient	Helium, 50	856, 857

Pyrolysis products of nitro-alkanes:

Compounds	Stationary phase	Column dimensions	Temperature	Carrier gas	Ref.
H_2, O_2, N_2, CO, CO_2, H_2S, NH_3, H_2O, C_1–C_5 hydrocarbons	Porapak Q + molecular sieve 5A	1.6 m × 4 mm 3.2 m × 4 mm	25 50	Argon, 25 Argon, 50	858
H_2, N_2, O_2, Ar, CO, CO_2, C_2H_4, C_2H_6, H_2O	Porapak Q + molecular sieve 5A	6 m × 6 mm 1 m × 6 mm	38–150 12°/min	Helium, 35	859
He, Ne, Ar, O_2, N_2, N_2O, CO_2, CF_4, H_2O	Poparak Q + molecular sieve 5A	1.8 m × 5 mm 2.7 m × 6 mm	Ambient	Oxygen, 40	860
CO_2 and H_2O in atmosphere	Poparak Q	1.8 m × 3 mm	40	20	861
HCl, Cl_2, SO_2, NO, N_2O, H_2O, CO_2	Porapak Q	1.8 m × 3 mm	110	25	861

Radiolysis products of methane:

Compounds	Stationary phase	Column dimensions	Temperature	Carrier gas	Ref.
CH_4, C_2H_2, C_2H_4, C_2H_6, C_3H_8, C_4H_{10}, iso-C_4H_{10}, iso-C_5H_{12}	Porapak Q + Polypak-1	0.9 m × 3 mm 3.6 m × 3 mm	−15–150	35	862
C_2H_2, C_2H_6, H_2O	Porapak S + Porapak S	0.6 m × 3 mm 1.5 m × 6 mm	20 50		863

stability and to introduce into their molecules different groups able to exhibit specific interactions with molecules to be separated.

Tranchant[889] showed that good separations could be achieved on a 2,6-dimethyl-*p*-phenyl oxide polymer, commercially available as PPO (poly-phenylene oxide), which is a linear polymer:

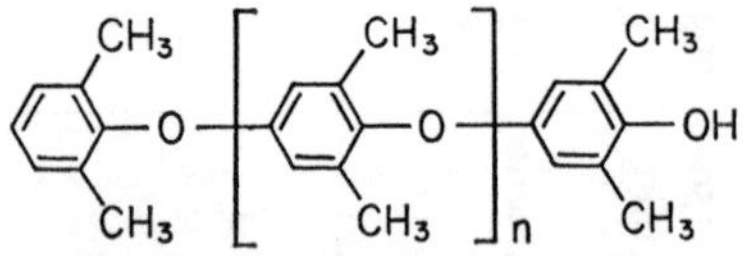

The column packed with this compound has high permeability and high efficiency, and the peaks are symmetrical for polar substances as well as for non-polar ones. Hydrocarbons, water, alcohols, and ketones are thus completely eluted, giving symmetrical peaks from columns of poly-phenylene oxide. The temperature behaviour of the polymer above 180°C was not reported.

Instead of PPO itself, van Wijk[890–892] has studied the behaviour of poly(2,6-diphenyl-*p*-phenylene oxide) as a stationary phase.

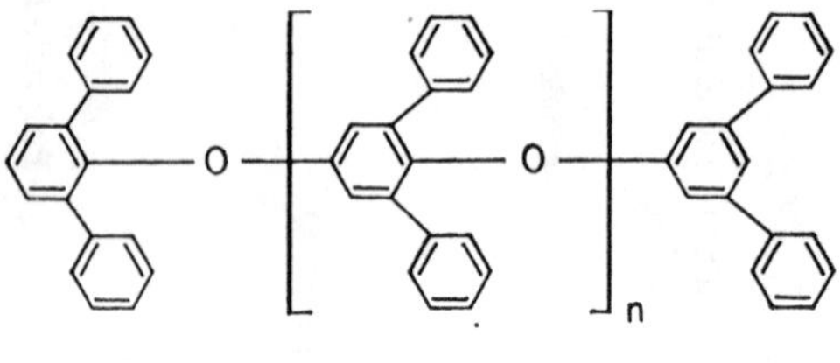

This material is a high molecular weight (5×10^5–10^6) linear polymer in the form of small granules. Thermogravimetric analysis of 300 mg of the polymer under a flow of nitrogen, showed no significant weight loss at 400°C, a small loss (0.3%/h) at 430°C, and a very pronounced loss at 450°C (3%/h). The separation behaviour of poly(2,6-diphenyl-*p*-phenylene oxide) is generally comparable with that of Porapak Q, but the latter requires higher temperatures and back pressures. Particularly polar components such as water, alcohols, amines, glycols, and the lower fatty acids behave well on such columns. Figure 42 shows the separation of glycols.

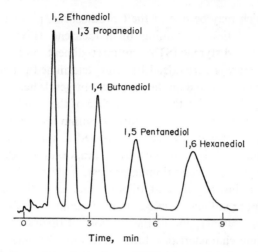

FIG. 42. Separation of glycols at 180°C on a stainless steel column (1.6×3 mm o.d.) packed with polydiphenyl phenylene oxide (35–60 mesh); carrier gas flow rate (nitrogen): 30 ml/min.[890]

The polymer was used untreated except for crude sieving. At temperatures up to 400°C, higher homologues of many classes of component can be separated and determined. Because of the good solubility of the polymer in lower, chlorinated hydrocarbons, the coating of a normal solid support with the polymer was investigated. However, the coating of normal solid supports was of no use. Presumably the porous character of the polymer is destroyed during the coating procedure. Such a material is available under the name Tenax GC (Applied Science Lab. Inc., Shimadzu Seisakusho Ltd.) and is recommended for the separation of high-boiling polar compounds such as alcohols, poly(ethylene glycol) compounds, diols, phenols, methyl esters of dicarboxylic acids, amines, diamines, ethanolamines, amides, aldehydes, and ketones.

Fuller[893] has described the preparation of porous polyaromatic copolymers formed directly on and physically bonded to the surface of suitable solid support materials, and has presented the initial results of experiments investigating their utility as gas chromatographic stationary phases and their surface properties. Basically, the technique consists of first coating a support (which may be one of the Chromosorbs, porous glass beads, etc.) with a solution containing divinylbenzene (DVB), ethylvinylbenzene (EVB), and styrene (STY), an inert diluent, and a polymerization initiator (lauroyl peroxide, LP) in predetermined proportions. The monomers are then reacted on the adsorbent by gentle heating to produce the porous copolymer directly on the surface of the support. Because the extensively cross-linked polymer is formed within the surface pore structure, it becomes permanently fixed or bonded to the support. The final step, solvent removal, is accomplished by evaporation under vacuum or else during the column conditioning process.

Based on an investigation of dilution and cross-linking effects on surface area, pore volume, average pore size and pore size distribution of the resulting polymer, and of the relations of these surface properties to actual column characteristics, broad limits for the practical ranges of initial dilution and cross-linking (approximately 30–95 wt.% n-heptane and DVB 30 or greater mole % of monomer reactants) have been formulated.

To illustrate the performance of the support-bonded polymer packings, the separation of some aromatic hydrocarbons up to butylbenzene is shown in Fig. 43. The column packing used was prepared on Chromo-

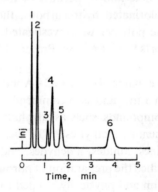

Fig. 43. Separation of aromatic hydrocarbons on a 6 ft DVB-EVB-STY support-bonded column at 180°C; helium carrier gas: 60 ml/min ≈0.02 μl sample. (1) Benzene, (2) toluene, (3) ethylbenzene, (4) o-xylene, (5) isopropylbenzene, (6) n-butylbenzene.[893]

sorb P (80–100 mesh) from a formulation consisting of 38 mole% DVB, 35 mole% STY, and 27 mole% EVB, caculated as mole% of reactant monomers with 50 wt.% of the total n-heptane as initial diluent and about 1 wt.% LP as initiator. The polymer formed on the support resulted in approximately 7% w/w loading after a reaction time of about 18 h.

The advantages of the support-bonded polymer phases, as compared to the polymer beads, include greater column efficiency, more rapid elution, and the capability of adjusting capacity ratios over a fairly wide range by varying the polymer loading. Because capacity ratios are normally lower, analyses can be extended to higher molecular weight materials. The author considers that although no investigation was attempted on other diluent systems or on the addition of other monomers such as acrylonitrile and methyl methacrylate to alter the chemical nature of the copolymers, there is no apparent reason why properties closely duplicating any of the polymer bead types could not be produced in support-bonded polymers as well as other entirely new types.

Sakodynsky et al.[894] investigated the possibility of using two types of polymer on a polyimide base. They were (1) pyromellitic dianhydride and diamine diphenyl ether (Polysorbimide-1), and (2) benzophenone-tetra carboxylic acid dianhydride and diaminodiphenyl ether (Polysorbimide-2). These polyimides are irregular yellow particles, soluble in concentrated nitric and sulphuric acids but insoluble in organic solvents. These compounds were synthesized by polyimerization in an inert diluent.[895]

They have a large total pore volume and a large average pore radius. The total pore volume of polymides is commensurable with the total pore volume of porous polymers on a styrene and divinylbenzene base. The thermogravimetric curves for Polysorbimides shows that they have a high thermal stability, and can be used in a gas chromatograph up to at least 400°C.

The principal retention characteristics of these stationary phases may be summarized as follows. The retention of unsaturated compounds depends not only on their boiling point and molecular weight but also on the presence and number of double bonds in the molecule: 2-pentane and 1,3-pentadiene elute after pentane; 1-hexene elutes after hexane.

It is also characteristic that cyclohexene and benzene, the boiling points of which are lower than that of cyclohexane, are more strongly retained on Polysorbimides than the latter.

The retention of polar molecules depends on the value of the dipole moment and on the ability of the solutes to form hydrogen bonds with the sorbent surface. Therefore, nitromethane is retained more strongly on polymides than heptane in spite of the significantly lower molecular weight. Likewise, acetonitrile is retained more strongly than cyclohexane, and ethyl acetate more strongly than n-hexane. Similarly, for

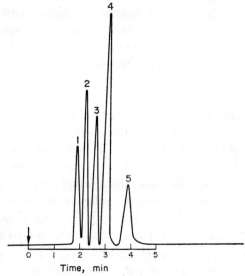

Fig. 44. Separation of esters on a Polysorbimide-2 column (1.5 m×4 mm i.d.) at 330°C. Helium flow rate: 30 ml/min. (1) Hexyl phenyl acetate, (2) heptyl phenyl acetate, (3) octyl phenyl acetate, (4) nonyl phenyl acetate, (5) decyl phenyl acetate.[894]

the chloromethanes, chloroform elutes later than the higher molecular weight and higher boiling carbon tetrachloride, while dichloromethane is retained two or three times more strongly than n-pentane.

The retention of alcohols and acids on Polysorbimides is considerably greater than on styrene-DVB copolymers.

It should be noted that on the Polysorbimides, a greater retention of fatty acids capable of forming stronger hydrogen bonds is observed as compared to alcohols with the same number of carbon atoms.

The retention of isomeric alcohols is determined mainly by the difference in the boiling points and volatilities of the components to be resolved. A stronger retention of molecules of normal structure is observed compared to that of branched alcohols.

Polymer sorbents on a polyimide base have a selective molecular interaction which can be assumed to arise from the imide and carbonyl groups on the surface.

These stationary phases are suitable for the separation of high boiling polar compounds such as alcohols, esters, aromatic hydrocarbons, pyrrolidones, aldehydes, and ketones. Figure 44 shows the separation of some esters on a column of Polysorbimide-2 at 330°C.

High maximum operating temperature ($< 400°C$) and a stable base line after a short conditioning time (3 h) are advantages of Polysorbimides.

The wide use of porous polymers as stationary phases in gas chromatography is justified by their following properties:

    (i) Porous polymers have a remarkable mechanical resistance which is not inferior to that of the majority of diatomitic supports currently used in gas chromatography.

    (ii) They can be used both at lower temperatures ($-190°C$) and higher ones (at least 200°C),[896] and thus to solve a wide range of analytical problems, such as the analysis of permanent gases and the analysis of compounds with very high boiling points.

    (iii) They give highly efficient separations, the number of theoretical plates for a 1 m column being between 1300–2600. The height equivalent of the theoretical plate is thus between 0.8–0.4 mm.[386]

    (iv) They can be used as supports for stationary liquid phases. Such modification of the porous polymers can improve separation efficiency and, depending on the polarity of the deposited liquid phase, the elution characteristics may be selected as appropriate for different classes of organic substances. The highest efficiency is obtained by depositing 2–7% of liquid phase. Up to 30% may be deposited.

(v) Their use assures good reproducibility and reliability of retention data obtained with the same column after some few months of use.[848] Their suitability for use with high sensitivity detectors also allows the analysis of trace components.

(vi) The introduction of different functional groups into porous polymers can suitably enlarge their applicability as stationary phases via specific molecular interactions with some components of the mixture to be separated.

4.2.6 Specific solid stationary phases

Molecular sieves

Molecular sieves are aluminosilicates in which the crystalline lattice is unusually open, so that it contains holes several Å across. Though they occur naturally by hydrothermal action of solutions in basaltic rocks and also as large sedimetary deposits, they may also be made synthetically, without any known natural conterparts.[897]

The molecular sieves of interest occur naturally and synthetically. Their general formula is $M_{2/m}O \cdot Al_2O_3 \cdot nSiO_2 \cdot xH_2O$. The structural units of the crystalline lattice of the aluminosilicates consist of tetrahedral anions $[SiO_4]^{4-}$ and $[AlO_4]^{5-}$ bound through the oxygen atoms. The excess of negativ echarge arising from the replacement of silicon by aluminium is compensated by the alkali or alkaline earth metal ions. Lamellar, filiform, and tridimensional structures are known.[898] The zeolites with a tridimensional structure are the only ones used as adsorbents.

Figure 45 shows schematically the structure of the synthetic zeolite A (faujasite type) which has a cubic structure.[899, 900]

The cavity is formed by the bonding of the octahedra in cycles of four structural units via the oxygen atoms (Fig. 45b). This gives a large cavity of 11 Å diameter (Fig. 45c), surrounded by eight octahedra. The cavity is accessible through six openings (corresponding to the surfaces of type 1, which, in their turn are constructed from eight oxygen atoms. Also, there is a smaller cavity of 7 Å diameter in each of the eight octahedra. Each of these small cavities is connected by a ring of six oxygen atoms (of 2 Å diameter, corresponding to the surface of type 2) to the greater cavity.

It should be emphasized that in the adsorption process the access ways to the small cavities of the synthetic zeolites A are much too narrow to allow the adsorption of the usually adsorbed molecules, so that they enter the cavities only through the larger holes. These large holes, composed of six or eight structural units (corresponding to surfaces 2 or 1), are connected in such a way that a very regular system of cavities is formed.

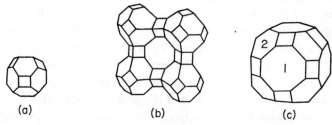

(a)   (b)   (c)

Fig. 45. Schematic structure of the synthetic zeolite A (faujasite type).
(899, 900)

The structure is microporous because the cavity (= pore) diameter falls in the range of small molecular dimensions. The micropores of the zeolites are characterized by two parameters—the diameter of the hole and the cavity diameter. The values assumed by these parameters can be evaluated from X-ray crystallographic data. From the adsorption measurements only the total volume of the pores may be obtained.

The same remarks concerning zeolite A may be extended to other synthetic zeolites. The differences between zeolite A and other synthetic zeolites lies in the binding of the polyhedra, i.e. in the lattice constants of the unit cell.

The zeolites, therefore, are porous bodies which have cavities or holes. The diameter of the hole may be varied by ion exchange.[901, 902] For example the 4 Å molecular sieve is the sodium form of zeolite A. By replacing the sodium ions by potassium ions, a 3 Å molecular sieve is obtained, and replacement with calcium ions leads to a 5 Å molecular sieve. Barrer[903] classifies the zeolites, according to the diameters of the access holes in the cavity, into five classes: type 3 A (3 Å), 4 A (4 Å), 5 A (5 Å), 10 X (8 Å), and 13 X (10 Å). The volume of the cavities connected to each other through a system of channels may be up to 56% of the total volume of the zeolites.[904]

In the anionic lattice of zeolites, large cavities filled with water of crystallization occur. They may be removed by heating without destruction of the crystalline zeolite lattice. Fresh material is usually activated by heating at 350–400°C for about 2 h.

The influence of the hydration on the ability of a zeolite column to separate the permanent gases has been discussed.[899, 905–907] Janak et al.[905] have shown that the shape of the chromatographic spectrum of the permanent gases (H_2—O_2—N_2—CH_4—CO) depends on the water content of the molecular sieve. The retention time of carbon monoxide depends on the hydration state of the adsorbent; for a water content higher than 9%, the elution order of the peaks corresponding to methane and carbon monoxide is reversed. The retention time of the gases generally decreases as the degree of hydration increases. This decrease is, however, more marked for the common gases H_2, O_2, N_2, CH_4, and CO, than for the rare gases Kr and Xe.[906] It is possible to explain the displacement of the peaks of the different gases as a function of the degree of hydration of the adsorbent by assuming that it partly depends on the diminution of the active surface area of the adsorbent by the increase in the degree of hydration.[907]

The selectivity of the sorption effect is considered to arise mainly because the cavities adsorb only those molecules which can penetrate into the cavities, i.e. those compounds with a cross-sectional molecular diameter equal to or smaller than the diameter of the channels which bind the cavities. For example, when a mixture of hydrocarbons is brought into contact with or passed over a bed of molecular sieve 5 Å, the only components of the mixture which can penetrate into the crystal lattice of such a calcium A-type zeolite are those whose molecular cross-sectional diameters are less than 5 Å, namely, n-alkanes.

The behaviour of various homologous series and compounds of general interest on columns run at 100°C packed with molecular sieve 5 Å has been investigated by Brenner et al.[908] The results are summarized in

Tables 23 and 24. Table 23 gives the materials which pass through molecular sieve 5 Å without adsorptive loss, and Table 24 shows the materials which were found to be completely adsorbed.

TABLE 23. COMPONENTS NOT ADSORBED BY A COLUMN OF MOLECULAR SIEVE 5Å[908]

Group	Compounds tested
Iso-alkanes	*Iso*butane, *iso*pentane, 2,3-dimethylbutane
Aromatic hydrocarbons	Benzene, toluene, *m*-xylene
Cyclo-alkanes	Cyclopentane, cyclohexane
Iso-olefins	*Iso*butylene, 2-methylbutadiene-1,3
Esters	Amyl formate, ethyl acetate, ethyl propionate
Ketones	Acetone, methyl ethyl ketone, mesityl oxide
Halogenated hydrocarbons	Methylene chloride, chloroform
Iso-alcohols	*Iso*propyl alcohol, methyl butyl alcohol
Ethers	Diethyl ether, di-*iso*propyl ether
Other compounds	Carbon monoxide, oxygen, nitrogen, rare gases, methane, nitromethane, carbon disulphide, dimethyl sulphide, thiophene

TABLE 24. COMPONENTS COMPLETELY ADSORBED ON A COLUMN OF MOLECULAR SIEVE 5 Å[908]

Group	Compounds tested
n-Alkanes (except methane)	Propane, n-butane, n-hexane
n-Olefins	Ethylene, propylene, hexene-2
n-Alcohols	Methyl alcohol, ethyl alcohol, n-butyl alcohol
Aldehydes	Acetaldehyde, propionaldehyde, isovaleraldehyde
Acids	Formic acid, propionic acid

The other types of molecular sieves do not behave similarly because their pore openings are different. Molecular sieve 4 Å (pore opening 4 Å diameter) adsorbs onlyethane of the n-alkanes. Propane and higher homologues are eluted. Of the n-olefins, ethylene and propylene are adsorbed on molecular sieve 4 Å but the butenes and higher homologues are not. The situation is similar with oxygenated compounds: e.g. the n-alcohols from n-butyl alcohol upwards pass through the column because their molecules are larger than 4 Å.

Because of the very large pore openings, molecular sieve type 13 X adsorbs practically all the organic compounds which are given in Table 23. This illustrates that it is possible to select a given type of molecular sieve to retain individual compounds in a sample, while the others pass through the column.

Whereas the important factor is the dimension and shape of the sorbed molecule relative to those of the interstitial cavities, it should be noted that the diffusion rates are modified by the subdivision of the particles. Barrer and Ibbitson[909] have shown that the sorption rate is modified by the fineness of subdivision of the mineral. The presence of gases in the interstitial channels (or outside the walls) may have some effect on the sorption rate. It is assumed that these gases may be sorbed only extremely slowly (or be present at the surface of the sieve) and may block the entrance to the interstitial channels and prevent uptake of species which in the absence of these gases would have been more freely occluded.[909, 910]

Oberholtzer and Rogers[911] attributed the anomalous behaviour in the variation of the height equivalent of a theoretical plate with temperature for methane and ethane on two different zeolites (types 4 Å and 5 Å) to the effects of slow intraparticle diffusion. They also suggested that the changes in retention volume with flow rate observed arose from the same cause. Habgood measured the heats of adsorption of methane and ethane on a wide-pore zeolite; the disparity between those values and the values measured on a narrow-pore zeolite were rationalized as being caused by restricted diffusion in the narrow-pore material.[912, 913]

The elution behaviour of CH_4, C_2H_6, n-C_3H_8, iso-C_4H_{10}, Ar, O_2, N_2, and SF_6 were studied in A-type zeolites containing K^+, Na^+, Li^+, Sr^{2+}, Ca^{2+}, and Mg^{2+} and on a 10X zeolite by Moreland and Rogers.[914] Temperature- and flow-dependent tailing of peaks, and changes in retention volume with flow rate, were found for those adsorbates able only partially to penetrate the pores. When diffusivities were large enough to allow total access to the pores, molecular size and the degree of adsorption for the adsorbate had significant effects on height equivalent of theoretical plate values.

In many analytical applications, these effects of micropores will be important. To minimize these effects the lowest temperature and highest flow rate possible should be used in order to minimize the penetration. For larger pores, the large coefficient of resistance to mass transfer, arising from the porosity, dictates that the flow rate should be as low as possible in order to preserve column efficiency. This indicates why an adsorbent having a wide range of pore sizes poses a problem. Obviously, compromises must be made and the optimum conditions arrived at empirically.

A number of common gases and vapours have permanent dipoles, so that with ionic, heteropolar adsorbbents such as zeolites, specific electrostatic interactions may be involved.[915] Thus in the molecular sieve zeolites, because of their anionic frameworks and mobile cations, the physical bonds for sorbed molecules having permanent electric moments (N_2, CO, CO_2, NH_3, etc.) may be much enhanced compared with molecules like the inert gases or methane. On the other hand, the possibility of modifying the zeolites by cation exchange allows their fine structure to be changed while maintaining a practically constant aluminosilicate framework. This creates favourable conditions for the study of the character of adsorption interactions during the chromatograpic separation of different substances on zeolites, depending on the nature of the cation.

For adsorbents where concentrated positive charges are present, as in the zeolites, when the molecules of adsorbed substances are characterized by the presence of π-bonds and dipole and quadrupole moments, the role of electrostatic interactions is especially important. In addition, in changing from one cation to another, interactions caused by the presence of chemically bound oxygen atoms in the zeolite framework should practically be unaffected.

Tsitsishvily and Andronikashvili[242] studied type X zeolites containing the following cations: Na^+, Li^+, K^+, Rb^+, Cs^+, Ag^+, Ca^{2+}, Mg^{2+}, Ba^{2+}, Sr^{2+} and Cd^{2+}. A mixture of C_1–C_4 hydrocarbon gases, carbon monoxide, and hydrogen was used as a model mixture. It was observed that the heats of adsorption, i.e. the energies of interaction of saturated hydrocarbons, characterized only by the presence of σ-bonds, with zeolites containing singly charged cations, increased for the heavier cations and had the highest value on silver-containing specimens. This agrees with the fact that the polarizability of cations increases with the increase in number of electrons in an atom, and hence the dispersion interaction with adsorbed molecules of saturated hydrocarbons increases. For unsaturated compounds and carbon monoxide, which are characterized by the presence of π-bonds and dipole and quadrupole moments, the reverse picture is observed in the sense that the interaction of these compounds is most pronounced on zeolites containing cations with small radii.

For zeolites with doubly charged cations, the influence of conditions of thermal activation on the values of adsorption heats is more pronounced. This influence depends to a great extent on the value of the energy of hydration of the cation. Thus greater thermal activation does not cause an appreciable increase in the heats of adsorption for magnesium-containing zeolites, but with zeolites containing calcium it leads to an appreciable increase in the heats of adsorption of the C_1–C_4 hydrocarbon gases and carbon monoxide.

The chromatographic properties of the silver and cadmium zeolites are especially noteworthy. For instance, if on a NaX zeolite at a column temperature of 180°C, there is a possibility of separating a mixture consisting of hydrogen, carbon monoxide (methane), ethane, ethylene, propane, butane, and propylene, then hydrogen, carbon monoxide, and unsaturated hydrocarbons can be strongly fixed on specimens containing silver, whereas the saturated hydrocarbons are readily eluted and are separated into different components.

A certain amount of selectivity, not so sharply pronounced as in the above compounds, is also shown by zeolites containing cadmium. Strong binding of hydrogen on the zeolite containing silver was explained by the formation of the hydrogen form of the zeolite on reduction of the silver ion by the hydrogen. Cadmium-substituted zeolites react similarly to those containing silver, but only at a high column temperature (300°C), whereas the silver form reacts readily at room temperature.

Carbon monoxide adsorption is assumed, on the basis of spectral data, to occur by complex formation with the silver or cadmium ions.[916] In the chromatographic separation of unsaturated hydrocarbons on such zeolites, complex formation with the adsorbent cations takes place. The silver complexes are more stable than those of cadmium. As is shown in Fig. 46, it is possible that bonding takes place by overlapping of the filled π-orbital of ethylene with the vacant $5sp$ orbital of the silver ion and by overlapping of the filled $4d$ orbital of the silver ion with a vacant π^* orbital of ethylene. For the cadmium zeolite, interaction only occurs

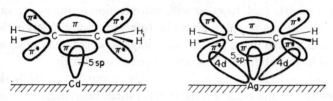

FIG. 46. Scheme of bond formation of ethylene molecules with Ag^+ and Cd^{2+} in a type X zeolite.[242]

by overlapping of π and $5sp$ orbitals; the molecule may rotate freely around this bond.

It has been also observed that the chromatographic separation of a methane–carbon monoxide mixture depends on the cation and the degree of its incorporation into the zeolite.[917] The presence of dipole and quadrupole moments is characteristic of carbon monoxide but not of methane. [918] Zeolites containing sodium or lithium ions display a certain selectivity with respect to carbon monoxide, which is eluted later than methane. A zeolite with a greater content of lithium retains carbon monoxide more than the sodium form. On zeolites containing potassium, rubidium, or cesium ions, carbon monoxide elutes before methane, even with a small amount of sodium ion substitution. Such behaviour occurs because the electrostatic interactions caused by the dipole and quadrupole moments of carbon monoxide are more important for cations with small radii (Na^+ and Li^+) than for the larger cations (K^+, Rb^+, and Cs^+).

The molecular sieves are the stationary phases most utilized for the separation of permenent gases.[919-927] These materials are the only commonly used chromatographic packing that will separate oxygen from nitrogen at room temperatures, the latter having more than twice the retention.

Of those molecular sieves commonly available, most work has been done with 5 Å variety. Rapid separations can be obtained on columns only a few feet long and, provided always that the sample material is not composed very largely of hydrogen, good separations of hydrogen, oxygen, nitrogen, methane, and carbon monoxide can be obtained, the gases being eluted in that order.[920, 926] Carbon dioxide is retained by the column packing material. This, together with the retention of the waper vapour present in most gas samples, gives rise to a progressive deterioration of the molecular sieve material. This deterioration is generally observed as a gradual loss of ability to separate oxygen from nitrogen.

For most applications the molecular sieve material is carefully sieved, and a fraction in the range 40–100 mesh is used. Bombaugh[928] has described the use of molecular sieve powder for the separation of hydrogen, oxygen, nitrogen, methane, and carbon monoxide. One part by weight of a molecular sieve 5 Å fraction passing a 200-mesh sieve was mixed with two parts by weight of Chromosorb W (60–80 mesh), mixed in a rotating cylinder for 18 h and screened. The fine material was rejected and the 60–80 mesh fraction was used to pack the chromatographic column. Of the original 33% of molecular sieve added, 28% was retained on the Chromosorb. Using such a column, Bombaugh obtained a sixfold increase in efficiency and an almost twofold increase in peak height compared with a column just packed with 60–80 mesh material.

Molecular sieve powder has been used successfully by Farre-Rius and Guiochon[929] to separate oxygen, nitrogen, methane, and carbon monoxide in 25 s. These authors obtained a powder size of 315–400 μm by washing the molecular sieve with water and decanting the fine particles with the supernatant liquid. The powder was dried and activated at 400°C and the column packed in the usual way. The 25 s separation was obtained on a 2 m column at 100°C, using hydrogen as carrier gas.

When a 5 Å molecular sieve column is used routinely for the analysis of mixtures of permanent gases and light hydrocarbons, occasional peak shift is observed. This may arise from the presence of *iso*butane,[237] the molecules of which are too large to pass through the screening pores, so that the adsorption and desorption of *iso*butane can only occur on the external surface of the molecular sieve. The difference in surface areas ($1-3$ m^2/g for the external area vs. $700-800$ m^2/g for the internal surface area) can account for the small retention volume of *iso*butane relative to the other compounds. Thus if *iso*butane is present, care must be taken to see that it is not reported as one of the permanent gases.

A rapid and simple method for the determination of oxygen and argon mixtures at room temperature on specially treated molecular sieves has been developed by Karlsson.[930] The molecular sieves were heated in an argon purge at a humidity of 100 ppm, at $450-500°C$ for 30 h, followed by slow cooling for 8 h. A column, 2.2 m long packed with molecular sieve 5 Å (70–80 mesh) treated in this way separates argon, oxygen, and nitrogen in 5 min at 23°C, if helium is used as carrier gas at a flow rate of 90 ml/min. Quite recently, carbon dioxide and nitrous oxide[931] were separated at 250°C on a column (6 ft$\times\frac{1}{4}$ in. i.d. (1.8 m$\times$6.3 mm)) packed with molecular sieve 5 Å (60–80 mesh), using helium as carrier gas with a flow rate of 60 ml/min.

In addition to the use of molecular sieves for the analysis pf permanent gases, owing to their well-defined pore size and structure, many separations are possible because of the exclusion of larger molecules from the large internal surface area. Thus linear and branched hydrocarbons may be separated as well as aromatic and cyclic hydrocarbons from aliphatic species.[932–945, 823]

More recently, a combination of 5 Å and 13X molecular sieve columns, which will analyse C_5-C_{10} cyclic, *iso*-, and n-alkanes on the basis of the number of carbon atoms, has been described by Peterson and Rodgers.[946] Aromatic and other unsaturated compounds, if present, are hydrogenated on a pre-column immediately after injection. The analytical process has two main stages. In the first, the sample is injected and the experimental conditions so arranged that a small fraction of the n-heptane and all of the C_8-C_{10} n-alkanes are retained on the molecular sieve 5 Å pre-column. All the other alkanes and all naphthenes (this includes all the aromatic compounds which are hydrogenated immediately after injection) elute from the pre-column during this period. They are separated into alkane and naphthane groups on the basis of their number of carbon atoms on the molecular sieve 13X main column in a temperature programmed procedure. The pre-column retards the C_8-C_{10} n-alkanes just sufficiently to resolve them on final elution from the 13X main column.

When elution of *iso* decane is complete, the oven containing the main column is cooled and the second stage starts with the desorption of the rest of the n-heptane and all the C_8-C_{10} n-alkanes from the pre-column by raising its temperature. This process "injects" these compounds onto the main column on which they are analysed by a second temperature programme cycle. The aromatic compounds are hydrogenated to prevent them interfering with the separation of the alkanes and naphthanes by their number of carbon atoms on the main column. The aromatic compounds are determined by a separate procedure and their contribution (as naphthenes) to the saturated compounds on the main column can be determined by a straightforward calculation on completion of the analysis. The method has been applied to the analysis of cyclic hydrocarbons and n- and *iso*alkanes in naphthas, and experience has shown that this procedure is no less accurate than more complicated methods which involve a physical on-line separation of the aromatic compound.

Carbon molecular sieves have sorption properties comparable with those of their inorganic counterparts. Their behaviour as stationary phases in gas chromatography has been investigated by Kaiser[947]. He studied the carbon molecular sieve type B obtained by the decomposition of poly(vinylidene chloride) at 180°C:

$$(C_2H_2Cl_2)_n \rightarrow 2nC + 2nHCl$$

To obtain an optimum stationary phase, the reaction must proced in a certain way, and further intensive processing is necessary later in order to obtain a homogeneous carbon molecular sieve from the standpoint of its adsorption properties, as well being chemically inactive. A solid foam of carbon is formed, mainly having two types of pores. These are micropores, open for rapid transport of substances, with a mean diameter of 1 μm, and micropores open for sorption with a mean diameter of 0.16 μm. Under certain conditions of synthesis, a dehalogenated carbon is obtained, with an adequate form and the highest purity.

Carbon molecular sieve type B shows very pronounced retention properties for the CH_2 group (water is eluted before methane on this stationary phase). Elution of water, alcohols, and formaldehyde to give symmetrical peaks is considered to result from the high purity of the stationary phase. The lack of any traces of inorganic oxides or of certain functional groups containing oxygen which could prove useful in all other types of active carbon is the origin of the inertness and uniformity of sorption of the material. The carbon molecular sieve can also behave as a useful stationary phase for the analysis of inorganic gases or even of the very polar gases with high chemical reactivity.

Clathrates

As a result of their ability to distinguish between molecular structures, gas–solid chromatography on clathrate forming stationary phases has been reported.[948-951] Mailen et al.[948] reported that thiourea, but not urea, is a good substrate in gas chromatography for the separation of the unsaturated compounds perfluoro-2-methylpentene-2 and cis- and trans-perfluoro-2-methylpentene-3 from perfluorohexanes and from one another.

The use of urea and of thiourea was an attempt to make use of the "channel" complexes that these chemicals form with numerous compounds.[952] If clathratrion should occur in a chromatographic column, a separation of isomers would be possible resulting in the elution of the most highly branched structure first and the n-isomer last. With both of the above substrates, the saturated perfluorohexanes appear in an order which is the reverse of that found on n-hexadecane,[953] on which the isomers appear strictly in order of increasing boiling point (which is the order of increasing complexity of molecular chain branching in the saturated perfluorohexanes). On urea and on the n-hexadecane substrates, the unsaturated molecules are not separated from the saturated ones.

The elution order from a 15.2 m×6.3 mm copper tube containing 31.2 g of thiourea on 124.6 g Chromosorb P (35–80 mesh) at 25°C with helium carrier flowing at 29.9 ml/min is: trans-perfluoro-2-methylpentene-3; perfluoro-2,3-dimethylbutane; perfluoro-2-methylpentane, cis-perfluoro-2-methylpentene-3; perfluoro-3-methylpentene; perfluoro-n-hexane; perfluoro-2-methylpentene-2. The peaks were symmetrical with no tailing for 1–2 ml samples. Hydrogen reacts with thiourea and destroys the separating ability of the column.

Bhattacharjee and Basu[949] investigated the selectivity of potassium benzene sulphonate (BK), sodium benzene sulphonate (BS), and rubidium benzene sulphonate (BR) as stationary phases for polar compounds. Stainless-steel columns (6 ft×⅛ in. o.d.) packed with 40% w/w of the freshly prepared salts on Chromosorb P (60–80 mesh) were used. Nitrogen was the carrier gas at a flow rate of 30 ml/min. The retention of compounds on the columns was found to be governed by three important factors: hydrogen bonding, clathration, and interactions with the lone pair of electrons on the metal ion.

The authors showed that clathration of alcohols on these phases is governed primarily by hydrogen bonding. Thus the retention of methyl alcohol is high compared with the other alcohols on BK and BR but not on BS, showing the deficiency in the sorption properties of the BS phase. The retention of n-alcohols decreased with an increase in chain length, except for n-butyl alcohol. For isomeric alcohols, the retention decreased in the order p- > s- > t-. The lower retention of *iso*alcohols is a function of their molecular shape.

The elution of di*iso*propyl ether and diethyl ether from columns of BR and BK clearly indicates the presence of holes in the crystal lattice of the sulphonates. Di*iso*propyl ether (b.p. 69.0°C) has a bulky structure and cannot enter the holes, so it is eluted earlier than diethyl ether (b.p. 34.7°C).

The order of elution of n-hexane, n-heptane, n-octane, *iso*octane, and cyclohexane followed the order of their boiling points on all the columns. The same result was also obtained for aromatic hydrocarbons.

On keeping the BR column at 160°C for 1 h and then lowering the temperature to that required for the elution of alcohols and ethers (80°C), the retentions followed the order of the boiling points, contrary to the behaviour of the non-thermally treated column, indicating the collapse of the holes or channels inside the crystal lattice.

These columns have excellent possibilities for the separation of phenols. The elution of phenols is governed by two factors—hydrogen bonding and the interaction of the metal ion with the lone pairs of electrons on the phenolic oxygen atom. A compound is probably retained by the combined effects of these two interactions. It is obvious that elution does not occur through clathration holes or channels because after the collapse of the channels the order of elution of the phenols remains the same. In o-substituted phenols, the higher the molecular volume of the alkyl group, the lower is the retention, owing to the steric hindrance by the alkyl group to hydrogen bonding and metal ion interactions. In phenols with n- and p-groups the decreased hydrogen bonding and increased metal ion interaction can be attributed to electronic effects.

The most significant effect on such columns is the metal ion–lone pair interaction, which is a maximum with the BR column, on which excellent separations of mixtures that are otherwise usually difficult to separate, such as 2,4- and 2,5-xylenol and m- and p-cresol, have been achieved. In these two pairs of compounds, the higher retentions of p-cresol and 2,4-xylenol arise from the higher electron density at the phenolic oxygen atom, caused by the methyl group in the 4-position.

The p-cresol/m-cresol retention ratio shows that the interaction decreases in the order BR > BK > BS. This may be because the ionic radii of the metals also decrease in the order Rb > K > Na.

In general, the peaks for alcohols and phenols are tailed. Tailing is minimal on the BR column and maximal on the BS column.

Six complexes of general composition $MB_4(NCS)_2$, where B is either 4-methyl- or 4-ethyl-pyridine and M is Fe, Co or Ni, and seven complexes of the type dithiocyanato-tetrakis(1-arylalkylamine)nickel(II) having different substituents in the aromatic ring, have been studied as stationary phases by Bhattacharyya and Bhattacharjee[950, 951] All these Werner complexes are known for their ability to clathrate various organic molecules.[954, 955] Stainless steel columns (180 cm×4 mm i.d.) packed with 30% w/w of the complex on Chromosorb P (60–80 mesh) were used, with a nitrogen flow rate of 60 ml/min.

Figure 47 shows the order of elution at 80°C of some aromatic hydrocarbons from columns containing $M(4\text{-MePy})_4(NCS)_2$ as stationary phases (M = Fe, Co, Ni).

The retentions in Fig. 47 are strictly governed by the shape of the "guest" molecules (except for 1,2,3-trimethylbenzene on cobalt- and iron-containing columns) and it appears that clathrate formation itself is not dependent on electronic interactions. The order of elution from these columns may be explained by the presence of permanent clathration "holes" in the host lattices. Linear molecules which fit in such holes are subjected to strong van der Waals forces inside, leading to their greater retentions over the non-linear molecules. This high selectivity toward

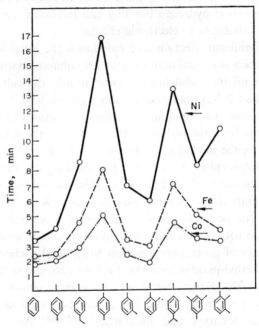

Fig. 47. Retention times of some aromatic hydrocarbons on M(4-MePy)$_4$(NCS)$_2$.[951] (M = Ni, Co, or Fe.)

molecular shape is primarily responsible for the broad, asymmetrical peak of the *p*-isomer; the *m*-xylene peak is of normal width.

The degree of asymmetry and broadness of the *p*-isomer peak increase in the order Fe > Co > Ni. Thus, from the nickel complex, the *p*-xylene peak is nearly symmetrical, from the cobalt complex it is much broader, and it is almost semicircular from the iron complex. This is consistent with the fact that the iron(II) complex has the highest capacity to accept foreign molecules into its lattice.

Increasing the length of the alkyl chain by one methylene group in passing from 4-methyl- to 4-ethyl pyridine results in the total loss of selectivity towards steric features under gas chromatographic conditions. This loss of selectivity is difficult to explain. The authors assume that an increase in one dimension of the cavity by placing an ethyl group in the 4-position of the pyridine ring possibly leads to a larger diameter hole in the host lattice so that only molecules having a cross-section larger than 1,3,5-trimethylbenzene would be preferentially retained on a gas chromatographic column.

For complexes of the type dithiocyanato-tetrakis(1-arylalkylamine)-nickel(II):

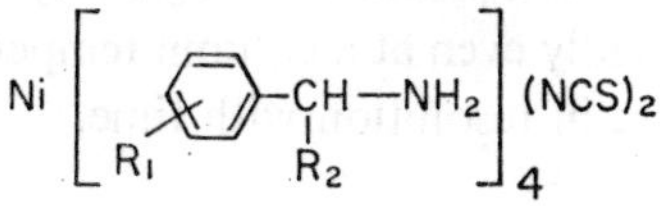

$$Ni \left[\underset{R_1}{\underset{\displaystyle |}{\bigotimes}} \underset{R_2}{\overset{\displaystyle |}{-CH}} -NH_2 \right]_4 (NCS)_2$$

de Radzitsky and Hanotier,[954] on the basis of exhaustive chemical and physical studies, have established that charge-transfer interactions between the aromatic nuclei of the host and the guest are primarily responsible for clathrate formation. Thus the elution order of some compounds studied on a $Ni(NCS)_2(1\text{-phenylethylamine})_4$ column are as follows:[950]

(a) cyclohexane < cyclohexene < benzene;
(b) methylcyclohexane < toluene;
(c) benzene < toluene < chlorobenzene < (m- + p-)xylenes < o-xylene < 1,3,5-trimethylbenzene < 1, 2, 3-trimethylbenzene;
(d) n-hexane < n-heptane < 2,2,4-trimethylpentane.

The column temperature for series (a), (b), and (d) was 30°C and for series (c), 60°C; the nitrogen flow rate was 30 ml/min in all instances. The elution orders of various compounds as shown in series (a), (b), and (c) agree with this explanation.

However, based on purely electronic interactions, ethylbenzene and

p-xylene (π-donors) should have emerged later than chlorobenzene and p-dichlorobenzene (π-acceptors), respectively, from the column containing the complex with $R_1{=}NO_2$ and $R_2{=}CH_3$ (π-acceptors). Obviously, non-charge-transfer factors like vapour pressure and diffusion override the weak charge-transfer effect in the gas chromatographic column.

In the columns having π-donor systems in the host complex ($R_1{=} R_2{=}CH_3$, and the trimethylphenylethylamine complex) the general order of elution of various guests, although seeming to conform to the charge-transfer effect (p-xylene is eluted earlier than p-dichlorobenzene), cannot be taken as unambiguous proof of this effect because in these examples the vapour pressure of the guest and electronic interactions operate in the same direction.

Finally, these Werner complexes, although very selective as stationary phases, decompose rapidly even at a column temperature of 90°C, resulting in a linear decrease in resolution with time.

Other compounds

Duffield and Rogers[956] show that silver nitrate-impregnated Chromosorb W and P columns should prove useful for the separation of aromatic compounds and olefins. Such packings should greatly simplify analyses for unsaturated compounds in complex mixtures because saturated hydrocarbons are not retained. Although larger amounts of silver nitrate on the support increased the distribution ratio, the effectiveness per milliequivalent of silver decreased. This may be the result of the formation of microscopic crystals or of blocking of pores. The values for 0.680 : 10 silver nitrate on Chromosorb W (80–100 mesh) appear to be anomalously high. When samples of 10^{-8} to 10^{-10} mole were used, the eluted peaks were quite symmetrical, whereas larger samples gave skewed elution peaks.

In all instances the *trans* isomer always eluted before the *cis* isomer, a result expected from steric effects.

Guha and Janak[229] note that because the experiments were carried out at relatively high temperatures (140–220°C), it is doubtful whether silver(I) charge transfer plays any role in the selectivity of such columns.

Cook and Givand[957] investigated the use of silver(I) complexes of various heterocyclic amines as column substrates. The general formula of the complex was AgL_2NO_3, where L represents the ligand. One exception was the 3-methylpyridine complex, AgL_3NO_3. The ligands used in preparing complexes were pyridine, 2-, 3-, and 4-methylpyridine, 2,6-dimethylpyridine, quinoline, isoquinoline, 2,2'-bipyridine, 2,2'-biquinoline, and 1,10-phenanthroline. Copper tubing $(8 \text{ ft} \times \frac{1}{8} \text{ in. o.d.})$ was packed with 10–15% of the complex on Chromosorb W, DMCS treated and acid washed (80–100 mesh). Silver(I) complexes prepared from pyridine and 2-, 3-, and 4-methylpyridine showed preferential adsorption for olefins, whereas those prepared from the other amines showed no interaction. This lack of interaction of the complexes prepared from the larger ligands may be explained by steric hindrance. The column temperature was limited to 40°C to ensure that no thermal breakdown of the complexes occurred.

The interaction of the adsorbents with unsaturated compounds decreased in the order Ag-2-Py > Ag-Py > Ag-4-Py.

Inconsistencies in retention volumes were observed for the Ag-3-Py complex column, so that it was not possible to place it in the above list. The presence of the third ligand molecule on this complex may be the cause of the inconsistencies.

Altenau et al.[958, 959] showed that complexes of 1,10-phenanthroline and 2,2'-bipyridine with copper(II) permitted the separation of a wide range of compounds. The compounds have good thermal stability and can be used as stationary phases up to 180°C. A column containing $Cu(Phen)_2(NO_3)_2$, 84 cm long, gives a good separation of aliphatic hydrocarbons, aromatic compounds, alcohols, esters, ethers, and ketones.[958] At a column temperature of 160°C, compounds with boiling points between 250–300°C have been eluted in 1–6 min.

The separation of a number of seven-ring polynuclear aromatic hydrocarbons derived from naphthalene and tetracene was achieved on a column containing $Cu(Bipy)_2(NO_3)_2$.

If a support is used for the copper–amine complexes, the problem of fine particles clogging the column may be eliminated because the support particle may be chosen for its size, and any "fines" produced can be removed by sieving. Moreover, particles of a support coated with a copper–amine complex display much more resistance to fracture or crumbling than do particles of a copper–amine complex alone.[959]

The dependence of the retention volumes of various organic ligands on the environment surrounding a transition metal ion was investigated by Pflaum and Cook.[960] Columns (6 ft × $\frac{1}{4}$ in. o.d. (1.8 m × 3 mm)) using bis(dimethylglyoximato)nickel(II) (Ni-DMG), bis(1,2-cyclohexanedio-nedioximato)nickel(II) (Ni-CHD), bis(salicylaldimino)nickel(II) (Ni-SAL), or nickel perchlorate as stationary phases on Gas-Chrom P (100–120 mesh) were prepared. Thermogravimetric analysis of the solid packing materials showed absolute thermal stability of all supports up to 220°C in a nitrogen atmosphere. The retention volumes for different compounds measured at the same temperature (50°C) were observed to pass through a minimum as the amount of nickel complex on the solid support increased. Peaks obtained on the nickel perchlorate column were less well-defined than peaks obtained for the same compounds on the columns of the nickel complexes. Only aliphatic and aromatic hydrocarbons were eluted from this column at 50°C. A good separation of C_3–C_7 aldehydes was observed using a 26.8% Ni-DMG column.

Of the metal complexes used as stationary phases, tailing increased in the order: Ni-DMG < Ni-SAL < Ni-CHD. Tailing of the solutes decreased as the stationary phase loading increased, and was found to follow the order: aliphatic hydrocarbons < ethers < esters < ketones < aldehydes < alcohols = amines. The order of decreasing heats of adsorption and thus an indication of increased interaction between adsorbents and adsorbates is: Ni-DMG = Ni-SAL < Ni-CHD.

Baiulescu and Ilie[385] showed that useful separations of olefins and aromatic hydrocarbons can be achieved on columns of thallium(I) tetraphenylborate. Table 25 gives the retentions of some aromatic hydrocarbons and their derivatives, relative to n-decane, at 129°C, on columns with $TlB(C_6H_5)_4$ comparated with those on $NaB(C_6H_5)_4$. The retention data were determined on thermoresistant glass columns (1.4 m × 3.6 mm) packed with 25% w/w stationary phase on hexamethylsilanized Chromosorb W (30–60 mesh). The substances generally elute from the column of sodium tetraphenylborate in the order of increasing boiling points.

TABLE 25. RETENTIONS (RELATIVE TO n-DECANE AT 129°C) ON COLUMNS CONTAINING $NaB(C_6H_5)_4$ AND $TlB(C_6H_5)_4$ [385]

Compound	Boiling point (°C)	Relative retention	
		$NaB(C_6H_5)_4$	$TlB(C_6H_5)_4$
Benzene	80.08	0.06	0.18
Cyclohexane	81.0	0.06	0.06
Toluene	110.6	0.24	0.41
Ethylbenzene	135.5	0.43	0.70
o-Xylene	144.1	0.47	0.94
m-Xylene	139.3	0.46	0.83
p-Xylene	138.0	0.46	0.76
Styrene	145.0	0.47	0.75
Phenylacetylene	142–144	0.56	0.60
isoPropylbenzene	152.0	0.57	1.01
Mesitylene	164.8	0 86	1.57
t-Butylbenzene	168.0	0.91	1.59
1-Methyl-4-isopropylbenzene	177.0	1.11	1.71
n-Decane	174.0	1.00	1.00
Napthalene	218.0	3.74	5.98
Phenyl-t-butylacetylene	72° (2 mm)	2.56	8.59
Benzo-t-butylcyclobutene	57° (1 mm)	1.90	3.77
p-Di-t-butylbenzene	236.5	6.69	17.23
n-Tridecane	234.0	5.90	7.66
1-Methylnaphthalene	241.0	5.65	10.30
2-Methylnaphthalene	240–242	6.74	10.76
Hexamethylbenzene	264.0	9.04	18.69
Diphenyl	254–255	14.93	17.38
n-Pentadecane	270.5	20.81	32.20

Larger differences in the relative retentions of the aromatic hydro-carbons on the column of thallium(I) tetraphenylborate should be a consequence of the interaction of the thallium(I) ions with the π-orbitals of the aromatic hydrocarbons. The presence of the electroposi-tive ethyne radical in phenylacetylene causes the decreased aromatic character of the molecule. This explains its relative retention, which is practically the same on both chromatographic columns. The introduction of the t-butyl radical into the molecule of phenylacetylene greatly increas-es the relative retention on the two chromatographic columns. The condensation of the benzene rings produces the diminished aromatic character of naphthalene, which explains the relatively small difference between the retention of naphthalene on columns containing sodium and thallium tetraphenylborates. Moreover, the heats of adsorption of the aromatic compounds and their derivatives are greater for $TlB(C_6H_5)_4$ than for $NaB(C_6H_5)_4$. This may be explained by the interaction of the thallium ions, which contain low-lying d-orbitals which can form a partial covalent bond with the π-orbitals of the adsorbate molecules.

REFERENCES

1. KAISER, R., *Chromatographia* **2**, 219 (1969).
2. KAISER, R., *ibid.* **3**, 443 (1970).
3. CRAMERS, C. A., LUYTEN, J. A., and RIJKS, J. A., *ibid.*, p. 441.
4. VERSINO, B., *ibid.*, p. 231.
5. PETERSON, M. L., and HIRSCH, J., *J. Lipid Res.* **1**, 132 (1959).
6. GOLD, H. J., *Analyt. Chem.* **34**, 174 (1962).
7. HANSEN, L. H., and ANDRESEN, K., *J. Chromat.* **34**, 246 (1968).
8. PALM, E., in E. SZ. KOVÁTS, ed., *Column Chromatography, Lausanne, 1969*, Sauerlaender AG, Aarau, 1970, p. 163.
9. ROHRSCHNEIDER, L., *Chromatographia* **2**, 437 (1969).
10. ROHRSCHNEIDER, L., ref. 8, p. 136.
11. JAMES, A. T., and MARTIN, A. J. P., *Biochem. J.* **50**, 679 (1952).
12. WICAROVA, O., NOVÁK, J., and JANAK, J., *J. Chromat.* **51**, 3 (1970).
13. GOEDERT, M., and GUIOCHON, G., *Analyt. Chem.* **42**, 962 (1970).
14. CRUBRYT, J. J., GESSER, H. D., and BOCK, E., *J. Chromat.* **53**, 439 (1970).
15. POSHKUS, D. P., and AFREIMOVITCH, A. J., *ibid.* **58**, 55 (1971).
16. ROHRSCHNEIDER, L., *J. Chromat. Sci.* **8**, 105 (1970).
17. MCREYNOLDS, W. D., *Gas Chromatographic Retention Data*, Preston Technical Abstracts Co., Evanston, Ill., 1966.
18. HARTKOPF, A., *J. Chromat. Sci.* **10**, 145 (1972).
19. KOVÁTS, SZ. E., *Helv. chim. Acta* **41**, 1915 (1958).
20. VAN DEN HEUVEL, W. J. A., GARDINER, W. L., and HORNING, E. C., *Analyt. Chem.* **36**, 1550 (1964).
21. MIWA, T. K., MIKOLAJCZAK, K. L., EARLE, F. R., and WOLFF, A. I., *ibid.* **32**, 1739 (1960).
22. VAN DEEMTER, J. J., ZUIDERWEG, F. J., and KLINKENBERG, A., *Chem. Engng Sci.* **5**, 271 (1956).
23. GOLAY, M. J. E., in D. H. DESTY, ed., *Gas Chromatography 1958*, Butterworths, London, 1958, p. 36.
24. JONES, W. L., *Analyt. Chem.* **33**, 829 (1961).
25. GIDDINGS, J. C., *Nature* **184**, 357 (1959); *Analyt. Chem.* **35**, 1338 (1963).
26. GIDDINGS, J. C., and SCHETTLER, P. D., *ibid.* **36**, 1483 (1964).
27. SAHA, N. C., and GIDDINGS, J. C., *ibid.* **37**, 822 (1965).
28. KNOX, J. H., and SALEEM, M., *J. Chromat. Sci.* **10**, 80 (1972).
29. GUIOCHON, G., and LANDAULT, C., *Chromatographia* **1**, 227 (1968).

30. PERRETT, R. H., and PURNELL, J. H., *Analyt. Chem.* **34**, 1336 (1962).
31. LITTLEWOOD, A. B., *in* A. Goldup, ed., *Gas Chromatography 1964*, Inst. of Petroleum, London, 1965, p. 77.
32. DAL NOGARE, S., and CHIU, J., *Analyt. Chem.* **34**, 890 (1962).
33. DE FORD, D. D., LOYD, R. J., and AYERS, B. O., *ibid.* **35**, 426 (1963).
34. JANAK, J., and STASZEWSKI, R., *J. Gas Chromat.* **2**, 47 (1964).
35. GIDDINGS, J. C., *J. Chromat.* **5**, 46 (1961).
36. GIDDINGS, J. C., *Analyt. Chem.* **34**, 1186 (1962).
37. SAHA, N. C., and GIDDINGS, J. C., *ibid.* **37**, 830 (1965).
38. PERRETT, R. H., and PURNELL, J. H., *ibid.* **35**, 430 (1963).
39. KNOX, J. H., and PARCHER, J. E., *ibid.* **41**, 1599 (1969).
40. HUBER, J. F. L., and HULSMAN, J. A. R., *Anal. chim. Acta* **38**, 305 (1967).
41. GLUECKAUF, E., *Trans. Faraday Soc.* **51**, 34 (1955).
42. GOLAY, M. J. E., *Nature* **182**, 1146 (1958).
43. PURNELL, J. H., *ibid.* **184**, 2009 (1959); *J. Chem. Soc.* 1268 (1960).
44. HALÁSZ, I., *J. Gas Chromat.* **5**, 51 (1967).
45. GIDDINGS, J. C., *in* J. C. GIDDINGS and R. A. KELLER, eds., *Dynamics of Chromatography*, Part I, Marcel Dekker, New York, 1965.
46. LITTLEWOOD, A. B., *Analyt. Chem.* **38**, 2 (1966).
47. AMBROSE, D., JAMES, A. T., KEULEMANS, A. I. M., and KOVÁTS, SZ. E., *in* R. P. W. SCOTT, ed., *Gas Chromatography 1960*, Butterworths, London, 1960, p. 429.
48. PURNELL, J. H., and QUINN, C. P., ref. 47, p. 184.
49. HALÁSZ, I., HARTMANN, K., and HEINE, E., ref. 31, p. 38.
50. KARGER, B. L., and COOKE, W. D., *Analyt. Chem.* **36**, 985 and 991 (1964).
51. GUIOCHON, G., *ibid.* **38**, 1020 (1966).
52. GUIOCHON, G., *in* J. C. GIDDINGS and R. A. KELLER, eds., *Advances in Chromatography*, vol. 8, Marcel Dekker, New York, 1969, p. 179.
53. GRUSHKA, E., *Analyt. Chem.* **42**, 1142 (1970).
54. GIDDINGS, J. C., *ibid.* **39**, 1027 (1967).
55. GIDDINGS, J. C., *Separ. Sci.* **4**, 181 (1969).
56. GRUSHKA, E., *Analyt. Chem.* **43**, 766 (1971).
57. GRUSHKA, E., YEPES-BARAYA, M., and COOKE, W. D., *J. Chromat. Sci.* **9**, 653 (1971).
58. GUIOCHON, G., in ref. 8, p. 250.
59. ZHUKHOVITSKII, A. A., SAZANOV, M. L., LUNSKII, M. KH., and YUSFIN, V., *J. Chromat.* **58**, 87 (1971).
60. TESARIK, K., and NECASOVA, M., *J. Chromat.* **65**, 39 (1972).
61. SCOTT, R. P. W., *in* J. C. GIDDINGS and R. A. KELLER, eds., *Advances in Chromatography*, vol. 9, Marcel Dekker, New York, 1970, p. 193.
62. HALÁSZ, I., and HEINE, E., *in* J. H. PURNELL, ed., *Progress in Gas Chromatography*, Interscience, New York, 1968, p. 153.
63. CALLEAR, A. B., and CVETANOVITS, R. J., *Can. J. Chem.* **33**, 1256 (1955).
64. HISTHA, C., and BOMSTEIN, J., *J. Gas Chromat.* **5**, 395 (1967).
65. HISTHA, C., and BOMSTEIN, J., ref. 61, p. 215.
66. HALÁSZ, I., and HORVÁTH, C., *Analyt. Chem.* **36**, 1178 (1964).
67. HALÁSZ, I., and HOLDINGHAUSEN, F., *J. Gas Chromat.* **5**, 385 (1967).
68. HALÁSZ, I., and HEINE, E., *Nature* **194**, 971 (1962).

69. HALÁSZ, I., and HEINE, E., *Analyt. Chem.* **37**, 495 (1965).
70. LANDAULT, C., and GUIOCHON, G., ref. 31, p. 121.
71. LANDAULT, C., and GUIOCHON, G., *Analyt. Chem.* **39**, 713 (1967).
72. LANDAULT, C., GUIOCHON, G., and GANANSIA, J., *Bull. Soc. chim. Fr.* 3985 (1967).
73. ADLARD, E. R., and ROBERTS, G. W., *J. Inst. Petrol.* **51**, 376 (1965).
74. ETTRE, L. S., PURCELL, J. E., and BILLEB, K., *Separ. Sci.* **1**, 777 (1966).
75. CRAMERS, C. A., RIJKS, J., and BOCEK, P., *J. Chromat.* **65**, 29 (1972).
76. CARTER, H. V., *Nature* **197**, 684 (1963).
77. VIRUS, W., *J. Chromat.* **12**, 406 (1963).
78. VIGDERGAUZ, M. S., and ANDREJEV, L. S., *Khimiya Tekhnol. Topl. Masel* **9**, 64 (1964).
79. WILHITE, W. F., *J. Gas Chromat.* **4**, 47 (1966).
80. SCOTT, R. P. W., and HAZELDDAN, G. S. F., ref. 47, p. 144.
81. DESTY, D. H., HARESNAPE, J. N., and WHYMAN, B. H. F., *Analyt. Chem.* **32**, 302 (1960).
82. HALÁSZ, I., and HORVÁTH, C., *ibid.* **35**, 499 (1963).
83. BOHEMEN, J., and PURNELL, J. H., *J. Chem. Soc.* 360 (1961).
84. ROHRSCHNEIDER, L., *Chromatographia* **9**, 431 (1970).
85. MARTIN, A. J. P., *Analyst, Lond.* **81**, 52 (1956).
86. YOUNG, C. L., *Chromat. Rev.* **10**, 129 (1968).
87. CONDER, J. R., ref. 62, p. 209.
88. VIGDERGAUZ, M. S., and IZMAILOV, R. I., *Application of Gas Cromatography to the Determination of the Physico-Chemical Characteristics of Substances*, Izd. Nauka, Moscow, 1970.
89. GOLOVNYA, R. V., and ARSENYEV, YU. N., *Chromatographia* **4**, 250 (1971).
90. LITTLEWOOD, A. B., *Gas Chromatography*, Academic Press, New York, 1962, p. 50.
91. WINDSOR, M. L., and YOUNG, C. L., *J. Chromat.* **27**, 355 (1967).
92. DESTY, D. H., GOLDUP, A., LUCHURST, G. R., and SWANTON, W. T., in M. VAN SWAAY, ed., *Gas Chromatography 1962*, Butterworths, London, 1962, p. 67.
93. ADLARD, E. R., KHAN, M. A., and WHITHAM, B. T., ref. 47, paper No. 17.
94. *Handbook of Chemistry and Physics*, Chemical Rubber Publ., Cleveland, Ohio (published annually).
95. JORDAN, T. E., *Vapor Presure of Organic Compounds*, Interscience, New York, 1954.
96. TIMMERMANS, J., *Physico-Chemical Constants of Pure Organic Compounds*, Elsevier, Amsterdam, 1950.
97. STULL, D. R., *Ind. Engng Chem.* **39**, 517 (1947).
98. DREISBACH, R. R., *Physical Properties of Chemical Compounds* (Advances in Chemistry Series), American Chemical Society, Washington, DC: (a) No. 15, 1955; (b) No. 22, 1959; (c) No. 29, 1961.
99. HALA, E., PICK, J., FRIED, V., and VILIM, O., *Vapour–Liquid Equilibrum*, Pergamon Press, Oxford, 1960.
100. ILIE, V., OPREA, M., and BAIULESCU, G., *Revistă Chim.* **20**, 761 (1969).
101. PORTER, P. E., DEAL, C. H., and STROSS, F. H., *J. Am. Chem. Soc.* **78**, 2999 (1956).
102. MUHS, M. A., and WEISS, F. T., *ibid.* **84**, 4697 (1962).
103. GIL-AV, E., and HERLING, J., *J. Phys. Chem.* **66**, 1208 (1962).

104. CVETANOVIC, R. J., DUNCAN, F. J., FALCONER, W. E., and IRWIN, R. S., *J. Am. Chem. Soc.* **87**, 1827 (1965).
105. PURNELL, J. H., in A. B. LITTLEWOOD, ed., *Gas Chromatography 1966*, Elsevier, Amsterdam, 1967, p. 3.
106. ROSSOTTI, F. J. C. and ROSSOTTI, H., *The Determination of Stability Constants*, McGraw–Hill, New York, 1961.
107. DE LIGNY, C. L., *J. Chromat.* **69**, 243 (1972).
108. DE LIGNY, C. L., VAN'T VERLAAT, T., and KARTHAUS, F., *ibid.* **76**, 115 (1973).
109. VITT, S. V., BONDAREV, V. B., PASKANOVA, E. A., SHCHERBINA, T. N., and BEZRUKOV, M. G., *Izv. Akad. Nauk SSSR*, Ser. Khim. 441 (1972).
110. MEEN, D. L., MORRIS, F., and PURNELL, J. H., *J. Chromat. Sci.* **9**, 281 (1971).
111. EON, C., POMMIER, C., and GUIOCHON, G., *J. Phys. Chem.* **75**, 2632 (1971).
112. EON, C., and KARGER, B. L., *J. Chromat. Sci.* **10**, 140 (1972).
113. CVETANOVIC, R. J., DUNCAN, F. J., FALCONER, W. E., and SUNDER, W. A., *J. Am. Chem. Soc.* **88**, 1602 (1966).
114. FALCONER, W. E., and CVETANOVIC, R. J., *J. Chromat.* **27**, 20 (1967).
115. CADOGAN, D. F., and PURNELL, J. H., *J. Chem. Soc.* A, 2133 (1968).
116. JUVET, R. S., Jr., SHAW, V. R., and ASLAM KHAN, M., *J. Am. Chem. Soc.* **91**, 3788 (1969).
117. SCHNECKO, H., *Analyt. Chem.* **40**, 1391 (1968).
118. CLEVER, H. L., BAKER, E. R., and HALE, W. R., *J. Chem. Engng Data* **15**, 411 (1970).
119. CASTELLS, R. C., and CATOGGIO, J. A., *Analyt. Chem.* **42**, 1268 (1970).
120. CADOGAN, D. F., and SAWYER, D. T., *ibid.* **43**, 941 (1971).
121. LITTLEWOOD, A. B., PHILLIPS, C. S. G., and PRICE, D. T., *J. Chem. Soc.* 1480 (1955).
122. MARTIRE, D. E., in L. FOWLER, ed., *Gas Chromatography*, Academic Press, New York, 1963, p. 33.
123. MARTIRE, D. E., and POLLARA, L. Z., in J. C. GIDDINGS and R. A. KELLER, eds., *Advances in Chromatography*, vol. 1, Marcel Dekker, New York, 1966, p. 335.
124. HIRSCHFELDER, J. O., CURTISS, C. F., and BIRD, R., *Molecular Theory of Gases and Liquids*, Wiley, New York, 1954, p. 252.
125. LAMBERT, J. D., ROBERTS, G. A. H., ROWLINSON, J. S., and WILKINSON, V. J. *Proc. R. Soc.* A, **196**, 113 (1949).
126. GUGGENHEIM, E. A., and McGLASHAN, M. L., *ibid.* **206**, 448 (1951).
127. McGLASHAN, M. L., and POTTER, D. J. B., *ibid.* **267**, 478 (1962).
128. GUGGENHEIM, E. A., and WORMALD, C. J., *J. Chem. Phys.* **42**, 3775 (1965).
129. KOBE, K. A., and LYNN, R. E., Jr., *Chem. Rev.* **52**, 117 (1953).
130. HERZOG, R., *Ind. Engng Chem.* **36**, 997 (1944).
131. EVERED, S., and POLLARD, F. H., *J. Chromat.* **4**, 451 (1960).
132. EVERETT, D. H., and STODDARD, C. T. H., *Trans. Faraday Soc.* **57**, 746 (1961).
133. DESTY, D. H., and SWANTON, W. T., *J. Phys. Chem.* **65**, 766 (1961).
134. EVERETT, D. H., and MUNN, R. J., *Trans. Faraday Soc.* **60**, 1951 (1964).
135. PECSAR, R. E., and MARTIN, J. J., *Analyt. Chem.* **38**, 1661 (1966).
136. MARTIRE, D. E., ref. 105, p. 21.
137. BARKER, P. E., and HILMI, A. K., *J. Gas Chromat.* **5**, 119 (1967).
138. EVERETT, D. H., GAINEY, B. W., and YOUNG, C. L., *Trans. Faraday Soc.* **64**, 2667 (1968).

139. HICKS, C. P., and YOUNG, C. L., *ibid.*, p. 2675.

140. PETSEV, N., and DIMITROV, C., *J. Chromat.* **42**, 311 (1969).

141. ZIELINSKI, W. L., Jr., FREEMAN, D. H., MARTIRE, D. E., and CHOW, L. C., *Analyt. Chem.* **42**, 176 (1970).

142. SMILEY, H. M., *J. Chem. Engng Data* **15**, 413 (1970).

143. MECHAEVA, M. A., VIGDERGAUZ, M. S., and MUHAMEDOVA, L. A., *Neftekhimiya* **10**, 289 (1970).

144. TEWARI, Y. B., MARTIRE, D. E., and SHERIDAN, J. P., *J. Phys. Chem.* **74**, 2345 (1970).

145. BLU, G., JACOB, L., and GUIOCHON, G., *J. Chromatog.* **50**, 1 (1970).

146. KUCHHAL, R. K., and MALLIK, K. L., *J. Chem. Engng Data* **17**, 49 (1972).

147. FREEGUARD, G. F., and STOCK, R., ref. 92, p. 102; *Trans. Faraday Soc.* **59**, 1655 (1963).

148. SEWELL, P. A., and STOCK, R., *J. Chromat.* **50**, 10 (1970).

149. ASHWORTH, A. J., and EVERETT, D. H., *Trans. Faraday Soc.* **56**, 1609 (1960).

150. LONGUET–HIGGINS, H. C., *Discuss. Faraday Soc.* **15**, 73 (1953).

151. BARKER, J. A., *J. Chem. Phys.* **20**, 1526 (1952); **21**, 1391 (1953).

152. CRUICKSHANK, A. J. B., GAINEY, B. W., and YOUNG, C. L., *Trans. Faraday Soc.* **64**, 337 (1968).

153. BHATTACHARYYA, S. N., and MUKHERJEE, A., *J. Phys. Chem.* **72**, 56, 63 (1968).

154. YOUNG, C. L., *Trans. Faraday Soc.* **64**, 1537 (1968).

155. GAINEY, B. W., and YOUNG, C. L., *ibid.* p. 349.

156. HILDEBRAND, J. H., and SCOTT, R. L., *Regular Solutions*, Prentice-Hall, Englewood Cliffs, 1962.

157. GUGGENHEIM, E. A., *Trans. Faraday Soc.* **44**, 1007 (1948).

158. GAINEY, B. W., and PECSOK, R. L., *J. Phys. Chem.* **74**, 2548 (1970).

159. LUCKHURST, G. R., and MARTIRE, D. E., *Trans. Faraday Soc.* **65**, 1248 (1969).

160. LONGUET-HIGGINS, H. C., *Proc. R. Soc. A*, **205**, 247 (1951).

161. KREGLEWSKI, A., *J. Phys. Chem.* **71**, 2860 (1967); **72**, 1897 (1968).

162. PIEROTTI, G. J., DEAL, C. H., DERR, E. L., and PORTER, P. E., *J. Am. Chem. Soc.* **78**, 2989 (1956).

163. PIEROTTI, G. J., DEAL, C. H., and DERR, E. L., *Ind. Engng Chem.* **51**, 95 (1959).

164. DEAL, C. H., DERR, E. L., and PAPDOPOULOS, M. N., *Ind. Engng Chem. Fundamentals* **1**, 17 (1962).

165. WILSON, G. M., and DEAL, C. H., *ibid.*, p. 20.

166. MARTIN, R. L., *Analyt. Chem.* **33**, 347 (1961).

167. PECSOK, R. L., DE YLLANA, A., and ABDUL–KARIM, A., *ibid.* **36**, 452 (1964).

168. MARTIRE, D. E., *ibid.* **38**, 244 (1966).

169. MARTIRE, D. E., ref. 62, p. 93.

170. BEREZKIN, V. G., PAKHOMOV, V. P., and BEREZKINA, L. G., *Gazovaya Kromatografiya*, Proc. 3rd USSR Conf. Gas Chromatography, Dzerzhinskii Filial OKBA, Dzerzinski, 1966.

171. BEREZKIN, V. G., PAKHOMOV, V. P., STAROBINETS, L. L., and BEREZKINA, L. G., *Neftekhimiya* **5**, 438 (1965).

172. BEREZKIN, V. G., and FATEEVA, V. M., *Chromatographia* **4**, 19 (1971).

173. BEREZKIN, V. G., and FATEEVA, V. M., *J. Chromat.* **58**, 73 (1971).

174. BEREZKIN, V. G., *ibid.* **65**, 227 (1972).

175. CONDER, J. R., LOCKE, D. C., and PURNELL, J. H., *J. Phys. Chem.* **73**, 700 (1969).
176. CADOGAN, D. F., CONDER, J. R., LOCKE, D. C., and PURNELL, J. H., *ibid.*, p. 708.
177. EON, C., CHATTERJEE, A. K., and KARGER, B. L., *Chromatographia* **5**, 28 (1972).
178. SUPRYNOWICZ, Z., WAKSMUNDZKI, A., and RUDZINSKI, W., *J. Chromat.* **67**, 21 (1972).
179. MARTIRE, D. E., PECSOK, R. I., and PURNELL, J. H., *Nature* **203**, 1279 (1964).
180. MARTIRE, D. E., PECSOK, R. L., and PURNELL, J. H., *Trans. Faraday Soc.* **61**, 2496 (1965).
181. URONE, P., and PARCHER, J. F., *Analyt. Chem.* **38**, 270 (1966).
182. ROGOZINSKI, M., and KAUFMAN, I., *J. Gas Chromat.* **4**, 413 (1966).
183. HABGOOD, H. W., and HANLAN, J. V., *Can. J. Chem.* **37**, 843 (1959).
184. GALE, R. L., and BEEBE, R. A., *J. Phys. Chem.* **68**, 555 (1964).
185. EBERLY, P. E., *J. Appl. Chem.* **14**, 330 (1964).
186. KING, J., Jr., and BENSON, S. W., *Analyt. Chem.* **38**, 261 (1966).
187. PETOV, G. M., and SHCHERBAKOVA, K. D., ref. 105, p. 50.
188. GURAN, B. T., and ROGERS, L. B., *Analyt. Chem.* **39**, 632 (1967).
189. DE BOER, J. H., and KRUYER, S., *Proc. Acad. Sci. Amsterdam* **55b**, 451 (1952).
190. SAWYER, D. T., and BROOKMAN, D. J., *Analyt. Chem.* **40**, 1847 (1968).
191. BROOKMAN, D. J., and SAWYER, D. T., *ibid.* p. 1368.
192. HARGROVE, G. L., and SAWYER, D. T., *ibid.*, p. 409; **39**, 945 (1967).
193. WHITE, D., and COWAN, C. T., *Trans. Faraday Soc.* **55**, 557 (1958).
194. CREMER, E., *Z. Analyt. Chem.* **170**, 219 (1959).
195. BASMADJIAN, D., *Can. J. Chem.* **38**, 149 (1960).
196. OKAMURA, J. P., and SAWYER, D. T., *Analyt. Chem.* **43**, 1730 (1971).
197. BAIULESCU, G. E., and ILIE, V. A., *ibid.* **44**, 1490 (1972).
198. ILIE, V. A., Doctoral Thesis, University of Bucharest, 1973.
199. JAMES, A. T., *Biochem. J.* **52**, 252 (1952).
200. JAMES, A. T., and MARTIN, A. J. P., *Analyst, Lond.* **77**, 915 (1952).
201. JAMES, A. T., MARTIN, A. J. P., and SMITH, G. H., *Biochem. J.* **52**, 238 (1952).
202. LANGER, S. H., and SHEEHAN, R. J., ref. 62, p. 289.
203. HIRSCHFELDER, J. O., CURTISS, C. F., and BIRD, R. B., *The Molecular Theory of Gases and Liquids*, 2nd edn., Wiley, New York, 1964.
204. A. D. McLACHLAN, *Discuss. Faraday Soc.* **40**, 239 (1965).
205. LONDON, F., *Trans. Faraday Soc.* **33**, 8 (1937).
206. LITTLEWOOD, A. B., *J. Gas Chromat.* **1**, 16 (1963).
207. DYSON, N., and LITTLEWOOD, A. B., *Trans. Faraday Soc.* **63**, 1895 (1967).
208. KEESOM, W. H., *Phys. Z.* **22**, 129 (1921); **23**, 225 (1922).
209. DEBYE, P., *ibid.* **21**, 178 (1920).
210. MEYER, E. F., and ROSS, R. A., *J. Phys. Chem.* **75**, 831 (1971).
211. WEIMER, R. F., and PRAUSNITZ, J. M., *Hydrocarbon Process* **44**, 237 (1965).
212. MEYER, E. F., and WAGNER, R. E., *J. Phys. Chem.* **70**, 3162 (1966).
213. MEYER, E. F., RENNER, T. A., and STEC, K. S., *ibid.* **75**, 642 (1971).
214. MEATH, W. J., and HIRSCHFELDER, J. O., *J. Chem. Phys.* **44**, 3197 and 3210 (1966).
215. COOPER, A. R., CROWNE, C. W. P., and FARRELL, P. G., *Trans. Faraday Soc.* **62**, 2725 (1966); **63**, 447 (1967).
216. NORMAN, R. O. C., *Proc. Chem. Soc.* 151 (1958).
217. CROWNE, C. W. P., and HARPER, M. F., *J. Chromat.* **61**, 7 (1971).

218. HADZI, D., and THOMPSON, H. W., *Hydrogen Bonding*, Pergamon Press, Oxford, 1959.
219. PIMENTEL, G. C., and MCCLELLAN, A. L., *The Hydrogen Bond*, Freeman, San Francisco, 1960.
220. SOKOLOV, N. D., *Annls. Chimie* **10**, 497 (1965).
221. BRATOZ, S., in P. O. LOWDIN, ed., *Advances in Quantum Chemistry*, vol. 3, Academic Press, New York, 1967.
222. AMBROSE, D., and AMBROSE, B. A., *Gas Chromatography*, Newnes, London, 1961, p. 132.
223. PURNELL, J. H., *Gas Chromatography*, Wiley, New York, 1962, p. 336.
224. DAL NOGARE, S., and JUVET, R. S., *Gas–Liquid Chromatography, Theory and Practice*, Interscience, New York, 1962, p. 107.
225. IOGANSEN, A. V., KURKCHI, G. A., and LEVINA, O. V., in ref. 105, p. 35.
226. VERNON, F., *J. Chromat.* **63**, 249 (1971).
227. DEWAR, M. J. S., *Bull. Soc. chim. Fr.* **18**, C71 (1951).
228. ILIE, V., *Revistă Chim.* **20**, 43 (1969).
229. GUHA, O. K., and JANAK, J., *J. Chromat.* **68**, 325 (1972).
230. PEARSON, R. G., *Science* **151**, 172 (1966).
231. PEARSON, R. G., and SONGSTAD, J., *J. Am. Chem. Soc.* **89**, 1827 (1967).
232. KISELEV, A. V., *in* J. C. GIDDINGS and R. A. KELLER, eds., *Advances in Gas Chromatography*, vol. 4, Marcel Dekker, New York, 1967, p. 113.
233. KISELEV, A. V., *Discuss. Faraday Soc.* **40**, 205 (1965).
234. LARD, E. W., and HORN, R. C., *Analyt. Chem.* **32**, 878 (1960).
235. JONES, K., and HALFORD, P., *Nature* **202**, 1003 (1964).
236. HERSH, C. K., *Molecular Sieves*, Reinhold, New York, 1961, Chapter 4.
237. SPENCER, C. F., *J. Chromat.* **11**, 108 (1963).
238. MCBAIN, J. W., *Colloid Symp. Monogr.* **4**, 7 (1926).
239. BARRER, R. M., *Nature* **181**, 176 (1958).
240. BACHMANN, L., BECHTOLD, E., and CREMER, E., *J. Catalysis* **1**, 113 (1949).
241. BARRER, R. M., and GIBBONS, R. M., *Trans. Faraday Soc.* **59**, 2569 (1963).
242. TSITSISHVILI, G. V., and ANDRONIKASHVILI, T. G., *J. Chromat.* **58**, 39 (1971).
243. GANT, P. L., and YANG, K., *J. Am. Chem. Soc.* **86**, 5063 (1964).
244. CROWELL, A. D., *in* E. A. FLOOD, ed., *The Solid–Gas Interface*, vol. 1, Marcel Dekker, New York, 1967, p. 175.
245. STEELE, W. A., *ibid.*, p. 307.
246. POSHKUS, D. P., *Discuss. Faraday Soc.* **40**, 195 (1965).
247. KISELEV, A. V., *Zh. fiz. Khim.* **38**, 2753 (1964); **41**, 2470 (1967).
248. AVGUL, N. N., and KISELEV, A. V., *in* P. L. WALKER, ed., *Chemistry and Physics of Carbon*, vol. 6, Marcel Dekker, New York, 1970, p. 1.
249. KISELEV, A. V., *J. Chromat.* **49**, 84 (1970).
250. CROWELL, A. D., and STEELE, R. B., *J. Chem. Phys.* **34**, 1347 (1961).
251. CROWELL, A. D., and CHAI OK-CHANG, *ibid.* **38**, 2584 (1963).
252. CURTHOYS, G., and ELKINGTON, P. A., *J. Phys. Chem.* **71**, 1477 (1967).
253. ROSS, S., and OLIVIER, J. P., *On Physical Adsorption*, Interscience, New York, 1964.
254. KOUZNETSOV, A. V., and SHCHERBAKOVA, K. D., *J. Chromat.* **49**, 21 (1970).
255. KISELEV, A. V., and POSHKUS, D. P., *Zh. fiz. Khim.* **32**, 2824 (1958).

256. CROWELL, A. D., *J. Chem. Phys.* **49**, 892 (1968).
257. BARRER, R. M., *Discuss. Faraday Soc.* **40**, 231 (1965).
258. GREGG, S. J., and SING, K. S. W., *Adsorption Surface Area and Porosity*, Academic Press, New York, 1967.
259. TIMOFEJEW, D. P., *Adsorption Kinetik*, VEB Deutscher Verlag für Grundstoffindustrie, Leipzig, 1967.
260. VON UNGER, K., *Angew. Chem.* **84**, 331 (1972).
261. EVERETT, D. H., *in* E. A. FLOOD, ed., *The Solid–Gas Interface*, vol. 2, Marcel Dekker, New York, 1967, p. 1055.
262. DUBININ, M. M., *J. Colloid Interface Sci.* **23**, 487 (1967); *Advan. Colloid Interface Sci.* **2**, 217 (1968).
263. BRUNAUER, S., EMMETT, P. H., and TELLER, E., *J. Am. Chem. Soc.* **60**, 309 (1938).
264. KISELEV, A. V., *Usp. Khim.* **14**, 367 (1945).
265. HARKINS, W. D., and JURA, G., *J. Am. Chem. Soc.* **66**, 1366 (1944).
266. BERING, B. P., DUBININ, M. M., and SERPINSKY, V. V., *J. Colloid Interface Sci.* **21**, 378 (1966).
267. DUBININ, M. M., *Chem. Rev.* **60**, 235 (1960).
268. BRUNAUER, S., *Z. phys. Chem.* **64**, 54 (1969).
269. BRUNAUER, S., KIKHAIL, R. SH., and BODOR, E. E., *J. Colloid Interface Sci.* **24**, 451 (1967).
270. KISELEV, A. V., *Usp. Khim.* **25**, 705 (1956).
271. VYAKHIREV, D. A., CHERNYAYEV, N. P., and BRUK, A. I., *Zh. fiz. Khim.* **34**, 1096 (1960).
272. KISELEV, A. V., PETROVA, R. S., and SHCHERBAKOVA, K. D., *Kinetica i Kataliz* **5**, 526 (1964).
273. KISELEV, A. V., and YASHIN, YA. I., *Neftekhimiya* **4**, 634 (1964).
274. ROHRSCHNEIDER, L., *Z. analyt. Chem.* **211**, 18 (1965).
275. SCHOMBURG, G. J., *Analyt. chim. Acta* **38**, 45 (1967).
276. KOVÁTS, SZ. E., *Chimia* **22**, 459 (1968).
277. WALRAVEN, J. J., LADON, A. W., and KEULEMANS, A. I. M., *Chromatographia* **1**, 195 (1968).
278. VANDEN HEUVEL, F. A., and COURT, A. S., *J. Chromat.* **38**, 439 (1968).
279. ALTENBURG, K., *in* H. G. STRUPPE, ed., *Gas Chromatographie 1968*, Akademie-Verlag, Berlin, 1968, p. 1.
280. PIRINGER, O., ref. 8, p. 173.
281. TAKÁCS, J., SZITA, C., and TARJÁN, G., *J. Chromat.* **56**, 1 (1971).
282. MATUKUMA, A., *in* C. L. A. HARBOURN and R. STOCK, eds., *Gas Chromatography 1968*, Elsevier, Amsterdam, 1969, p. 55.
283. BAUMANN, F., STRAUS, A. E., and JOHNSON, J. F., *J. Chromat.* **20**, 1 (1965).
284. WILLIS, D. E., *ibid.* **30**, 86 (1967).
285. SOJAK, L., KRUPCIK, J., TESARIK, K., and JANAK, J., *ibid.* **55**, 93 (1970).
286. CASTELLO, G., and D'AMATO, G., *ibid.* **58**, 127 (1971).
287. MAIER, H. J., and KÁRPÁTHY, O. C., *ibid.* **8**, 308 (1962).
288. BROWN, I., *ibid.* **10**, 284 (1963).
289. ROHRSCHNEIDER, L., *Z. analyt. Chem.* **211**, 18 (1965).
290. CHOVIN, P., and LEBBE, J., *Separation Immediate et Chromatographie 1961*, GAMS, Paris, 1961, p. 90.

291. LITTLEWOOD, A. B., *J. Gas Chromat.* **1**, 16 (1963).
292. LAZARRE, F., and ROUMAZEILLES, S., *Bull. Soc. chim. Fr.* 3371 (1965).
293. CHOVIN, P., and LEBBE, J., *J. Gas Chromat.* **4**, 37 (1966).
294. BENASTRE, J., and GRENIER, P., *Bull. Soc. chim. Fr.* 1395 (1967); 118 (1968).
295. WEHRLI, A., and KOVÁTS, SZ. E., *Helv. chim. Acta* **42**, 2709 (1959).
296. KOVÁTS, SZ. E., *Z. analyt. Chem.* **181**, 351 (1961).
297. KOVÁTS, SZ. E., ref. 123, p. 229.
298. SCHOMBURG, G., *J. Chromat.* **14**, 157 (1964).
299. PULLIN, I. A., and WERNER, R. L., *Nature* **206**, 393 (1965).
300. ROHRSCHNEIDER, L., *J. Chromat.* **17**, 1 (1965); **22**, 6 (1966); **39**, 383 (1969).
301. TAKÁCS, J., SZENTIRMAI, ZS., MOLNÁR, E. B., and KRALIK, D., *ibid.* **65**, 121 (1972).
302. TAKÁCS, J., TÁLAS, ZS., BERNÁTH, I., CZAKÓ, GY., and FISCHER, A., *ibid.* **67**, 203 (1972).
303. KÁPLÁR, L., SZITA, C., TAKÁCS, J., and TARJÁN, G., *ibid.* **65**, 115 (1972).
304. WEINER, P. H., DACK, C. J., and HOWERY, D. G., *ibid.* **69**, 249 (1972).
305. WEINER, P. H., and HOWERY, D. G., *Analyt. Chem.* **44**, 1189 (1972).
306. McREYNOLDS, O. W., *J. Chromat. Sci.* **8**, 685 (1970).
307. ECKNIG, W., KREIGSMANN, H., and ROTZSCHE, H., *Ber. Bunsenges. Physik. Chem.* **71**, 587 (1967).
308. ROBINSON, P. G., and ODELL, A. L., *J. Chromat.* **57**, 1 (1971).
309. PULLIN, I. A., and WERNER, R. L., *Spectrochim. Acta* **21**, 1257 (1965).
310. SUPINA, W. R., and ROSE, L. P., *J. Chromat. Sci.* **7**, 192 (1969); **8**, 214 (1970).
311. PRESTON, S. T., Jr., *ibid.* **8** (12), 18 A (1970); **9** (5), 12 A (1971); **9** (6), 12 A (1971).
312. SLADKOV, A. Z., *Zav. Lab.* **36**, 1017 (1970).
313. SUPELCO, Inc., Bellefonte, Pa., *Product Catalog*, 1971, p. 3.
314. SAKODYNSKY, K. I., and PANINA, L. I., *J. Chromat.* **58**, 61 (1971).
315. TENNEY, H. M., *Analyt. Chem.* **30**, 2 (1958).
316. EVANS, M. B., and SMITH, J. F., *Nature* **190**, 905 (1961).
317. SCOTT, R. P. W., and PHILLIPS, C. S. G., ref. 31, p. 266.
318. JENTZSCH, D. and BERGMANN, G., *Z. analyt. Chem.* **170**, 239 (1959).
319. LANGER, S. H., *Analyt. Chem.* **39**, 524 (1967).
320. ILIE, V., OPREA, M., and BAIULESCU, G., *Proc. 3rd Conference on Analytical Chemistry, Budapest, August 1970*, vol. 1, p. 291.
321. PEPPARD, D. E., and FERRARO, J. R., *J. Inorg. Nucl. Chem.* **10**, 275 (1959).
322. LANGER, S. H., ZAHN, C., and PANTAZOPOLOS, G., *Chemy. Ind.* 1145 (1958); *J. Chromat.* **3**, 154 (1960).
323. GIL-AV, E., FEIBUSH, B., and CHARLES-SINGLER, R., *Tetrahedron Lett.* **10**, 1009 (1966).
324. GIL-AV, E., FEIBUSH, B., and CHARLES-SINGLER, R., ref. 105, p. 227.
325. KARGER, B. L., *Analyt. Chem.* **39** (8), 24A (1967).
326. KARGER, B. L., STERN, R. L., ROSE, H. C., and KEANE, W., ref. 105, p. 240.
327. KARGER, B. L., STERN, R. L., KEANE, W., HALPERN, B., and WESTLEY, J. W., *Analyt. Chem.* **39**, 228 (1967).
328. MOSHIER, R. W., and SIEVERS, R. E., *Gas Chromatography of Metal Chelates*, Pergamon Press, Oxford, 1965.
329. SOKOLOV, D. N., *Zh. analit. Khim.* **27**, 993 (1972).

330. BELCHER, R., MARTIN, R. J., STEPHEN, W. I., HENDERSON, D. E., KAMALIZAD, A., and UDEN, P. C., *Analyt. Chem.* **45**, 1197 (1973).

331. KARGER, B. L., ELMEHRIK, Y., and STERN, R. L., *ibid.* **40**, 1227 (1968).

332. KARGER, B. L., ELMEHRIK, Y., and ANDRADE, W., *J. Chromat. Sci.* **7**, 209 (1969).

333. KREMSER, M., JERNEJCIC, M., and PREMRU, L., *J. Chromat.* **65**, 129 (1972).

334. HAMMETT, L. P., *J. Chem. Phys.* **4**, 613 (1936).

335. JAFFE, H. H., *Chem. Rev.* **53**, 191 (1953).

336. PURNELL, J. H., and SPENCER, M. S., *Nature* **175**, 988 (1955).

337. PHIFER, L. H., and PLUMMER, H. K., Jr., *Analyt. Chem.* **38**, 1652 (1966).

338. KARGER, B. L., and HARTKOPF, A., *ibid.* **40**, 215 (1968).

339. KARGER, B. L., HARTKOPF, A., and POSMANTER, H., *J. Chromat. Sci.* **7**, 315 (1969).

340. NOVAK, J., and JANAK, J., *Analyt. Chem.* **38**, 265 (1966).

341. KWANTES, A., and RIJNDERS, G. W. A., *in* D. H. DESTY, ed., *Gas Chromatography 1958*, Butterworths, London, 1958, p. 125.

342. MARTIRE, D. E., *Analyt. Chem.* **33**, 1143 (1961).

343. MARTIRE, D. E., ref. 105, p. 21.

344. ROHRSCHEIDER, L., *Gas Chromat.* **6**, 5 (1968).

345. WAKSMUNDZKI, A., SUPRYNOWICZ, Z., and NIEDZIELSKA, K., *Chem. Anal., Warsaw* **14**, 611 (1969).

346. LANGER, S. H., and PURNELL, J. H., *J. Phys. Chem.* **67**, 263 (1963).

347. CLARK, R. K., and SCHMIDT, H. H., *ibid.* **69**, 3682 (1965).

348. GRAY, G. W., *Molecular Structure and the Properties of Liquid Crystals*, Academic Press, New York, 1962.

349. WULF, A., and DE ROCCO, A. G., *in* T. F. JOHNSON and R. S. PORTER, eds., *Liquid Crystals and Ordered Fluids*, Plenum Press, New York, 1970.

350. DEWAR, M. J. S., and SCHROEDER, J. P., *J. Am. Chem. Soc.* **86**, 5235 (1964).

351. KELKER, H., and WINTERSCHEIDT, H., *Z. Analyt. Chem.* **220**, 1 (1966).

352. KELKER, H., and VERHEIST, A., *J. Chromat. Sci.* **7**, 79 (1969).

353. PRICE, F. P., and WENDORFF, J. H., *J. Phys. Chem.* **76**, 2605 (1972).

354. KELKER, H., and VON SCHIVIZHOFFEN, E., *in* J. C. GIDDINGS, and R. A. KELLER, eds., *Advances in Chromatography*, vol. 6, Marcel Dekker, New York, 1968, p. 247.

355. CHOW, L. C., and MARTIRE, D. E., *J. Phys. Chem.* **75**, 2005 (1971).

356. FREDERICKS, E. M., and BROOKS, F. R., *Analyt. Chem.* **28**, 297 (1956).

357. CHOVIN, P., *Bull. Soc. chim. Fr.* 104 (1964).

358. HILDEBRAND, G. P., and REILLEY, C. N., *Analyt. Chem.* **35**, 47 (1964).

359. BARNARD, J. A., and HUGHES, H. W. D., *Nature* **183**, 250 (1959).

360. MCFADDEN, W. H., *Anal. Chem.* **30**, 479 (1958).

361. ROHRSCHNEIDER, L., *Z. analyt. Chem.* **170**, 256 (1959).

362. WAKSMUNDZKI, A., and SUPRYNOWICZ, Z., *J. Chromat.* **18**, 232 (1965).

363. GASCO-SANCHEZ, L., and BURRIEL-MARTI, F., *Analyt. Chim. Acta* **36**, 460 (1966).

364. SCHLUNEGGER, U. P., *J. Chromat.* **26**, 1 (1967).

365. REZNIKOV, S. A., *Zh. fiz. Khim.* **42**, 1730 (1968).

366. WAKSMUNDZKI, A., SUPRYNOWICZ, Z., and MIEDZIAK, I., *Chem. Anal., Warsaw* **13**, 635 (1968).

367. NOVITSKAYA, R. N., and VIGDERGAUZ, M. S., *Izv. Akad. Nauk SSSR*, Ser. Khim. 2798 (1970).

368. WINDHAM, E. S., *J. Ass. Offic. Anal. Chem.* **52**, 1237 (1969).

369. ANNINO, R., and McCREA, P. F., *Analyt. Chem.* **42**, 1486 (1970).
370. KELLER, R. A., and STEWART, G. H., *ibid.* **36**, 1186 (1964).
371. ZHUKOVITSKII, A. A., SELENKINA, M. S., and TURKEL'TAUB, N. M., *Zh. fiz. Khim.* **36**, 519 (1962).
372. LITTLEWOOD, A B., and WILLMOTT, F. W., *Analyt. Chem.* **38**, 1031 (1966).
373. REZNIKOV, S. A., AGANOVA, I. A., and SIDOROV, R. I., *Zh. fiz. Khim.* **44**, 1267 (1970).
374. YOUNG, C. L., *J. Chromat. Sci.* **8**, 103 (1970).
375. BOGOSLOVSKY, YU. N., SAKHAROV, V. M., and SHEVCHUK, I. M., *J. Chromat.* **69**, 17 (1972).
376. MELNIKOV, A. S., and AIVAZOV, B. V., *Zh. analit. Khim.* **28**, 791 (1973).
377. PORTER, R. S., HINKINS, R. L., TORNHEIM, L., and JOHNSON, J. F., *Analyt. Chem.* **36**, 260 (1964).
378. LEWIS, J. S., *Compilation of Gas Chromatographic Data*, ASTM Spec. Tech. Publ. No. 343, American Society for Testing Materials, Philadelphia, Pa., 1963.
379. PURNELL, J. H., WASIK, S. P., and JUVET, R. S., *Acta chim. hung.* **50**, 201 (1966).
380. POPOV, N. M., KASATOCHKIN, V. I., and LUKYANOVICH, V. M., *Dokl. Akad. Nauk SSSR* **131**, 609 (1960).
381. GRAHAM, D., and KAY, W. S., *J. Colloid. Sci.* **16**, 182 (1961).
382. BELYAKOVA, L. D., KISELEV, A. V., and KOVALEVA, N. V., *Analyt. Chem.* **36**, 1517 (1964).
383. KISELEV, A. V., and YASHIN, YA. I., *Zh. fiz. Khim.* **40**, 429 (1966).
384. PETOV, G. M., and SHCHERBAKOVA, K. D., ref. 105, p. 50.
385. BAIULESCU, G. E., and ILIE, V. A., Paper presented at *IUPAC International Congress on Analytical Chemistry, Kyoto, 1972*, B1 507, p. 111.
386. HOLLIS, O. L., *Analyt. Chem.* **38**, 309 (1966).
387. JOHNSON, J. F., and BARRALL, E. M., *J. Chromat.* **31**, 547 (1967).
388. PATZELOVA, V., and VOLKOVA, J., *ibid.* **65**, 255 (1972).
389. ZADO, F. M., and FABECIC, J., *ibid.* **51**, 37 (1970).
390. DRESSLER, M., GUHA, O. K., and JANAK, J., *ibid.* **65**, 261 (1972).
391. GVOSDOVICH, T. N., and JASHIN, YA. I., *ibid.* **49**, 36 (1970).
392. HERTL, W., and NEUMANN, M. G., *ibid.* **60**, 319 (1971).
393. SAKODYNSKY, K. I., *Chromatographia* **1**, 483 (1968).
394. SAKODYNSKY, K. I., and PANINA, L. I., *Zh. analit. Khim.* **27**, 1024 (1972).
395. PHILLIPS, C. S. G., and SCOTT, C. G., ref. 62, p. 121.
396. SERPINET, J., *J. Chromat.* **68**, 9 (1972).
397. KISELEV, A. V., POSHKUS, D. P., and AFREIMOVITCH, A. JA., *Zh. fiz. Khim.* **44**, 981 (1970).
398. MORTIMER, J. V., and GENT, P. L., *Analyt. Chem.* **36**, 754 (1964).
399. DI CORCIA, A., and BRUNER, F., *J. Chromat.* **62**, 462 (1971).
400. LINDSAY-SMITH, J. R., and WADDINGTON, D. J., *Analyt. Chem.* **40**, 522 (1968).
401. DI CORCIA, A., FRITZ, D., and BRUNER, F., *ibid.* **42**, 1500 (1970).
402. DI CORCIA, A., FRITZ, D., and BRUNER, F., *J. Chromat.* **53**, 135 (1970).
403. CARTONI, G. P., LIBERTI, A., and PELA, A., *Analyt. Chem.* **39**, 1618 (1967).
404. PECSOK, R. L., and VARY, E. M., *ibid.* p. 289.
405. SAKHAROV, V. M., *J. Chromat.* **65**, 103 (1972).

406. HORVÁTH, C., in L. S. ETTRE and A. ZLATKIS, eds., The Practice of Gas Chromatography, Interscience, New York, 1967, p. 129.
407. CHEN, C., and GACKE, D., Analyt. Chem. 36, 72 (1964).
408. GERRARD, W., HAWKES, S. J., and MOONEY, E. F., ref. 47, p. 299.
409. BÉKÁSSY, S., LIPTAY, G., and TAKÁCS, J., ref. 279, p. 75.
410. MCKINNEY, R. W., LIGHT, J. F., and JORDAN, R. L., J. Gas Chromat. 6, 97 (1968).
411. HAWKES, S. J., and CARPENTER, D. J., Analyt. Chem. 39, 393 (1967).
412. ALTENAU, A. G., KRAMER, R. E., MCADOO, D. J., and MERRITT, C., Jr., J. Gas Chromat. 4, 96 (1966).
413. KOROL, A. N., Stationary Phases in Gas–Liquid Chromatography, Naukova Dumka, Kiev, 1969.
414. LYNN, T. R., Guide to Stationary Phases for Gas Chromatography, Analabs. Inc., Hamden, Conn., 1967.
415. PRESTON, S. T., Jr., A Guide to Selected Liquid Phases and Adsorbents Used in Gas Chromatography, Debton, Antwerp, 1969.
416. ROTZSCHE, H., and HOFMANN, M., in E. LEIBNITZ and H. G. STRUPPE, eds., Handbuch der Gas-Chromatographie, Academie-Verlag, Leipzig, 1970, p. 364.
417. LITTLEWOOD, A. B., Gas Chromatography: Principles, Techniques and Applications, 2nd edn., Academic Press, New York, 1970.
418. LITTLEWOOD, A. B., J. Gas Chromat. 1, 6 (1963).
419. DATA SUBCOMMITTEE OF THE GC DISCUSSION GROUP OF THE INSTITUTE OF PETROLEUM, J. Gas Chromat. 4, 1 (1966).
420. EGGERTSEN, F. T., and GROENNINGS, S., Analyt. Chem. 30, 20 (1958).
421. NELSON, K. H., HINES, W. J., GRIMES, M. D., and SMITH, D. E., ibid. 32, 10 (1960).
422. SULLIVAN, J. H., WALSH, J. T., and MERRITT, C., Jr., ibid. 31, 1826 (1959).
423. CHENEAU, M., and GUIOCHON, G., Bull. Soc. chim. Fr. 2601 (1963).
424. DESTY, D. H., and WHYMAN, B. H. F., Analyt. Chem. 29, 320 (1957).
425. CASTELLO, G., J. Chromat. 66, 213 (1972).
426. ASINGER, F., FELL, B., and COLLIN, G., Chem. Ber. 96, 716 (1963).
427. ADLARD, E. R., and WHITMAN, B. T., in D. H. DESTY, Gas Chromatography, Butterworths, London, 1958, p. 351.
428. IRVINE, L., and MITCHELL, T. J., J. Appl. Chem. 8, 3 (1958).
429. METCALFE, L. D., J. Gas Chromat. 1, 32 (1963).
430. LINK, W. E., MORRISSETTE, R. A., COOPER, A. D., and SMULLIN, C. F., J. Am. Oil. Chem. Soc. 37, 364 (1960).
431. BRAMLETT, C. L., Jr., J. Ass. Offic. Analyt. Chem. 49, 857 (1966).
432. PARKER, W. W., and HUDSON, R. L., Analyt. Chem. 35, 1334 (1963).
433. BLAY, N. J., WILLIAMS, J., and WILLIAMS, R. L., J. Chem. Soc. 424 (1960).
434. RIEDMANN, M., Z. analyt. Chem. 260, 341 (1972).
435. TRASH, C. R., J. Chromat. Sci. 11, 196 (1973).
436. COLEMAN, A. E., ibid. p. 198.
437. BOHEMEN, J., LANGER, S. H., PERRETT, R. H., and PURNELL, J. H., J. Chem. Soc. 2444 (1960).
438. SUPINA, W. R., HENLY, R. S., and KRUPPA, R. F., J. Am. Oil. Chem. Soc. 43, 202A (1966).
439. OTTENSTEIN, D. M., J. Gas Chromat. 6, 129 (1968).

440. ABEL, E. W., POLLARD, F. H., UDEN, P. C., and NICKLESS, G., *J. Chromat.* **22**, 23 (1966).

441. AUE, W. A., and HASTINGS, C. R., *ibid.* **42**, 319 (1969).

442. HASTINGS, C. R., AUE, W. A., and LARSEN, F. N., *ibid.* **60**, 329 (1971).

443. AUE, W. A., HASTINGS, C. R., AUGL, J. M., NORR, M. K., and LARSEN, J. V., *ibid.* **56**, 295 (1971).

444. AL-TAIAR, A. H., LINDSAY–SMITH, J. R., and WADDINGTON, D. J., *Analyt. Chem.* **42**, 935 (1970).

445. HASTINGS, C. R., and AUE, W. A., *J. Chromat.* **53**, 487 (1970).

446. GARDON, J. L., *J. Paint. Tech.* **38**, 43 (1966).

447. WADA, S., *Bull. Chem. Soc. Jap.* **35**, 707 (1962).

448. YOSHIDA, Z. and OSAWA, E., *J. Am. Chem. Soc.* **88**, 4019 (1966).

449. GILES, C., and NAKHWA, S. N., *J. Appl. Chem.* **11**, 197 (1961).

450. GREEN, J., FEIN, M. M., MAYES, N., DONOVAN, G., ISRAEL, M., and COHEN, M.S. *J. Polymer Sci.* B, 987 (1964).

451. BEROZA, M., and BOWMAN, M. C., *Analyt. Chem.* **43**, 808 (1971).

452. BURGETT, C. A., and FRITZ, J. S., *Talanta* **20**, 363 (1973).

453. ROTZSCHE, H., ref. 92, p. 111.

454. ROSS, W. D., and SIEVERS, R. E., ref. 105, p. 272.

455. GENTY, C., HOUIN, C., and SCHOTT, R., ref. 282, p. 142

456. BRACE, N. O., *J. Org. Chem.* **27**, 3027 (1962).

457. HANADA, Y., and KITAJINA, M., *J. Chem. Soc. Japan* **80**, 1272 (1959).

458. PRESTON, S. T., Jr., *A Guide to the Analysis of Pesticides by Gas Chromatography*, 2nd ed., Evanston, Ill., 1969.

459. ZWEIG, G., and SHERMA, J., *Analytical Methods for Pesticides and Plant Growth Regulators*, vol. 6 of *Gas Chromatographic Analysis* (G. Zweig, ed.), Academic Press, New York, 1972.

460. TULLOCK, A. P., *J. Am. Oil. Chem. Soc.* **41**, 833 (1964).

461. CLARKE, D. D., WILK, S., and GITLOW, S. E., *J. Gas Chromat.* **4**, 310 (1966).

462. VAN DEN HEUVEL, W. J. A., HAAHTI, E. O. A., and HORNING, E. C., *J. Am. Chem. Soc.* **83**, 1513 (1961).

463. BUSH, I. E., *The Chromatography of Steroids*, Pergamon Press, Oxford, 1961, p. 117.

464. LAU, H. L., *J. Gas Chromat.* **4**, 136 (1966).

465. VESTERGAARG, P., and RAABO, E., *ibid.*, p. 422.

466. RUCHELMAN, M. W., *ibid.*, p. 265.

467. VASILESCU, V., ref. 279, p. 585.

468. JANAK, J., and HRIVNAK, M., *J. Chromat.* **3**, 297 (1960).

469. JAMES, A. T., and MARTIN, A. J. P., *J. Appl. Chem.* **6**, 105 (1956).

470. HILDEBRAND, G., and LESCHNER, O., *Chem. tech.*, Berlin **18**, 424 (1966).

471. EBING, W., *J. Gas Chromat.* **6**, 79 (1968).

472. SALZWEDEL, M., *Z. analyt. Chem.* **209**, 299 (1965).

473. TA-CHUANG LO CHANG, and KARR, C., Jr., *Analyt. chim. Acta* **26**, 410 (1962).

474. HENBEST, H. B., REID, J. A. W., and STIRLING, C. J. M., *J. Chem. Soc.* 5239 (1961).

475. BEESON, J. H., and PECSAR, R. E., *Analyt. Chem.* **41**, 1678 (1969).

476. RICHARDSON, G. A., and BLAKE, E. S., *Ind. Engng Chem. Prod. Res. Develop.* **7**, 22 (1968).

477. VON RUDLOFF, E., *Can. J. Chem.* **38**, 631 (1960).

478. ETTRE, L. S., CIEPLINSKI, E. W., and KABOT, F. J., *J. Gas Chromat.* **1**, 38 (1963).

479. MUSAEV, I. A., SANIN, P. I., PAKHOMOV, V. P., and KURASHOVA, E. H., *Neftekhimiya* **9**, 914 (1969).

480. HAMMANN, W. C., SCHISLA, R. M., ADAMS, J. S., and KOCH, S. D., *J. Chem. Engng Data* **15**, 352 (1970).

481. HETPER, J., and PIELICHOWSKI, J., *Chem. Anal., Warsaw* **14**, 813 (1969).

482. NADEAU, H. G., and OAKS, D., Jr., *Analyt. Chem.* **33**, 1157 (1961).

483. RAY, N. H., *J. Appl. Chem.* **4**, 21 (1954).

484. JOFFE, B. V., and STOLJAROV, B. V., *Neftekhimiya* **2**, 911 (1962).

485. ZUBYK, W. J., and CONNER, A. Z., *Analyt. Chem.* **32**, 912 (1960).

486. WALLING, C., and PADWA, A., *J. Am. Chem. Soc.* **85**, 1593 (1963).

487. HING, F. S., and WECKEL, K. G., *J. Food Sci.* **29**, 149 (1964).

488. TYDEN, I., *Talanta*, **13**, 1353 (1966).

489. LIBERTI, A., *Analyt. Chim. Acta* **17**, 247 (1957).

490. YOUNG, J. R., *Chem. Ind.* 594 (1958).

491. LANGER, S. H., *Analyt. Chem.* **44**, 1915 (1972).

492. RYBA, M., *Chromatographia* **5**, 23 (1972).

493. CSICSERY, S. M., and PINES, H., *J. Chromat.* **9**, 34 (1962).

494. LICHTENFELS, D. H., FLECK, S. A., BUROW, F. H., and COGGESHALL, N. D., *Analyt. Chem.* **28**, 1376 (1956).

495. KOMAREK, K., CHURACEK, J., DUFKA, O., and VLCEK, J., *Chem. Prumysl.* **21**, 176 (1971).

496. SHKVYRYA, A. G., and VIGDERGAUZ, M. S., *Zh. analit. Khim.* **21**, 625 (1966).

497. LEWIS, J. S., PATTON, H. W., and KAYE, W. I., *Analyt. Chem.* **28**, 1370 (1956).

498. RYCE, S. A., and BRYCE, W. A., *ibid.* **29**, 925 (1957).

499. SINGLIAR, M., BOBAK, A., and BRIDA, J., *Chem. Zvesti* **14**, 209 (1960).

500. BROOKS, V. T., *Chem. Ind.* 1317 (1959).

501. LINDSAY-SMITH, J. R., NORMAN, R. O. C., and RADDA, G. K., *J. Gas Chromat.* **2**, 146 (1964).

502. ILIE, V., BOROANCA, M., and BAIULESCU, G., *Chim. Analit, Bucharest* **1**, 33 (1971).

503. JACKSON, R. B., *J. Chromat.* **22**, 251 (1966).

504. VIGDERGAUZ, M. S., GOLBERT, K. A., and AFANASIEV, M. I., *Khimiya Tekhnol. Topl. Mas.* **12**, 61 (1964).

505. BEREZKIN, V. G., KOLBANOVSKII, YU. A., KIAZIMOV, E. A., PAKHOMOV, V. P., and SCHABURKINA, V. I., *Neftekhimiya* **7**, 141 (1967).

506. LITTLEWOOD, A. B., and WILLMOTT, F. W., *Analyt. Chem.* **38**, 1076 (1966).

507. HICKERSON, J. F., *J. Inst. Petrol.* **45** (422), 29A (1959).

508. HOIGNE, J., WIDMER, H., and GAUMANN, T., *J. Chromat.* **11**, 459 (1963).

509. WIDMER, H., and GAUMANN, T., *Helv. chim. Acta* **46**, 2766 (1963).

510. KOVÁTS, SZ. E., *ibid.*, p. 2705.

511. ASCOLI, F., PISPISA, B., and SERVELLO, F., *J. Chromat.* **6**, 544 (1961).

512. DESTY, D. H., and HARBOURN, C. L. A., *Analyt. Chem.* **31**, 1965 (1959).

513. UMEH, E. O., *J. Chromat.* **51**, 147 (1970).

514. ZAREMBO, J. E., and LYSYJ, I., *Analyt. Chem.* **31**, 1833 (1959).

515. REED, T. M., III, *ibid.* **30,** 221 (1958).
516. ELLIS, J. F., FORREST, C. W., and ALLEN, P. L., *Analyt. chim. Acta* **22,** 27 (1960).
517. JUVET, R. S., and FISHER, R. L., *Analyt. Chem.* **38,** 1860 (1966).
518. RUNGE, H., *Z. analyt. Chem.* **189,** 111 (1962).
519. NOVOTNY, M., BEKTESH, S. L., DENSON, K. B., GROHMANN, K., and PARR, W., *Analyt. Chem.* **45,** 971 (1973).
520. HALÁSZ, I., and SEBESTIAN, I., *Angew. Chem.* **81,** 464 (1969).
521. MAHADEVAN, V., and DECKER, R., *J. Chromat. Sci.* **8,** 279 (1970).
522. LOCKE, D. C., SCHMERMUND, J. T., and BANNER, B., *Analyt. Chem.* **44,** 90 (1972).
523. KIRKLAND, J. J., and DeSTEFANO, J. J., *J. Chromat. Sci.* **8,** 309 (1970).
524. KIRKLAND, J. J., *ibid.* **9,** 206 (1971).
525. PARR, W., and GROHMANN, K., *Tetrahedron Lett.* 2633 (1971).
526. DAVISON, V. L., and MOORE, D. J., *J. Gas Chromat.* **6,** 540 (1968).
527. PRYDE, E. H., AWE, R. A., TCETER, H. M., and COWAN, J. C., *J. Polymer Sci.* **59,** 1 (1962).
528. NEFF, W. E., PRYDE, E. H., SELKE, E., and COWAN, J. C., *J. Chromat. Sci.* **10,** 512 (1972).
529. KARGER, B. L., *J. Chem. Educ.* **43,** 47 (1966).
530. KEULEMANS, A. I. M., KWANTES, A., and ZAAL, P., *Analyt. Chim. Acta* **13,** 357 (1955).
531. SMITH, B., *Acta chem. scand.* **13,** 480 (1959).
532. FUNASAKA, W., KOJIMA, T., and IGAKI, H., *Analyt. Chem.* **36,** 2214 (1964).
533. SAGERT, N. H., and LAIDLER, K. J., *Can. J. Chem.* **41,** 838 (1963).
534. PRESTON, S. T., Jr., *J. Gas Chromat.* **1,** 31 (1963).
535. KARCHMER, J. H., *Analyt. Chem.* **31,** 1377 (1959).
536. CHOVIN, P., and SANNIER, H., *J. Gas Chromat.* **2,** 83 (1964).
537. McNAIR, H. M., and DE VRIES, T., *Analyt. Chem.* **33,** 806 (1961).
538. FROBERGER, C. F., and McEWEN, D. J., *J. Org. Chem.* **27,** 1891 (1962).
539. LEMOINE, T. J., *J. Gas Chromat.* **3,** 323 (1965).
540. KUCHHAL, R. K., and MALLIK, K. L., *Chromatographia* **6,** 202 (1973).
541. BRUSON, H. A., and RIENER, T. W., *J. Am. Chem. Soc.* **65,** 23 (1943).
542. CLEMONS, C. A., LEACH, P. W., and ALTSHULLER, A. P., *Analyt. Chem.* **35,** 1548 (1963).
543. KEULEMANS, A. I. M., KWANTES, A., and ZAAL, P., *Analyt. Chim. Acta* **13,** 337 (1955).
544. SMITH, B., *Acta chem. scand.* **13,** 877 (1959).
545. PETRANEK, J., and SLOSAR, J., *Coll. Czech. Chem. Commun.* **26,** 2667 (1961).
546. CASTELLS, R. C., *Chromatographia* **6,** 57 (1973).
547. FABRIZIO, F. A., KING, R. W., CERATO, C. C., and LOVELAND, J. W., *Analyt. Chem.* **31,** 2060 (1959).
548. SZE, Y. L., BORKE, M. L., and OTTENSTEIN, D. M., *ibid.* **35,** 240 (1963).
549. BAHJAT, K. S., *J. Pharm. Sci.* **52,** 1109 (1963).
550. WINTER, L. N., and ALBRO, P. W., *J. Gas Chromat.* **2,** 1 (1964).
551. ECKNIG, W., and LENZ, E., *J. Chromat.* **64,** 7 (1972).
552. PODBIELNIAK, W. J., and PRESTON, S. T., *Petrol. Refiner* **35,** 215 (1956).
553. JANAK, J., and NOVÁK, J., *Coll. Czech. Chem. Commun.* **24,** 384 (1959).
554. PERRIN-DROWIN, M., *Bull. Soc. chim. Fr.* 1440 (1962).

555. MORROW, H. N., and BUCKLEY, K. B., *Petrol. Refiner* **36**, 157 (1957).
556. FRISONE, G. J., *Nature* **193**, 370 (1962).
557. GADSDEN, R. H., and McCORD, W. M., *J. Gas Chromat.* **2**, 7 (1964).
558. CHESKO, F. F., and AZEN, V. E., *Zh. Obshch. Khim.* **34**, 1843 (1964).
559. SPEARS, A. W., *Analyt. Chem.* **35**, 320 (1963).
560. MATHEWS, R. G., SCHWARTZ, R. D., STOUFFER, J. E., and PETTITT, B. C., *J. Chromat. Sci.* **8**, 508 (1970).
561. JENDEN, D. J., ROCH, M., and BOOTH, R., *ibid.* **10**, 151 (1972).
562. VAN DER WIEL, A., *Nature* **187**, 142 (1960).
563. BLOCH, M. G., *in* H. J. NOEBELS, R. F. WALL, and N. BRENNER, eds., *Gas Chromatography*, Academic Press, New York, 1961, p. 133.
564. KNIGHTS, B. A., *J. Gas Chromat.* **2**, 338 (1964).
565. PERRY, S., and HIBBERT, H., *J. Am. Chem. Soc.* **62**, 2599 (1940).
566. EVANS, M. B., and SMITH, J. F., *J. Chromat.* **36**, 489 (1968).
567. BIGHI, C., BETTI, A., SAGLIETTO, G., and DONDI, F., *ibid.* **35**, 309 (1968).
568. BYARE, B., and JORDAN, G., *J. Gas Chromat.* **2**, 304 (1964).
569. GROB, K., *J. Chromat.* **2**, 80 (1962).
570. ASSMANN, K., GEPPERT, G., STRUPPE, H. G., and SERFAS, O., ref. 279, p. 33.
571. DURRETT, L. R., *Analyt. Chem.* **32**, 1393 (1960).
572. HRIVNAK, J., SOJAK, L., BESKA, E., and JANAK, J., *J. Chromat.* **68**, 55 (1972).
573. FALK, F., and DIETRICH, P., *ibid.* **33**, 411 (1968).
574. FALK, F., and DIETRICH, P., *ibid.*, p. 422.
575. ASSMANN, K., SERFAS, O., and GEPPERT, G., *ibid.* **26**, 495 (1967).
576. ANDERSON, D. G., and ANSEL, R. E., *J. Chromat. Sci.* **11**, 192 (1973).
577. ORR, C. H., and CALLEN, J. E., *J. Am. Chem. Soc.* **80**, 246 (1958).
578. LIPSKY, S. R., and LANDOWNE, R. A., *Biochim. biophys. Acta* **27**, 666 (1959).
579. JANAK, J., and HAINOVA, O., *Chem. Prumysl.* **18**, 308 (1968).
580. DUFKA, O., VLCEK, J., KASTANEK, A., CHURACEK, J., and KOMAREK, K., *ibid.* **20**, 489 (1970).
581. BAPAT, B. V., CHATGE, B. B., and BHATTACHARYYA, S. C., *J. Chromat.* **18**, 308 (1965); **23**, 227 (1966).
582. GEHRKE, C. W., ZUMWALT, R. W., and WALL, L. L., *ibid.* **37**, 398 (1968).
583. METCALFE, L. D., *Nature* **188**, 142 (1960).
584. CRAIG, B. M., and MURTY, N. L., *J. Am. Oil. Chem. Soc.* **36**, 549 (1959); *Can. J. Chem.* **36**, 1297 (1958).
585. HAAHTI, E. O. A., VAN DEN HEUVEL, W. J. A., and HORNING, E. C., *J. Org. Chem.* **26**, 626 (1961).
586. SUPINA, W. R., *Analyt. Chem.* **35**, 1304 (1963).
587. HRIVNAC, M., *Chem. Ind.* 930 (1960).
588. ASPINAL, G. O., and ROSS, K. M., *J. Chem. Soc.* 1681 (1963).
589. LANDOWNE, R. A., and LIPSKY, S. R., *Nature* **199**, 141 (1963).
590. CRAIG, B. M., *Chem. Ind.* 1442 (1960).
591. SWEELY, C. C., and HORNING, E. C., *Nature* **187**, 144 (1960).
592. JONES, C. E. R., ref. 47, p. 401.
593. FALK, F., *J. Chromat.* **17**, 450 (1965).
594. BOTTCHER, C. J. F., WOODFORD, F. P., BOELSMA-VAN HOUTE, E., and VAN GENT, C. M., *Recl Trav. chim. Pays-Bas Bels.* **78**, 794 (1959).

595. ZEMAN, I., *J. Chromat.* **4**, 314 (1966).
596. SCHNELL, H., *Chemistry and Physics of Polycarbonates*, Interscience, New York, 1964.
597. SAND, D. M., and SCHLENK, H., *Analyt. Chem.* **33**, 1624 (1961).
598. SCHLENK, H., GELLERMAN, J. L., and SAND, D. M., *ibid.* **34**, 1529 (1962).
599. SCHLENK, H., GELLERMAN, J. L., AARONSON, S., and HAINES, T. H., *Ind. Chim. Belge*, **27**, 539 (1962).
600. SMITH, D. M., BARTLET, J. C., and LEVI, L., *Analyt. Chem.* **32**, 568 (1960).
601. HANACK, M., and KEBERLE, W., *Chem. Ber.* **96**, 2937 (1963).
602. LILIEDAHL, H., *J. Gas Chromat.* **3**, 263 (1965).
603. O'DONNELL, J. F., and MANN, C. K., *Analyt. Chem.* **36**, 2097 (1964).
604. LEE, H. L., and NEVILLE, K. O., *Epoxy Resins. A Practical Review of their Chemistry and Technology*, McGraw-Hill, New York, 1957.
605. WISNIEWSKI, J. V., and SPENCER, S. F., *J. Gas Chromat.* **2**, 34 (1964).
606. MATHEWS, R. G., SCHWARTZ, R. D., NOVOTNY, M., and ZLATKIS, A., *Analyt. Chem.* **43**, 1161 (1971).
607. GORDON, J. E., SELWYN, J. E., and THORNE, R. L., *J. Org. Chem.* **31**, 1925 (1966).
608. GORDON, J. E., *ibid.* **30**, 2760 (1965).
609. FREISER, H., *Analyt. Chem.* **31**, 1440 (1959).
610. JUVET, R. S., and WACHI, F. M., *ibid.* **32**, 290 (1960).
611. TADMOR, J., *ibid.* **36**, 1565 (1964).
612. ZADO, F. M., and JUVET, R. S., ref. 105, p. 283.
613. GUIOCHON, G., and POMMIER, C., *La Chromatographie en phase gazeuse en chimie inorganique*, Gauthier-Villars, Paris, 1971.
614. HANNEMAN, W. W., SPENCER, C. F., and JOHNSON, J. F., *Analyt. Chem.* **32**, 1386 (1960).
615. GEISS, F., VERSINO, B., and SCHLITT, H., *Chromatographia* **1**, 9 (1968).
616. GIL-AV, E., and SCHURIG, V., *Analyt. Chem.* **43**, 2030 (1971).
617. ONGEL, L. E., *An Introduction to Transition-metal Chemistry. Ligand Field Theory*, Methuen, London, 1960.
618. DAVIES, D. R., and WEBB, G. A., *Coord. Chem. Rev.* **6**, 95 (1971).
619. CARTONI, G. P., LOWRIE, R. S., PHILLIPS, C. S. G., and VENANZI, L. M., ref. 47, p. 273.
620. BASOLO, F., and PEARSON, R. G., *in* F. A. COTTON, ed., *Prog. Inorg. Chemistry*, vol. 4, Interscience, New York, 1962, p. 381.
621. BRADFORD, B. W., HARVEY, D., and CHAKLEY, D. E., *J. Inst. Petroleum* **41**, 80 (1955).
622. GIL-AV, E., HERLING, J., and SHABTAI, J., *Chem. Ind.* 1483 (1957).
623. GIL-AV, E., HERLING, J., and SHABTAI, J., *J. Chromat.* **1**, 508 (1958).
624. HERLING, J., SHABTAI, J., and GIL-AV, E., *ibid.* **8**, 349 (1962).
625. SHABTAI, J., HERLING, J., and GIL-AV, E., *ibid.* **2**, 406 (1959).
626. SHABTAI, J., *Israel J. Chem.* **1**, 300 (1963).
627. SHABTAI, J., HERLING, J., and GIL-AV, E., *J. Chromat.* **11**, 32 (1963).
628. SMITH, B., and OHLSON, R., *Acta chem. scand.* **13**, 1253 (1959).
629. SMITH, B., and OHLSON, R., *ibid.* **16**, 351 (1962).
630. ACHE, H. J., and WOLF, A. P., *J. Am. Chem. Soc.* **88**, 888 (1966), *Z. analyt. Chem.* **230**, 19 (1967).

631. ATKINSON, J. G., RUSSELL, A. A., and STUART, R. S., *Can. J. Chem.* **45**, 1963 (1967).
632. CVETANOVIC, R. J., DUNCAN, F. J., and FALCONER, W. E., *ibid.* **41**, 2095 (1963).
633. DUBRIN, J., MACKAY, C., and WOLFGANG, R., *J. Am. Chem. Soc.* **86**, 959 (1964).
634. KEULEMANS, A. I. M., *Gas Chromatography*, 2nd edn., Reinhold, New York, 1960, p. 205.
635. BEDNAS, M. E., and RUSSELL, D. S., *Can. J. Chem.* **36**, 1272 (1958).
636. SHABTAI, J., *J. Chromat.* **18**, 302 (1965).
637. ARMITAGE, F., *ibid.* **2**, 655 (1959).
638. VAN DE CRAATS, F., *Analyt. chim. Acta* **14**, 136 (1956).
639. ZLATKIS, A., *Chromatographia* **2**, 298 (1969).
640. COPE, A. C., LEVEL, N. A., LEE, H. H., and MOORE, W. R., *J. Am. Chem. Soc.* **79**, 4720 (1957).
641. COPE, A. C., and ACTON, E. M., *ibid.* **80**, 355 (1958).
642. SAUERS, R. R., *ibid.* **81**, 4873 (1959).
643. SCHURIG, V., CHANG, R. C., ZLATKIS, A., GIL-AV, E., and MIKES, F., *Chromatographia* **6**, 223 (1973).
644. WASIK, S. P., and TSANG, W., *J. Phys. Chem.* **74**, 2970 (1970).
645. WASIK, S. P., and TSANG, W., *Analyt. Chem.* **42**, 1648 (1970).
646. BANTHORPE, D. V., GATFORD, C., and HOLLEBONE, B. R., *J. Gas Chromat.* **6**, 61 (1968).
647. SACCONI, L., LOMBARDO, G., and PAOLETTI, P., *J. Chem. Soc.* 848 (1958).
648. BARBER, D. W., PHILLIPS, C. S. G., TUSA, G. F., and VERDIN, A., *ibid.* **18** (1959).
649. BAYER, E., ref. 341, p. 333.
650. CARTONI, G. P., LIBERTI, A., and PALOMBARI, R., *J. Chromat.* **20**, 278 (1965).
651. FEIBUSH, B., and GIL-AV, E., *J. Gas Chromat.* **5**, 257 (1967).
652. CORBIN, J. A., and ROGERS, L. B., *Analyt. Chem.* **42**, 974 (1970).
653. GIL-AV, E., and FEIBUSH, B., *Tetrahedron Lett.* 3345 (1967).
654. FEIBUSH, B., and GIL-AV, E., *Tetrahedron* **26**, 1361 (1970).
655. NAKAPARKSIN, S., BIRRELL, P., GIL-AV, E., and ORO, J., *J. Chromat. Sci.* **8**, 177 (1970).
656. KOENIG, W. A., PARR, W., LICHTENSTEIN, H. A., BAYER, E., and ORO, J., *ibid.*, p. 183.
657. PARR, W., PIETERSKI, J., YANG, C., and BAYER, E., *ibid.* **9**, 141 (1971).
658. PARR, W., YANG, C., BAYER, E., and GIL-AV, E., *ibid.* **8**, 591 (1970).
659. PARR, W., YANG, C., PIETERSKI, J., and BAYER, E., *J. Chromat.* **50**, 510 (1970).
660. PARR, W., and HOWARD, P., *Chromatographia* **4**, 162 (1971).
661. CORBIN, J. A., RHOAD, J. E., and ROGERS, L. B., *Analyt. Chem.* **43**, 327 (1971).
662. GROHMANN, K., and PARR, W., *Chromatographia* **5**, 18 (1972).
663. PARR, W., and HOWARD, P. Y., *J. Chromat.* **67**, 227 (1972).
664. LEHMANN, O., *Z. phys. Chem.* **5**, 427 (1890).
665. BROWN, G. H., and SHAW, W. G., *Chem. Rev.* **57**, 1049 (1957).
666. USOL'TSEVA, V. A., and CHISTYAKOV, I. G., *Russian Chem. Rev.* **32**, 495 (1963).
667. KELKER, H., *Z. analyt. Chem.* **198**, 254 (1963).
668. KELKER, H., *Ber. Bunsenges. Phys. Chem.* **67**, 698 (1963).
669. KELKER, H., and WINTERSCHEIDT, H., *Z. analyt. Chem.* **220**, 1 (1966).
670. DEWAR, M. J. S., and SCHROEDER, J. P., *J. Org. Chem.* **30**, 3485 (1965).

671. BARRALL E. M., II, PORTER, R. S., and JOHNSON, J. F., *J. Chromat.* **21,** 392 (1966).
672. VIGALOK, R. V., and VIGDERGAUZ, M. S., *Izv. Akad. Nauk SSSR,* Ser. Khim. 715 (1972).
673. TAYLOR, P. J., CULP, R. A., LOCHMULLER, C. H., ROGERS, L. B., and BARRALL, E. M., II, *Separ. Sci.* **6,** 841 (1971).
674. MACZEK, A. O. S., and PHILLIPS, C. S. G., ref. 47, p. 281.
675. MACZEK, A. O. S. and PHILLIPS, C. S. G., *J. Chromat.* **29,** 7 (1967).
676. GRANT, D. W., *Chromatographia* **6,** 239 (1973).
677. VERSINO, B., GEISS, F., and BARBERO, G., *Z. analyt. Chem.* **201,** 20 (1964).
678. MORTIMER, J. V., and GENT, P. L., *Nature* **197,** 789 (1963).
679. VERGNAUD, J. M., *J. Chromat.* **27,** 54 (1967).
680. MACZEK, A. O. S., and PHILLIPS, C. S. G., *ibid.* **29,** 15 (1967).
681. MANN, J. R., and PRESTON, S. T., Jr., *J. Chromat. Sci.* **11,** 216 (1973).
682. PRESTON, S. T., *ibid.* **8**(12), 18A, 1970.
683. RICHMOND, A. B., *ibid.* **9**(5), 12A (1971).
684. HAKEN, J. K., *ibid.*, p. 13A.
685. SCUPP, O. E., *ibid.* **9**(6), 12A (1971).
686. HAWKES, S. J., *ibid.*
687. PRESTON, S. T., *ibid.* **9**(8), 10A (1971).
688. CLUTTON, D. W., *ibid.* **10**(6), 14A (1972).
689. HENLY, R. S., *ibid.* **11,** 221 (1973).
690. KELLER, R. A., *ibid.*, p. 188.
691. JAMES, A. T., and MARTIN, A. J. P., *Biochem. J.* **52,** 242 (1952).
692. HESSE, G., *Ann. Chem.* **546,** 251 (1941).
693. CLAESSON, S., *Arkiv Kemi* A, 23 (1) (1946).
694. PHILLIPS, C. S. G., *Discuss. Faraday Soc.* **7,** 241 (1949).
695. CREMER, E., and PRIOR, F., *Z. Elektrochem.* **55,** 66 (1951).
696. GIDDINGS, J. C., *Analyt. Chem.* **36,** 1170 (1964).
697. KISELEV, A. V., ref. 31, p. 238.
698. KULEY, C. J., *Analyt. Chem.* **35,** 1472 (1963).
699. MADISON, J. J., *ibid.* **30,** 1858 (1958).
700. JANAK, J., *Coll. Czech. Chem. Commun.* **19,** 917 (1954).
701. GREENE, S. A., *Analyt. Chem.* **31,** 480 (1959).
702. WENCKE, K., *Chem. Tech. Berlin* **8,** 728 (1956); **9,** 404 (1957).
703. GRAVEN, W. M., *Analyt. Chem.* **31,** 1197 (1959).
704. FALCONER, J. W., and KNOX, J. H., *Proc. R. Soc.* A, **250,** 493 (1959).
705. TA-CHUANG LO CHANG, *J. Chromat.* **37,** 14 (1968).
706. GLUEKAUF, E., *Ann. NY Acad. Sci.* **72,** 562 (1959).
707. KYRYACOS, G., and BOORD, C. E., *Analyt. Chem.* **29,** 787 (1957).
708. KISELEV, A. V., POSHKUS, D. P., and AFREIMOVICH, A. J., *Zh. fiz. Khim.* **42,** 1201 (1968).
709. KISELEV, A. V., and YASHIN, YA. I., *ibid.* **40,** 603 (1966).
710. KISELEV, A. V., KOUZNETSOV, A. V., FILATOVA, I. YU., and SHCHERBAKOVA, K. D., *ibid.* **44,** 1272 (1970).
711. BORTNIKOV, G. N., KISELEV, A. V., VYASANKIN, N. S., and YASHIN, YA. I., *Chromatographia* **4,** 14 (1971).
712. ZANE, A., *J. Chromat.* **38,** 130 (1968).

713. FRYCKA, J., *ibid.* **65**, 341 and 432 (1972).
714. CANTUTI, V., CARTONI, G. P., LIBERTI, A., and TORRI, A. G., *ibid.* **17**, 60 (1965).
715. DAVIS, H. J., *Talanta* **16**, 621 (1969).
716. CHORTYK, O. T., SCHLOTZHAUER, W. S., and STEDMAN, R. L., *J. Gas Chromat.* **3**, 394 (1965).
717. GUMP, B. H., *J. Chromat. Sci.* **7**, 755 (1969).
718. GOUW, T. H., WHITTEMORE, I. M., and JENTOFT, R. E., *Analyt. Chem.* **42**, 1394 (1970).
719. VIDAL-MADJAR, C., GANANSIA, J., and GUIOCHON, G., in H. STOCK and S. G. PERRY, eds., *Gas Chromatography 1970*, Inst. of Petroleum, London, 1971, p. 20.
720. POPE, C. G., *Analyt. Chem.* **35**, 654 (1963).
721. GORETTI, G., LIBERTI, A., and NOTA, G., ref. 282, p. 22.
722. YASHIN, YA. I., ref. 105, p. 423.
723. GORETTI, G. C., LIBERTI, A., and NOTA, G., *J. Chromat.* **34**, 96 (1968).
724. DI CORCIA, A., and BRUNER, F., *Analyt. Chem.* **43**, 1634 (1971).
725. DI CORCIA, A., CICCIOLI, P., and BRUNER, F., *J. Chromat.* **62**, 128 (1971).
726. BELYAKOVA, L. D., and KISELEV, A. V., *Izv. Akad. Nauk. SSSR*, Ser. Khim. **4**, 638 (1966).
727. CROWELL, A. D., and CHANG, C. O., *J. Chem. Phys.* **43**, 4364 (1965).
728. FAVRE, J. A., and KALLENBACH, L. R., *Analyt. Chem.* **36**, 63 (1964).
729. SOLOMON, P. W., *ibid.*, p. 476.
730. ONUSKA, F., and JANAK, J., *Chem. Zvesti* **22**, 929 (1968).
731. GROB, R. L., WEINERT, G. W., and DRELICH, J. W., *J. Chromat.* **30**, 305 (1967).
732. HIRSCHMANN, R. P., SIMON, H. L., ANDERSON, L. R., and FOX, W. B., *ibid.* **50**, 118 (1970).
733. GROB, R. L., GONDEK, R. J., and SCALES, T. A., *ibid.* **53**, 477 (1970).
734. GROB, R. L., and MCGONIGLE, E. J., *ibid.* **59**, 13 (1971).
735. ROGERS, L. B., and ALTENAU, A. G., *Analyt. Chem.* **35**, 915 (1963).
736. ALTENAU, A. G., and ROGERS, L. B., *ibid.* **36**, 1726 (1964).
737. BRENNER, N., and ETTRE, L. S., *ibid.* **31**, 1815 (1959).
738. DONOHUE, J. A., and JONES, F. S., *ibid.* **38**, 1858 (1966).
739. TURKELTAUB, N. M., and ZHUKHOVITSKII, A. A., *Zavod. Lab.* **22**, 1032 (1956).
740. JEFFERY, P. G., and KIPPING, P. J., *Gas Analysis by Gas Chromatography*, 2nd edn., Pergamon Press, Oxford, 1972.
741. SZULCZEWSKI, D. H., and HIGUCHI, T., *Analyt. Chem.* **29**, 1541 (1957).
742. MARVILLET, L., and TRANCHANT, J., *Proc. 3rd Symp. Gas Chromat., Edinburgh, 1960*, Butterworths, London, 1960, p. 321.
743. SAKAIDA, R. R., RINKER, R. G., CUFFEL, R. F., and CORCORAN, W. H., *Analyt. Chem.* **33**, 32 (1961).
744. KISELEV, A. V., ref. 92, p. 34.
745. BOEHM, H. P., *Angew. Chem.* **78**, 617 (1966).
746. KISELEV, A. V., *Z. Chem.* **7**, 369 (1969).
747. BEBRIS, N. K., KISELEV, A. V., and NIKITIN, YU. S., *Kolloid. Zh.* **29**, 326 (1967).
748. BEBRIS, N. K., KISELEV, A. V., MOKEEV, B. YA., NIKITIN, YU. S., YASHIN, YA. I., and ZAIZEVA, G. E., *Chromatographia* **4**, 93 (1971).
749. ZHURAVLEV, L. T., KISELEV, A. V., NAIDINA, V. P., and POLYAKOV, A. L., *Zh. fiz. Khim.* **37**, 1113 (1963); **37**, 1216 (1963).

750. BASILA, M. R., J. Chem. Phys. 35, 1151 (1961).
751. PERI, J. B., J. Phys. Chem. 70, 2937 (1966).
752. HOCKEY, J. A., and PETHICA, B. A., Trans. Faraday Soc. 57, 2247 (1961).
753. ELKINGTON, P. A., and CURTHOYS, G., J. Phys. Chem. 72, 3475 (1968).
754. KISELEV, A. V., Dokl. fiz. Chem. 181, 914 (1968).
755. CADOGAN, D. F., and SAWYER, D. T., Analyt. Chem. 42, 190 (1970).
756. PICKETT, J. H., LOCHMULLER, C. H., and ROGERS, L. B., Separ. Sci. 5, 23 (1970).
757. ASH, S., KISELEV, A. V., KUZNETSOV, B. V., and NIKITIN, YU. S., Trans. Faraday Soc. 67, 3118 (1971).
758. CHAPMAN, I. D., and HAIR, M. L., ibid. 61, 1507 (1965).
759. BEBRIS, N. K., ZAIZEVA, G. E., KISELEV, A. V., NIKITIN, YU. S., and YASHIN, YA. I., Neftekhimiya 8, 481 (1968).
760. BEREZKIN, V. G., KISELEV, A. V., NIKITINA, N. S., and NIKITIN, YU. S., Izv. Akad. Nauk. SSSR, Ser. Khim. 1385 (1969).
761. DAVYDOV, Y. YA., KISELEV, A. V., and KUZNETSOV, B. V., Zh. fiz. Khim. 39, 2058 (1965).
762. BEBRIS, N. K., KISELEV, A. V., and NIKITIN, YU. S., Kolloid. Zh. 29, 326 (1967).
763. HALÁSZ, I., and GERLACH, H. O., Analyt. Chem. 38, 281 (1966).
764. BRUNNER, H., Pharm. Ind. 20, 581 (1958).
765. SCHWARTZ, R. D., BRASSEAUX, D. J., and MATHEWS, R. G., Analyt. Chem. 38, 303 (1966).
766. GUILLEMIN, C. L., LE PAGE, M., BEAU, R., and DE VRIES, A. J., ibid. 39, 940 (1967).
767. GUILLEMIN, C. L., DELEUIL, M., CIRENDINI, S., and VERMONT, J., ibid. 43, 2015 (1971).
768. FELTL, I., and SMOLKOVA, E., J. Chromat. 65, 249 (1972).
769. SNYDER, L. R., Principles of Adsorption Chromatography, Marcel Dekker, New York, 1968.
770. GUILLEMIN, C. L., LE PAGE, M., and DE VRIES, A. J., J. Chromat. Sci. 9, 470 (1971).
771. ZLATKIS, A., and KAUFMAN, H. R., in H. J. NOEBELS, R. F. WALL, and N. BRENNER, eds., Gas Chromatography, Academic Press, New York, 1961, p. 339.
772. CRESPI, V., and CEVOLANI, F., La Chimica e l'Industria 91, 215 (1959).
773. ZSHDANOV, S. P., Dokl. Akad. Nauk SSSR 82, 281 (1952).
774. ZSHDANOV, S. P., KISELEV, A. V., and YASHIN, YA. I., Neftekhimiya 3, 417 (1963).
775. YASHIN, YA. I., ZSHDANOV, S. P., and KISELEV, A. V., in H. P. ANGELE and H. G. STRUPPE, eds., Gas Chromatography 1963, Akademie-Verlag, Berlin, 1963, p. 402.
776. LYSYJ, I., and NEWTON, P. R., Analyt. Chem. 36, 2514 (1964).
777. BLANDENET, G., and ROBIN, J., Seance 1857 (1963).
778. HAIR, M. L., Infrared Spectroscopy in Surface Chemistry, Marcel Dekker, New York, 1967, pp. 141 ff.
779. LITTLE, L. H., KLAUSER, H. E., and AMBERG, C. H., Can. J. Chem. 39, 42 (1961).
780. GREENLER, R. G., J. Chem. Phys. 37, 2094 (1962).
781. NEUMANN, M. G., and HERTL, W., J. Chromatog. 65, 467 (1972).
782. KLEMM, L. H., and AIREE, S. K., ibid. 13, 40 (1964).
783. HALÁSZ, I., and HEINE, E., ref. 232, p. 207.
784. SCOTT, C. G., J. Inst. Petrol. 45, 118 (1959).

785. HOFFMANN, R. L., and EVANS, C. D., *Analyt. Chem.* **38,** 1309 (1966).
786. HOFFMANN, R. L., LIST, G. R., and EVANS, C. D., *Nature* **211,** 965 (1966).
787. LIST, G. R., HOFFMANN, R. L., and EVANS, C. D., *J. Am. Oil. Chem. Soc.* **42,** 1058 (1965).
788. HOFFMANN, R. L., LIST, G. R., and EVANS, C. D., *Nature* **206,** 823 (1965).
789. LIST, G. R., HOFFMANN, R. L., and EVANS, C. D., *ibid.* **213,** 380 (1967).
790. HOFFMANN, R. L., LIST, G. R., and EVANS, C. D., *J. Am. Oil. Chem. Soc.* **48,** 675 (1966).
791. VAN DER VLIST, E., and DE JONG, J. M., *J. Chromat.* **52,** 486 (1970).
792. KILARSKA, M., and ZIELINSKI, E., *Chem. Anal., Warsaw* **10,** 403 (1965).
793. BELYAKOVA, L. D., KISELEV, A. V., and SOLOYAN, G. A., *Chromatographia* **3,** 254 (1970).
794. BELYAKOVA, L. D., KISELEV, A. V., and SOLOYAN, G. A., *Zh. anal. Khim.* **27,** 1182 (1972).
795. EGGERTSEN, F. T., KNIGHT, H. S., and GROENNINGS, S., *Analyt. Chem.* **28,** 303 (1956).
796. EGGERTSEN, F. T., and KNIGHT, H. S., *ibid.* **30,** 15 (1958).
797. PURNELL, J. H., *Gas Chromatography*, Wiley, New York, 1962, p. 376.
798. DI CORCIA, A., *Analyt. Chem.* **45,** 492 (1973).
799. DI CORCIA, A., *J. Chromat.* **80,** 69 (1973).
800. DI CORCIA, A., LIBERTI, A., and SAMPERI, R., *Analyt. Chem.* **45,** 1228 (1973).
801. KISELEV, A. V., KOVALEVA, N. V., and NIKITIN, YU. S., *J. Chromat.* **58,** 19 (1971).
802. BRUNER, F., CICCIOLI, P., CRESCENTINI, G., and PISTOLESI, M. T., *Analyt. Chem.* **45,** 1851 (1973).
803. BRUNER, F., CARTONI, G. P., and LIBERTI, A., ref. 31, p. 301.
804. MACDONELL, H. L., NOONAN, J. M., and WILLIAMS, J. P., *Analyt. Chem.* **35,** 1253 (1963).
805. ROWAN, R., Jr., and SORRELL, J. B., *ibid.* **42,** 1716 (1970).
806. KIRKLAND, J. J., *ibid.* **35,** 1295 (1963).
807. MCKENNA, T. A., Jr. and IDLEMAN, J. A., *ibid.* **32,** 1290 (1960).
808. URBACH, G., *ibid.* **36,** 2368 (1964).
809. VIDAL-MADJAR, C., and GUIOCHON, G., *Nature* **215,** 1372 (1967); *Separ. Sci.* **2,** 155 (1967); *Bull. Soc. chim. Fr.* 1096 (1966).
810. GUIOCHON, G., and VIDAL-MADJAR, C., ref. 279, p. 295.
811. VIDAL-MADJAR, C., and GUIOCHON, G., *J. Chromat. Sci.* **9,** 664 (1971).
812. FRANKEN, J. J., VIDAL-MADJAR, C., and GUIOCHON, G., *Analyt. Chem.* **43,** 2034 (1971).
813. ALBERINI, G., BRUNER, F., and DEVITOFRANCESCO, G., *ibid.* **41,** 1940 (1969).
814. GAUMANN, T., PIRINGER, O., and WEBER, A., *Chimia* **24,** 112 (1970).
815. ISBELL, A. F., Jr., and SAWYER, D. T., *Analyt. Chem.* **41,** 1381 (1969).
816. BROOKMAN, D. J., and SAWYER, D. T., *ibid.* **40,** 106 (1968).
817. HUNT, P. P., and SMITH, H. A., *J. Phys. Chem.* **64,** 383 (1960); **65,** 87 (1961).
818. MOORE, W. R., and WARD, H. R., *ibid.* **64,** 832 (1960).
819. FURUYAMA, S., and KWAN, T., *ibid.* **65,** 190 (1961).
820. WEST, D. L., and MARSTON, A. L., *J. Am. Chem. Soc.* **86,** 4731 (1964).
821. GENTY, C., and SCHOTT, R., *Analyt. Chem.* **42,** 7 (1970).
822. VERNON, F., *J. Chromat.* **60,** 406 (1971).

823. BRUNNOCK, J. V., and LUKE, L. A., *ibid.* **39**, 502 (1969).
824. LLOYD, W. G., and ALFREY, T., *J. Polymer Sci.* **62**, 301 (1962).
825. MOORE, J. C., *ibid.* A, **2**, 835 (1964).
826. POSSANZINI, M., PELA, A., LIBERTI, A., and CARTONI, G. P., *J. Chromat.* **38**, 492 (1968).
827. BAYER, E., and NIKOLSON, G., *J. Chromat. Sci.* **8**, 467 (1970).
828. JANSSON, B. O., HALLGREN, K. C., and WIDMARK, G., *ibid.* p. 398.
829. SMITH, J. R. L., and WADDINGTON, D. J., *J. Chromat.* **36**, 145 (1968).
830. DUFKA, O., MALINSKY, J., CHURACEK, J., and KOMAREK, K., *ibid.* **51**, 111 (1970).
831. SMITH, J. R. L., and WADDINGTON, D. J., *Analyt. Chem.* **40**, 523 (1968).
832. BURGER, J. D., *J. Gas Chromat.* **6**, 177 (1968).
833. SAKODINSKY, K., and MOSEVA, L., *Chromatographia* **1**, 483 (1968).
834. GVOSDOVICH, T. N., KISELEV, A. V., and YASHIN, YA. I., *ibid.* **2**, 234 (1969).
835. HOLLIS, O. L., and HAYES, W. V., *J. Gas Chromat.* **4**, 235 (1966).
836. KLEIN, A., *Porapak Columns*, Varian Aerograph Technical Bull., 128 (1966).
837. *Choosing Porapak Applications (Bull. Waters Associates)* 4 (1968).
838. GVOSDOVICH, T. N., KISELEV, A. V., and YASHIN, YA. I., *Chromatographia* **6**, 179 (1973).
839. Chromosorb-104, Johns-Manville Celite Division, *Bulletin FF-189.*
840. YASHIN, YA. I., BEBRIS, N. K., and GVOSDOVICH, T. N., ref. 279, p. 333.
841. DUSEK, K., SEIDL, J., and MALINSKY, J., *Coll. Czech. Chem. Commun.* **32**, 2766 (1967).
842. DERGE, K., *Fette, Siefen, Anstrichmittel* **21**, 407 (1967).
843. BOGART, G., *Facts and Methods, Technical Bull. Hewlett Packard,* **7** (5) (1966).
844. HOLLIS, O. L., and HAYES, W. V., ref. 105, p. 57.
845. JONES, C. N., *Analyt. Chem.* **39**, 1858 (1967).
846. OBERMILLER, E. L., and CHARLIER, G. O., *J. Gas Chromat.* **6**, 446 (1968); *J. Chromat. Sci.* **7**, 580 (1969).
847. WILHITE, W. F., and HOLLIS, O. L., *J. Gas Chromat.* **6**, 84 (1968).
848. ZLATKIS, A., and KAUFMAN, H. R., *ibid.* **4**, 240 (1966).
849. CARLE, G. C., *J. Chromat. Sci.* **8**, 550 (1970).
850. DRENNAN, G. A., and MATULA, R. A., *J. Chromat.* **34**, 77 (1968).
851. JERMAN, R. I., and CARPENTER, L. R., *J. Gas Chromat.* **6**, 298 (1968).
852. PAPIC, M., *ibid.* p. 493.
853. JARIWALA, S. L., and WAKEMAN, I. B., *J. Chromat. Sci.* **8**, 612 (1970).
854. DI LORENZO, A., *ibid.* p. 224.
855. CROSS, R. A., *Nature* **211**, 409 (1966).
856. BENNETT, D., *J. Chromat.* **26**, 482 (1967).
857. SOLOMON, P., *ibid.* **30**, 593 (1967).
858. CRAWFORTH, C. G., and WADDINGTON, D. J., *J. Gas Chromat.* **6**, 103 (1968).
859. FORSEY, R. R., *ibid.*, p. 555.
860. IKELS, K. G., and NEVILLE, J. R., *ibid.*, p. 222.
861. GRICE, H. W., and DAVID, D. J., *J. Chromat. Sci.* **7**, 247 (1969).
862. SEELER, A. K., and CAHILL, R. W., *ibid.*, p. 159.
863. GALLIARD, T., and GREY, T. C., *J. Chromat.* **41**, 442 (1969).
864. GOUGH, T. A., and SIMPSON, C. F., *ibid.* **51**, 129 (1970); **68**, 31 (1972).

865. CASTELLO, G., and MUNARI, S., *ibid.* **31**, 202 (1967).
866. JONES, K., *J. Gas Chromat.* **5**, 432 (1967).
867. ONUSKA, F., JANAK, J., DURAS, S., and KRCHMAROVA, M., *J. Chromat.* **40**, 209 (1969).
868. THOMSON, T. B., *ibid.* **39**, 500 (1969).
869. MURACA, R. F., WILLIS, E., MARTIN, C. H., and CRUTCHFIELD, C. A., *Analyt. Chem.* **41**, 295 (1969).
870. BALLINGER, J. T., BARTELS, T. T., and TAYLOR, J. H., *J. Gas Chromat.* **6**, 295 (1968).
871. MAHADEVAN, V., and STENROOS, U., *Analyt. Chem.* **39**, 1652 (1967).
872. HENKEL, H. G., *J. Chromat.* **58**, 201 (1971).
873. PALFRAMANN, J. F., and WALTER, E. A., *Analyst Lond.* **92**, 71 (1967).
874. UMANSKAYA, A. A., MAKARENKOVA, R. M., SOKOLOV, N. M., and ZHAVORONKOV, N. M., *Zh. analit. Khim.* **25**, 1211 (1970).
875. SMITH, J. R. L., and WADDINGTON, D. J., *J. Chromat.* **42**, 105 (1969).
876. GVOSDOVICH, T. N., KOVALEVA, M. P., PETROVA, G. K., and YASHIN, YA I., *Neftekhimiya* **8**, 123 (1968).
877. PALO, V., and ILKOVA, H., *J. Chromat.* **53**, 363 (1970).
878. BAKER, R. N., ALENTY, A. L., and ZACK, I. F., *J. Chromat. Sci.* **7**, 312 (1969).
879. COLEHOUR, J. K., *Analyt. Chem.* **39**, 1190 (1967).
880. MAHADEVAN, J., and ZIEVE, L., *J. Lipid Res.* **10**, 338 (1969).
881. KIKUCHI, Y., KIKKAWA, T., and KATO, R., *J. Gas Chromat.* **5**, 261 (1967).
882. JENDEN, D. J., HANIN, J., and LAMB, S. J., *Analyt. Chem.* **40**, 125 (1968).
883. JOHNSTON, G. A. R., LLOYD, H. J., and DE MARCHI, W. J., *J. Chromat.* **47**, 482 (1970).
884. GRANMER, M. F., and PEOPLES, A., *ibid.* **57**, 365 (1971).
885. CHARRANSOL, G., and DESGREZ, P., *ibid.* **48**, 530 (1970).
886. HALVARSON, H., *ibid.* **57**, 406 (1971).
887. DARDENNE, G. A., SEVERIN, M., and MARLIER, M., *ibid.* **47**, 176 and 182 (1970).
888. SIE, S. T., BLEUMER, J. P. A., and RIJNDERS, G. W A., ref. 282, p. 235.
889. TRANCHANT, J., *Z. Analyt. Chem.* **236**, 137 (1968).
890. VAN WIJK, R., ref. 8, p. 254.
891. VAN WIJK, R., *in* A. ZLATKIS, ed., *Advances in Chromatography 1970*, Houston, Texas, 1970, p. 122.
892. VAN WIJK, R., *J. Chromat. Sci.* **8**, 418 (1970).
893. FULLER, E. N., *Analyt. Chem.* **44**, 1747 (1972).
894. SAKODYNSKY, K. I., KLINSKAYA, N. S., and PANINA, L. I., *ibid.* **45**, 1369 (1973).
895. ADROVA, N. A., BESSONOV, M. I., LAIUS, L. A., and RUDAKOV, A. P., *Polyimides– New Class of Thermally Stable Polymers*, Nauka, Leningrad, 1968.
896. D'AUBIGNE, J. M., and GUIOCHON, G., *Chromatographia* **3**, 153 (1970).
897. BARRER, R . M., *Ber. Bunsenges. Phys. Chem.* **69**, 786 (1965).
898. BARRER, R. M., *Brennstoff-Chem.* **35**, 325 (1954).
899. BRECK, D. W., EVERSOLE, W. G., MILTON, R. M., REED, T. B., and THOMAS, T. L. *J. Am. Chem. Soc.* **78**, 5963 (1956).
900. REED, T. B., and BRECK, D. W., *ibid.*, p. 5972.
901. BARRER, R. M., *J. Chem. Soc.*, 2342 (1950).
902. BARRER, R. M., and RILEY, D. W., *Trans. Faraday Soc.* **46**, 853 (1950).

903. BARRER, R. M., *Br. Chem. Engng* 436 (1959).
904. GRUBNER, O., RALEK, M., and JIRU, P., *Chem. Prumysl.* 11, 521 (1961).
905. JANAK, J., KREJCI, M., and DUBSKY, H. E., *Chem. Listy* 52, 1099 (1958); *Ann. NY Acad. Sci.* 72, 731 (1959).
906. AUBEAU, R., LEROY, J., and CHAMPEIX, L., *J. Chromat.* 19, 249 (1965).
907. PATZELOVA, V., *Chromatographia* 3, 170 (1970).
908. BRENNER, N., CIEPLINSKI, E., ETTRE, L. S., and COATES, V. J., *J. Chromat.* 3, 230 (1960).
909. BARRER, R. M., and IBBITSON, D. A., *Trans. Faraday Soc.* 40, 206 (1944).
910. BHATNAGAR, V. M., *J. Gas Chromat.* 5, 43 (1967).
911. OBERHOLTZER, J. E., and ROGERS, L. B., *Analyt. Chem.* 41, 1590 (1969).
912. HABGOOD, H. W., *Can. J. Chem.* 42, 2340 (1964).
913. HABGOOD, H. W., and MacDONALD, W. R., *Analyt. Chem.* 42, 543 (1970).
914. MORELAND, A. K., and ROGERS, L. B., *Separ. Sci.* 6, 1 (1971).
915. BARRER, R. M., *J. Colloid Interface Sci.* 21, 415 (1966).
916. ANGELL, C. L., and SCHAFFER, P. C., *J. Phys. Chem.* 70, 1413 (1966).
917. ANDRONIKASHVILI, T. G., TSITSISHVILI, G. V., and SABELASHVILI, SH. D., *J. Chromat.* 58, 47 (1971).
918. NEDDENRIEP, R. J., *J. Colloid Interface Sci.* 28, 293 (1968).
919. KYRIACOS, G., and BOARD, C. E., *Analyt. Chem.* 29, 787 (1957).
920. JANAK, J., KREJCI, M., and DUBSKY, H. E., *Coll. Czech. Chem. Commun.* 24, 1080 (1959).
921. KREJCI, M , *ibid.* 25, 2457 (1960).
922. BOSACEK, V., *ibid.* 29, 1797 (1964).
923. FURTIG, H., and WOLF, F., *Ber. Bunsengesellschaft* 69, 842 (1965).
924. THOROGOOD, R. M., WILSON, G. M., and GEIST, J. M., *J. Chem. Engng Data* 10, 269 (1965).
925. WALKER, J. A. J., *Nature* 209, 197 (1966).
926. GNAUCK, G., and TREUTLER, E., ref. 279, p. 231.
927. KISELEV, A. V., CHERNENKOVA, YU. L., and YASHIN, YA. I., *Neftekhimiya* 5, 141 (1965).
928. BOMBAUGH, K. J., *Nature* 197, 1102 (1963).
929. FARRE–RIUS, F., and GUIOCHON, G., *J. Gas Chromat.* 1, 33 (1963).
930. KARLSSON, B. M., *Analyt. Chem.* 38, 668 (1966).
931. McALLISTER, W. A., and SOUTHERLAND, W. V., *ibid.* 43, 1536 (1971).
932. BARRALL E. M., II, and BAUMAN, F., *J. Gas Chromat.* 2, 256 (1964).
933. MAIR, B. J., and SHAMAIENGER, M., *Analyt. Chem.* 30, 276 (1958).
934. BRENNER, N., and COATES, V. J., *Nature* 181, 1401 (1958).
935. WHITHAM, B. T., *ibid.* 182, 391 (1958).
936. ADLARD, E. R., and WHITHAM, B. T., *ibid.* 192, 966 (1961).
937. SCHENCK, P. A., and EISMA, E., *ibid.* 199, 170 (1963).
938. EGGERTSEN, F. T., and GROENNINGS, S., *Analyt. Chem.* 33, 1147 (1961).
939. ALBERT, D. K., *ibid.* 35, 1918 (1963).
940. BLYTAS, G. C., and PETERSON, D. L., *ibid.* 39, 1434 (1967).
941. BRUNNOCK, J. V., *ibid.* 38, 1648 (1966).
942. BRUNNOCK, J. V., and LUKE, L. A., *ibid.* 40, 2158 (1968); 41, 1126 (1969).
943. MORTIMER, J. V., and LUKE, L. A., *Analyt. chim. Acta* 38, 119 (1967).

944. KNIGHT, H. S., *Analyt. Chem.* **39,** 1452 (1967).
945. KISELEV, A. V., and YASHIN, YA I., *Zh. fiz. Khim.* **40,** 944 (1966).
946. PETERSON, R. M., and RODGERS, J., *Chromatographia* **5,** 13 (1972).
947. KAISER, R., *Chromatographia* **3,** 38 (1970).
948. MAILEN, J. C., REED, T. M., III, and YOUNG, J. A., *Analyt. Chem.* **36,** 1883 (1964).
949. BHATTACHARJEE, A., and BASU, A. N., *J. Chromat.* **71,** 534 (1972).
950. BHATTACHARYYA, A. C., and BHATTACHARJEE, A., *ibid.* **41,** 446 (1969).
951. BHATTACHARYYA, A. C., and BHATTACHARJEE, A., *Analyt. Chem.* **41,** 2055 (1969).
952. BROWN, J. F. Jr., *Sci. Am.* **207,** 82 (July 1962).
953. REED, T. M., III, *J. Chromat.* **9,** 419 (1962).
954. DE RADZITSKY, P., and HANOTIER, J., *Ind. Engng. Chem. Process Res. Develop.* **1,** 10 (1962).
955. SCHAEFFER, W. D., DORSEY, W. S., SKINNER, D. A., and CHRISTIAN, C. G., *J. Am. Chem. Soc.* **79,** 5870 (1957).
956. DUFFIELD, J. J., and ROGERS, L. B., *Analyt. Chem.* **34,** 1193 (1962).
957. COOK, L. E., and GIVAND, S. H., *J. Chromat.* **57,** 313 (1971).
958. ALTENAU, A. G., and ROGERS, L. B., *Analyt. Chem.* **37,** 1432 (1965).
959. ALTENAU, A. G., and MERRITT, C., Jr., *J. Gas Chromat.* **5,** 30 (1967).
960. PFLAUM, R. T., and COOK, L. E., *J. Chromat.* **50,** 120 (1970).

INDEX